Agricultural Adaptation to Climate Change in Africa

A changing climate is likely to have a drastic impact on crop yields in Africa. The purpose of this book is to document the effects of climate change on agriculture in Africa and to discuss strategies for adaptation to hotter weather and less predictable rainfall. These strategies include promoting opportunities for farmers to adopt technologies that produce optimal results in terms of crop yield and income under local agro-ecological and socioeconomic conditions.

The focus is on sub-Saharan Africa, an area that is already affected by changing patterns of heat and rainfall. Because of the high prevalence of subsistence farming, food insecurity, and extreme poverty in this region, there is a great need for practical adaptation strategies. The book includes empirical research in Ethiopia, Kenya, South Africa, Tanzania, and other sub-Saharan countries, and the conclusion summarizes policy-relevant findings from the chapters.

It is aimed at advanced students, researchers, extension and development practitioners, and officials of government agencies, NGOs, and funding agencies. It also will provide supplementary reading for courses in environment and development and in agricultural economics.

Cyndi Spindell Berck, M.P.P., J.D., is the principal of International Academic Editorial Services, based in Orinda, California, USA, and Managing Editor of the EfD Discussion Paper series and Research Brief series.

Peter Berck is Professor of Agricultural and Resource Economics and Policy and S.J. Hall Professor of Forestry, University of California, Berkeley, USA. He is Editor-in-Chief of the EfD Discussion Paper series.

Salvatore Di Falco is Professor of Environmental Economics and Director of the Institute of Economics and Econometrics, Geneva School of Economics and Management, University of Geneva, Switzerland.

About Resources for the Future *and* RFF Press

Resources for The Futures (RFF) improves environmental, energy, and natural resource decisions through impartial economic research and policy engagement.

Founded in 1952, RFF pioneered the application of economics as a tool for developing more effective policy about the use and conservation of natural resources. Its scholars continue to employ social science methods to analyze critical issues concerning pollution control, energy policy, land and water use, hazardous waste, climate change, biodiversity, and the environmental challenges of developing countries.

RFF Press supports the mission of RFF by publishing book-length works that present a broad range of approaches to the study of natural resources and the environment. Its authors and editors include RFF staff, researchers from the larger academic and policy communities, and journalists. Audiences for publications by RFF Press include all of the participants in the policymaking process—scholars, the media, advocacy groups, NGOs, professionals in business and government, and the public.

Resources for the Future
Board of Directors

Environment for Development Book Series
Series Editors: Thomas Sterner and Gunnar Köhlin

The Environment for Development (EfD) initiative (www.efdinitiative.org) supports poverty alleviation and sustainable development through the increased use of environmental economics in the policymaking process. EfD identifies the environment as an important resource for development rather than a constraint. The EfD initiative is a capacity-building program in environmental economics focusing on research, policy advice, and teaching in Central America, Chile, Colombia, China, Ethiopia, India, Kenya, South Africa, Tanzania, USA, Vietnam and Sweden. The eleven EfD centers are hosted by leading universities or academic, Vietnam institutions in their respective country/region.

The EfD is initiated and managed by the Environmental Economics Unit, University of Gothenburg, Sweden. The core funding for the EfD initiative is provided by Sida, the Swedish International Development Cooperation Agency.

Institutions interested in partnering with the EfD initiative, please contact the EfD secretariat at info@efdinitiative.org.

Agricultural Adaptation to Climate Change in Africa
Food Security in a Changing Environment
Edited by Cyndi Spindell Berck, Peter Berck and Salvatore Di Falco

Forest Tenure Reform in Asia and Africa
Local control for improved livelihoods, forest management and
carbon sequestration
Edited by Randall A. Bluffstone and Elizabeth J.Z. Robinson

Biodiversity Conservation in Latin America and the Caribbean
Prioritizing policies
Allen Blackman, Rebecca Epanchin-Niell, Juha Siikamäki and Daniel Velez-Lopez

Environmental Regulation and Public Disclosure
The case of PROPER in Indonesia
Shakeb Afsah, Allen Blackman, Jorge H. Garcia and Thomas Sterner

For more information on books in the Environment for Development Book series, please visit the book series page on the Routledge website: http://www.routledge.com/books/www.routledge.com/books/ series/ECEFD/.

Agricultural Adaptation to Climate Change in Africa

Food Security in a Changing Environment

Edited by
Cyndi Spindell Berck, Peter Berck, and Salvatore Di Falco

Routledge
Taylor & Francis Group

LONDON AND NEW YORK

First published 2018 by RFF Press

2 Park Square, Milton Park, Abingdon, Oxon OX14 4RN
605 Third Avenue, New York, NY 10017

Routledge is an imprint of the Taylor & Francis Group, an informal business

First issued in paperback 2021

British Library Cataloguing-in-Publication Data
A catalogue record for this book is available from the British Library

Library of Congress Cataloging-in-Publication Data
Names: Berck, Cyndi Spindell, editor. | Berck, Peter, 1950– editor. | Di Falco, Salvatore, editor.
Title: Agricultural adaptation to climate change in Africa : food security in a changing environment / edited by Cyndi Spindell Berck, Peter Berck, and Salvatore Di Falco.
Description: New York, NY : RFF Press, 2018. | Includes bibliographical references and index.
Identifiers: LCCN 2017047543 | ISBN 9781138555976 (hardback) | ISBN 9781315149776 (ebook)
Subjects: LCSH: Crops and climate—Africa, Sub-Saharan. | Climatic changes—Africa, Sub-Saharan. | Agriculture—Environmental aspects—Africa, Sub-Saharan. | Farms, Small—Africa, Sub-Saharan. | Food security—Africa, Sub-Saharan.
Classification: LCC S600.7.C54 A36 2018 | DDC 630.25150967—dc23
LC record available at https://lccn.loc.gov/2017047543

ISBN: 978-1-138-55597-6 (hbk)
ISBN: 978-0-367-50742-8 (pbk)

Typeset in Goudy
by Apex CoVantage, LLC

Contents

Acknowledgements

This book represents a collaboration across the Environment for Development Initiative and beyond. Researchers from EfD centers in Ethiopia, Kenya, Tanzania, South Africa, and Sweden generously contributed their original research or took the time to write chapters synthesizing earlier research. We thank those who were early contributors and waited patiently for completion of the book, and those who helped us push toward the finish line.

We can always count on Thomas Sterner and Gunnar Köhlin, the editors of the EfD Book Series, as friends and colleagues. We thank Tim Hardwick and Amy Louise Johnston, of Resources for the Future Press/Routledge/Taylor and Francis, for their support. We appreciate the editorial assistance of Anna Snow Berck, Poojan Thakrar, and Tyler N. Jacobson. Tyler and Poojan, who are undergraduates at the University of California, Berkeley, demonstrated writing ability and knowledge of environmental science and economics that earned them a much-appreciated role as co-authors. Poojan, who came on board when Cyndi and Peter Berck were unable to work because of a family health crisis, took the initiative to keep the Bercks' end of the project going and demonstrated impressive competence and good spirits.

We hope that this book will be of use to researchers, policymakers, extension agents, and, above all, the hard-working farm households of Africa.

Cyndi Spindell Berck
Berkeley, California

Peter Berck
Berkeley, California

Salvatore Di Falco
Geneva, Switzerland
August 2017

About the authors

Yonas Alem is the Research Director of the Environment for Development Initiative in Gothenburg, Sweden.

Tekie Alemu is a Senior Research Fellow at the Environment and Climate Research Center at the Ethiopian Development Research Institute and an Associate Professor at Addis Ababa University, Ethiopia.

Jonse Bane is a Lecturer at the Economics Department of Addis Ababa University and a PhD student at the Department of Economics of Addis Ababa University, Ethiopia.

Rahel Deribe Bekele is a PhD candidate at the Center for Development Research (ZEF), University of Bonn, Germany.

Peter Berck is Professor of Agricultural and Resource Economics and Policy and the S.J. Hall Professor of Forestry at the University of California, Berkeley, USA, and the Editor-in-Chief of the EfD Discussion Paper Series.

Cyndi Spindell Berck, MPP, JD, is the principal of International Academic Editorial Services, based in Orinda, California, USA, and the Managing Editor of the EfD Discussion Paper series and Research Brief series.

Mintewab Bezabih (Ayele) is a Senior Research Fellow and Research Director at the Environment and Climate Research Center, Addis Ababa University, Ethiopia.

Lars Bohlin is an Assistant Professor at Mälardalen University, Sweden.

Martina Bozzola is a Lecturer in the Economics of Agriculture, Food and Health at Queen's University, Belfast, UK, and a Research Associate in the Agricultural Economics and Policy Group at ETH Zurich, Switzerland.

Kerri Brick is a Research Fellow at the Environmental-Economics Policy Research Unit at the University of Cape Town, South Africa.

Martin J. Chegere is a Research Fellow at EfD-Tanzania and a Lecturer in the Department of Economics, University of Dar es Salaam, Tanzania.

Salvatore Di Falco is a Professor of Environmental Economics at the University of Geneva, Switzerland.

Hailu Elias is an Assistant Professor of Economics and Head of the Rural Development Unit of the Institute of Development Policy Research of Addis Ababa University, Ethiopia.

Tadele Ferede is an Assistant Professor in the Department of Economics, Addis Ababa University, Ethiopia.

Zenebe Gebreegziabher is an Associate Professor at the Department of Economics, Mekelle University, Ethiopia, and a Senior Research Fellow at the Environment and Climate Research Center of the Ethiopian Development Research Institute.

Tsegazeab Gebremariam (Yihdego) is a Lecturer and head of the Department of Economics at Assosa University, Ethiopia.

Fantu Guta is an Assistant Professor at the Department of Economics, Addis Ababa University, Ethiopia.

Richard Zadocky Jacob is a Junior Research Fellow at EfD, Tanzania.

Tyler N. Jacobson is a BS student in Environmental Economics and Policy at the University of California, Berkeley, USA.

Hafsah Jumare is the Executive Director of Solidarity Development Cooperative Ltd., a social enterprise that engages in capacity building for agricultural cooperatives in low-income communities in Nigeria. She has a master's degree from the University of Cape Town, South Africa, where she carried out agricultural research under the Environmental-Economics Policy Research Unit.

Millicent Kabara is a PhD student at the School of Economics, Kenyatta University, Kenya. She has worked as a Senior Economist in the Kenyan Ministry of Environment and Natural Resources, handling climate change issues.

Jane Kabubo-Mariara is Executive Director of the Partnership for Economic Policy and a Professor of Economics at the University of Nairobi, Kenya. She is immediate former Director of the School of Economics, University of Nairobi, and Director of EfD, Kenya.

Gunnar Köhlin is an Associate Professor at the Environmental Economics Unit, Department of Economics, University of Gothenburg, Sweden, and director of the Environment for Development initiative.

Coretha Komba is a Lecturer with the Economics Department at Mzumbe University in Tanzania. She obtained her PhD (Economics) at the University of Cape Town, South Africa.

Jörgen Levin is an Associate Professor at Örebro University School of Business, Sweden, and works as a Research Fellow at UNU-WIDER.

Razack Lokina is an Associate Professor in the Department of Economics and the Center Director at EfD, Tanzania, Department of Economics, University of Dar es Salaam, Tanzania.

Mathilde Maurel is the Research Director at the Centre d'Economie de la Sorbonne CNRS – Université Paris 1, France.

Alemu Mekonnen is an Associate Professor at the Department of Economics of Addis Ababa University, Ethiopia, and a Senior Research Fellow at the Environment and Climate Research Center of the Ethiopian Development Research Institute.

Katrin Millock is a CNRS Research Fellow at the Paris School of Economics, France.

Edwin Muchapondwa is a Professor at the University of Cape Town, South Africa, and a Research Fellow at EfD, South Africa.

Melinda Smale is a Professor of International Development and a member of the Food Security Group of the Agricultural, Food and Resource Economics Department, Michigan State University, USA.

Hailemariam Teklewold is a Research Fellow at the Environment and Climate Research Center at the Ethiopian Development Research Institute.

Poojan Thakrar is a BS student in Environmental Economics and Policy at the University of California, Berkeley, USA.

Byela Tibesigwa is a Senior Research Fellow in EfD-Tanzania and a Research Associate in the Environmental-Economics Policy Research Unit at the University of Cape Town's School of Economics, South Africa.

Martine Visser is a Professor in the School of Economics, University of Cape Town, South Africa, and Director of the Environmental-Economics Policy Research Unit.

Tesfamicheal Wossen is an impact evaluation economist at the International Institute of Tropical Agriculture, Nigeria.

Samuel Abera Zewdie is a statistician at the Environment and Climate Research Center of the Ethiopian Development Research Institute.

1   Introduction

Cyndi Spindell Berck, Peter Berck,
Salvatore Di Falco, and Poojan Thakrar

The purpose of this book is to explore ways that African farmers and governments can adapt to a changing climate. We take it as given that mitigation of climate change, through reduction of greenhouse gas (GHG) emissions, is essential.
However, farmers in Africa already are experiencing the effects of a changed climate, and the prognosis for effective global policies to reduce GHG is uncertain.
In addition, many of the policy prescriptions for agricultural adaptation to climate change, such as increased adoption of yield-enhancing technologies, would
benefit African farmers even without climate change. Therefore, responses to
climate change must proceed on both the mitigation and adaptation tracks.

About two out of three people in sub-Saharan Africa (SSA) make a living in
agriculture, mostly as smallholder subsistence farmers who depend directly on
rainfall. Hotter weather and less predictable rainfall are already having impacts
on their crop yields. These climate impacts are expected to hit SSA harder than
other regions of the world.

Increased food production is essential to improve food security and reduce poverty among the growing population of Africa. Increasing the productivity of land
already under cultivation is crucial for two reasons. First, much of the arable land
is already under cultivation in many parts of Africa. Second, clearing more
land for agriculture would compound the problem of GHG emissions by removing
carbon sinks.

Agricultural production takes place in a context of overall economic development. In addition to discussing strategies to increase crop yield, the studies in this
book consider a number of income-producing responses to climate change, such
as the crop-livestock mix and migration for off-farm work opportunities.

This book seeks to describe some of the challenges facing African farmers
and to suggest practical responses. The chapters include both original research
and syntheses of existing research. Many of the contributors are part of the
Environment for Development Initiative research network. This is reflected in
the focus on the four countries that host EfD centers: Ethiopia, Kenya, South
Africa, and Tanzania. However, empirical findings for other SSA countries are
included as well.

This book is divided into five parts, after this introduction. The first part considers climate science in relation to the agronomic and agroecological factors that

influence farm productivity. The second part looks at on-farm practices to increase productivity of food crops, and the third part at on-farm practices other than those related to food crop yield. The fourth part considers gender, and the fifth discusses agricultural adaptation in a broader development context. The concluding chapter summarizes policy recommendations.

Part I: climate science, agronomic, and agroecological factors

Chapters 2 through 4 explore the possible effects of climate change on agriculture in light of the diverse agroecological conditions in sub-Saharan Africa. The chapters examine important indicators, such as food security, crop yield, and geographical adaptability, to shed light on the consequences of climate change. Institutions such as social capital are also considered in light of their relationship with biophysical factors.

Chapter 2 outlines the processes by which climate change is studied, as well as potential adaptations to alleviate the expected negative consequences. Peter Berck first explains five methods of obtaining data, with their associated practicality, benefits, and drawbacks. Compared to other regions around the globe, sub-Saharan Africa suffers from a lack of quality data. This leads to climate models that can be imprecise and inaccurate. Most of the models in this book will be regression models, which use the covariance of weather, yield, and other important factors to make predictions of the effects of climate change.

This level of detail is important because the optimal adaptation strategy for a given region depends on the precise changes in climate experienced in that area. For example, in response to increased temperatures, farmers might switch from temperature-sensitive maize to heat-resistant sorghum. To combat decreased expected rainfall and increased rainfall variability, farmers would benefit from wells and irrigation, but most smallholder farmers cannot finance such projects without an increase in access to credit. By studying climate change and formulating adaptation strategies that are carefully tailored to specific agroecological conditions, some of the productivity lost to climate change can be restored.

Previous climate change literature has often addressed countries as single units. In Chapter 3, "Mapping Vulnerability to Climate Change of the Farming Sector in the Nile Basin of Ethiopia," Zenebe Gebreegziabher et al. extend the analysis to local ecological and geographic variation. The chapter analyzes four different regions in Ethiopia to discover which specific regions are most vulnerable to climate change. According to the latest IPCC report, a region's vulnerability is a function of both its exposure and its sensitivity to climate change effects, as well as the region's adaptive capability. Sensitivity can be further broken down into factors such as climate extremes or population density. Meanwhile, adaptive capability can be broken down into social capital, human capital, financial capital, and physical capital.

To assess a region's vulnerability to climate change, Gebreegziabher et al. construct a vulnerability index. The authors use an indicator method in which the index is calculated by combining the observables using different weights, calculated using a principal component analysis. Using this method, the authors are able to quantify the exposure, sensitivity, and adaptive capability for four different regions. When all three were taken into account, the two regions with the most vulnerability were the humid lowlands and the drought-prone highlands.

In Chapter 4, "Climate Change and Food Security in Kenya," Jane Kabubo-Mariara and Millicent Kabara analyze the linkages between climate change and food security, proxied by food crop yields, in Kenya. The authors use county-level data collected over three decades for maize, sorghum, beans, and millet. Using this, they were able to find relationships between yield and climate variables. Next, the authors simulated the expected impact of future climate change on food insecurity based on ten climate and emission scenarios.

Most of the variables exhibited nonlinear U or inverse U-shaped relationships with crop yields. For example, while adequate rainfall is obviously crucial for crop productivity, too much rain can cause flooding, waterlogging, and rotting. Summer temperatures exhibit a U-shaped relationship for three of the four crops, while winter temperatures exhibit a hill-shaped relationship. Population density also has significant, positive effects on maize and beans. Population density can be a proxy for agricultural adaptation options and can also capture availability of farm labor.

As for the simulation results, nine of the ten simulations predict declining crop yields due to climate change, with the largest prediction being a 69% decline in yields. The simulations also predict that, if there are no changes, 73% of Kenya's population will be maize-insecure by 2100. Because maize is the main food crop, those who are insecure in maize are likely food-insecure in general.

Part II: on-farm practices related to food crop productivity

The next five chapters examine the relationship between climate change and farmers' decisions related to increasing the yields from food crops. These chapters examine the adaptive measures a farmer can take and the driving forces that determine whether farmers adopt such adaptive techniques. These driving forces include behavioral considerations, such as risk aversion, which in turn are influenced by the relative wealth or poverty of individual farmers. These five chapters go to the heart of the question of how African farmers can grow enough food in a changing climate.

Chapter 5, "Adaptation to Climate Change in Sub-Saharan Agriculture: Assessing the Evidence and Rethinking the Drivers," identifies levels of climate change perception while also evaluating how climate variables can affect risk aversion. According to previous literature, the two steps to climate change adaptation are the perception of change and the adoption of adaptation strategies. Perception does not always lead to adoption. For instance, while 83% of Ethiopian farmers perceive variations in temperature, only 44% of them use adaptation strategies. The main barriers to adoption include a lack of education and the absence of credit and capital. When examining the effectiveness of adaptation strategies, Salvatore Di Falco noted that strategies adopted in combination with other strategies rather than in isolation are more effective. Adaptation is, therefore, more effective when it is composed of a portfolio of actions rather than a single action.

Additionally, Di Falco considers the behavioral factors of risk aversion. Specifically, he asks whether farmers who are exposed to more variable rainfall are more likely to display higher risk aversion. Di Falco determines whether an individual is risk-averse, using an experiment where a farmer chooses between two different scenarios with different income means and variances. The author found that higher rainfall is negatively correlated with the probability of being risk-averse, while the coefficient of variation of rainfall is positively correlated with the probability of being risk-averse.

Chapter 6, "Climate, Shocks, Weather and Maize Intensification Decisions in Rural Kenya," aims to measure the effects of climate change and climatic shocks in Kenya on a small farmer's decision to allocate part of the farm to hybrid maize seeds in order to intensify productivity. In this sense, the term "climatic shocks" refers to the number of serious droughts in the past decade. Secondly, Bozzola et al. investigate whether adoption of improved maize seeds, which is known to increase farmers' mean income, also increases the income variability, thus exposing the farmer to more downside risk. The authors assume that farmers choose the proportion of their farm dedicated to hybrid seeds based on both the incentive of increased mean income and the disincentive of increased income variance. Thus, the proportion of the farm dedicated to hybrid seeds is a function of each individual farmer's tolerance for risk and aversion to a negatively skewed income distribution.

The results showed that many explanatory variables had statistically significant effects on both adoption of improved seeds and farmers' income. Factors such as the number of educated men, soil quality, and population density were shown to have positive effects on adoption. As for climatic variables, droughts had a negative and statistically significant impact on adoption, while rainfall had a significantly positive effect on adoption. However, the regression results showed no correlation between adoption of improved seeds and either variance or skewness of the distribution of crop income.

In Chapter 7, "Adaptation to Climate Change by Smallholder Farmers in Tanzania," Coretha Komba and Edwin Muchapondwa study climate change perception and adaptation levels. The authors survey whether farmers perceive climate change, whether they adapt their agricultural activities, and what factors influence their choice of adaptation methods. Komba and Muchapondwa used a probit to model the binary choice of whether a farmer adapts and a multinomial probit to model the particular adaptation method.

When surveyed, 98.9% of farmers said they perceived changes in both the mean and variance of both temperature and rainfall from the 1990s to the 2000s. This shows that there is overwhelming evidence that farmers in Tanzania have perceived climate change.

While almost all the farmers perceived climate change, only a fraction implemented one or more adaptation methods. Komba and Muchapondwa found a variety of factors that increase a farmer's probability of adapting to climate change, including education and proximity to markets. Those who did not adopt methods cited a lack of water, a lack of funds, poor planning, and a shortage of seeds as barriers to adaptation. The authors urge the government to address these factors to alleviate farmers' vulnerability to climate change.

Next, Komba and Muchapondwa investigated how farmers choose their primary method of adaptation, such as changing planting dates or increased irrigation. Farmers with access to media are 7.7% more likely to change their planting date, presumably because they have access to weather forecasts and information. Meanwhile, the inability to access credit increases the probability of changing planting dates as the primary method of adaptation by 9.9%. This is because changing planting dates is less capital-intensive than methods such as increased irrigation.

The poverty trap theory suggests that low-income individuals make choices that keep them in poverty, due to factors such as risk intolerance. Chapter 8, "Risk Preferences and the Poverty Trap: A Look at Farm Technology Uptake among Smallholder Farmers in the Matzikama Municipality," studies this theory in South Africa. Hafsah Jumare et al. study the effects on farm technology uptake of risk aversion and loss aversion (greater sensitivity to the possibility of loss than to the possibility of gain). They also considered how farmers' decisions are affected when their decision-making processes give too much weight to low-probability events, or too little weight to high-probability events. The authors looked at the decision to use naturally occurring drought-resistant seeds and the decision to purchase improved seeds. The latter produce higher mean yields but with more variance; because they are more expensive, they carry more risk for the farmer, who would have to repay a loan even in the case of a failed harvest.

The authors measured risk tolerance and loss aversion using a set of hypothetical lottery scenarios. Subjects were presented with two scenarios with different expected payouts and variances and were instructed to choose between the two. Education and household income were among the factors correlated with lower loss aversion. This suggests that less educated and lower-income households, all else equal, are more likely to remain in a poverty trap because they are so sensitive to the risk of loss that they avoid choices that offer possible gain.

Next, Jumare et al. model the decision to use either drought-resistant crops or improved seeds. Looking at the risk preference parameters, the authors find no evidence to suggest that risk or loss aversion influences the uptake of either option. However, a farmer who gives too much weight to a low-probability event is more likely to adopt drought-resistant crop varieties. This may be because fear of a drought causes farmers to overestimate its likelihood and thus switch to more drought-resistant crops. In addition, higher education, higher income, and being male increase the likelihood of one or both of the adaptation methods. Again, these findings support the poverty trap theory.

Agricultural production can be intensified by practices and inputs that are neutral, negative, or positive in their environmental impacts. Purchased inputs such as high-yielding seeds and fertilizer have been shown to increase crop yield, especially when these methods are used as part of a package. Under some circumstances, including a combination of sustainable intensification practices in a farming system can increase productivity while reducing the impact that external inputs have on the environment. Chapter 9, "Good Things Come in Packages: Sustainable Intensification Systems in Smallholder Agriculture," explores systems of sustainable intensification practices. This chapter summarizes some previous research on smallholder farmers' intensification practices that have the potential to increase productivity, while considering their environmental effects.

Examples of intensification practices include improved varieties of crops, such as those with improved drought resistance or pest tolerance. For example, a study in Malawi found a positive correlation between improved maize varieties and own maize consumption, income, and asset holdings. Diversification strategies, such as intercropping and crop rotation, restore nitrogen to the soil, as well as allowing farmers to spread their risk across different crops.

Adoption of these different intensification practices in combination usually improves productivity more than an individual practice. A study in Ethiopia examined three intensification practices: modern seeds, inorganic fertilizers, and agricultural water management. The results showed that adoption of any of the three practices resulted in higher net crop income compared to non-adoption. However, the three practices were found to be complements and the highest net income was obtained when all three were jointly adopted.

A wide range of factors influence a farmer's decision to adopt these practices. For example, each additional year of education for the spouse of a head of household increases the probability of adopting more than two intensification practices by 12%. Farmers who relied on walking to markets are 9% less likely to adopt more than two sustainable intensification practices, indicating the importance of overall development (including road development), as further discussed in Chapter 17. Farmers' confidence in Ethiopia's Productive Safety Net Program, which allows farmers to work for food or cash in case their crops fail, increases the chance of adopting more than two practices by 20%. These studies highlight the need for increased education, availability of resources, and social programs to mitigate downside risk in order for smallholder farmers to efficiently manage their crops.

Part III: on-farm practices other than those related to food crop productivity

Besides adjustments made to crop production, farms have several other dimensions for adapting to climate change. The next three chapters examine livestock preferences, the role of biofuels in agriculture, and postharvest crop losses. These three chapters provide insights into often overlooked options for adapting to climate change.

Many farmers in SSA, particularly in Ethiopia, engage in mixed crop-livestock farming. Chapter 10, "Climate Change Adaptation and Livestock Activity Choice in the Nile Basin of Ethiopia," considers the impact of climate change on whether Ethiopian farmers manage livestock and, if so, what types of animals. Tsegazeab Gebremariam and Zenebe Gebreegziabher began by modeling the decision of whether to keep livestock. Out of 1,000 sample households, 95% of them held livestock. The authors reported their findings as the marginal effects for the average farmer. For example, a summer temperature increase of 1°C would increase the probability of the average farmer holding livestock by 4.8%, while a 1°C degree increase in the winter temperature would decrease the probability by 6.7%. As a whole, the decision to hold livestock was found to be more sensitive to temperature variations than to rainfall variations. When demographic variables were included, factors such as household size and education were found to increase the probability of owning livestock.

Next, to determine how climate affected which livestock were chosen, the authors used a multinomial logit model to predict the primary animal chosen. The results suggest that, with increasing warming, farmers will reduce milk cattle in preference for draught oxen. In addition, goats and sheep are less responsive to climate change than are dairy cattle and oxen. Based on the results of the multivariate probit model, the probability of choosing goats increases, while the probability of sheep-rearing decreases. The corresponding impacts of climate-related decrease in rainfall also lead to substitution of dairy cattle in place of chickens, and goats instead of sheep. The authors recommend that adaptation policies should be based on comprehensive knowledge of the structure and dynamics of livestock production systems.

Biofuels crops are another option by which farmers can earn income, but these crops can compete for land with food crops and livestock. Chapter 11, "The Distributive Effect and Food Security Implications of Biofuels Investment in Ethiopia: A CGE Analysis," examines whether there will be positive or negative impacts on food security and on the income of smallholder farmers as more land is used for biofuels production. This chapter uses a dynamic computable general equilibrium (CGE) model, which simulates the entire economy using a set of simultaneous equations.

Using the CGE, Zenebe Gebreegziabher et al. produced scenarios that hypothetically increase the production of four biofuel crops – sugarcane, jatropha, palm oil, and castor bean – relative to a baseline. Three of the scenarios include "spillover effects," which can help smallholder farmers by improving infrastructure, farming techniques, or access to agricultural inputs. Depending on the agroecological zone and the particular crop examined, the spillover effects can increase the production of food cereals without increasing cereal prices. When spillover effects are considered, biofuel investment tends to improve the welfare of most rural poor households. Urban households benefit from returns to labor under some scenarios. The authors emphasize the role of road construction, so that biofuels crops will be raised on land that is currently underutilized, rather than by displacing smallholder farmers.

A third potential area for improving food security in the face of climate change is to reduce postharvest losses. Postharvest losses are an often overlooked part of food insecurity. Chapter 12, "Climate Change and Post-Harvest Agriculture," presents estimates that between 10% and 20% of crops in sub-Saharan Africa suffer from postharvest losses, equating to billions of dollars' worth of food. Next, the authors look at factors that influence postharvest losses. For example, warm weather can support pests and accelerate biological deterioration of crops. Unseasonal rains can dampen the matured crop before harvesting and result in mold growth. Poor harvesting and handling practices also expose crops to postharvest losses. Harvesting crops before maturity encourages fungus growth. If farmers pile their moist harvest to dry, infections from a small population of the harvest can quickly spread to the entire harvest. According to a study from Tanzania, the mean postharvest loss is about 11.7%, with most of that occurring during the storage phase. The study also found strong correlations between poor harvesting practices and postharvest losses, suggesting a role for educating farmers about good handling and storage practices.

Part IV: gender issues

The gender gap between male-headed households (MHHs) and female-headed households (FHHs) is intensified due to climate change. While there is already a difference in welfare in the two categories of households, future effects of climate change will only increase the disparity. Because of disadvantages in income, resources, and education, FHHs are projected to be more vulnerable to the effects of climate change. The next two chapters will further explore this disparity.

Chapter 13, "Contribution of Smallholder Agriculture to Daily Calories, Macronutrients, Minerals and Vitamins in Male- and Female-Headed Farm Households in Sub-Saharan Africa," aims to shed light on the nutritional gender gaps in several countries across sub-Saharan Africa. The UN's Food and Agriculture Organization estimates that removing barriers facing female heads of farm households could raise productivity by 20%–30%, increase food production by 2.5%–4%, and reduce food insecurity by 12%–17%. Some of the reasons for the gender gap include a lack of income, land, and resources, as well as limited education and agricultural knowledge. In particular, the study looks at the differences among male-headed households, de jure female-headed households, and de facto female-headed households. The de facto heads are female heads of households who are married, but their spouses are away; de jure female-headed households are headed by women who are divorced, widowed, or never married.

To study these gender gaps, Byela Tibesigwa et al. use data from Ethiopia, Malawi, Nigeria, South Africa, Tanzania, and Uganda. They begin by observing that crop farming provides an average of 1034 kilocalories (kcal) per household member per day, representing 39% of the daily energy requirement. MHHs have much higher average kcal per household member per day than either type of FHHs. Consumption gaps are larger between male- and de facto female-headed households than they are between male- and de jure female-headed households. Additionally, de jure female-headed households have a higher kcal average than de facto female-headed households, suggesting that de jure female-headed households benefit more than de facto female-headed households from crop farming. In rural areas, the gender gap pertaining to calories, macronutrients, minerals, and vitamins is greater for de facto female-headed households than for de jure female-headed households. In urban areas, the gender gap is similar regardless of de facto and de jure status, except for vitamins. The results reflect, in part, the different crops grown by male and female farmers, and suggest the importance of considering gender, household structure, rural versus urban location, and the types of crops and nutritional deficiencies when developing policy responses to food insecurity in the face of a changing climate.

In Chapter 14, "Gender-Differentiated Impacts of Climate Variability in Ethiopia," Tesfamicheal Wossen addressed the question of whether climate variability affects the welfare of male-headed households and female-headed households differently. The author also examined the role of adaptation methods and policies in mitigating the detrimental effects of climate change on both male- and female-headed households. Wossen employs a micro-simulation approach that captures farm-level impacts of climate variability.

Wossen found a significant difference in the well-being of MHHs and FHHs. MHHs enjoyed more education, increased initial endowment, and decreased dependence on government aid and safety nets. These differences then translated into differences in climate vulnerability. In the simulation, the author first established a base scenario without any climate variability. In the scenario with climate variability, FHHs saw an average of 12.4% decrease in household income, while MHHs saw a 5.7% average decrease. The author then added adaptation policies into his simulation to examine any gender differences. Each of three policies examined – short-term production credit, information about new crop varieties, and fertilizer subsidy – offsets some of the effects of climate variability, with similar or slightly greater benefits for FHHs.

Part V: the broader development context

The last group of chapters examines climate change through the lens of broader development trends. These chapters examine adaptation strategies that are related to large-scale changes in policies and economic activity. The chapters look at land certification, migration, and growth trends.

Chapter 15, "The Land Certification Program in Ethiopia: A Review of Achievements, Constraints and Opportunities," aims to analyze, synthesize, and critique studies that looked into the land certification program in Ethiopia, one of the largest in the world. Before the program, Ethiopia faced frequent land redistribution and a ban on land market activities, which discouraged investment and had negative effects on productivity. To solve this problem, the government created the current land certification program. The current program grants farmers essentially perpetual user rights to the land, notably giving women equal legal rights.

The resulting increase in tenure security allowed for improvement on a variety of parameters. The program increased perceived levels of tenure security and trust in the regional and national governments. Certification reduced border and inheritance disputes, in part thanks to women placing their names and pictures on the land certificates. The program increased the likelihood that farmers would plant trees as well as invest in soil and water conservation. The official certification of land also led to increased land productivity by allowing farmers to benefit from long-term investments, such as perennial crops or irrigation technologies. Lastly, the program saw an increase in the allocative efficiency of land because of the legalization of renting out land. Because female-headed households had less tenure security prior to the program, female-headed households saw a significantly greater increase in many of these parameters when compared to male-headed households. Overall, the findings show that the program has substantially improved tenure security as well as gender equality.

Tenure security is part of a structural labor transformation toward off-farm employment. One effect of tenure security is that farmers can leave the land for off-farm employment without fear of losing land rights. Off-farm employment often entails migration. The ability to migrate gives farmers another option for responding to climate change. To mitigate risks from climate change, smallholder farm households often send at least one household member away as a migrant. In Chapter 16, "Migration as an Adaptation Strategy to Weather Variability: An Instrumental Variables Probit Analysis," Yonas Alem et al. investigate the impact of climate variables on the decision by households in Ethiopia to send away a member as a migrant.

This study uses exogenous, objective measures of rainfall, and applies an instrumental variables approach to control for endogeneity of income. The authors separate rainfall variability, as a determinant of ex ante migration decisions, and rainfall levels/shocks, as a determinant of ex post migration decisions. Ex ante, prior to observing the current year's rainfall, a household may make the decision to send a migrant, based on past observations of rainfall variability. Meanwhile, after observing the current year's rainfall, a household may make an ex post decision to send a migrant.

Results suggest that households in rural Ethiopia adapt to weather variability through migration. Ex ante, smallholder farm households that live in places with higher rainfall variability are more likely to send away a household member as a migrant. This effect is economically important and statistically significant at the 1% level. The authors also find suggestive evidence of an ex post impact of low rainfall in the year preceding migration, but this result is not statistically significant.

Finally, in the context of a structural transformation, Chapter 17 looks to the future. Adoption of modern technologies, whether in farming or other sectors, is constrained by inadequacies in three categories: human capital, infrastructure, and institutions. In Chapter 17, we suppose that East Africa continues to grow, educate, and urbanize at its recent rates. We concentrate on education, infrastructure, and institutions because there are policy levers that work directly on them. We think that agricultural adaptation to climate change will be driven (or held back) by the same factors that promote or retard development more generally. The policy implication is that investing in schools, roads, and other infrastructure and institutions will reap benefits in adapting to climate change.

Part VI: conclusion and policy implications

The technologies that can increase farm production in response to climate change are, to a great extent, the same technologies that would increase farm productivity even if climate change were not an issue. There are a number of policy levers that can promote adoption of such technologies. In addition, there are institutional changes that can promote agricultural development in the context of overall development. Chapter 18 concludes this book by discussing several themes for agricultural development, climate change adaptation, and overall development in SSA.

Part I

Climate science, agronomic, and agroecological factors

2 Understanding adaptation to climate change

Peter Berck

What changes in agricultural systems will be adaptive to climate change? That is the major question that agriculturists and pastoralists will have to face in the coming decades. To understand responses or potential responses to climate change, it is best to begin at the beginning and discuss what is meant by climate change and what is meant by adaptation. Given that the bulk of climate change hasn't happened yet, we then discuss how we model adaptation. First, we model climate change and then we take a few variables from those models, such as temperature or precipitation, and use them in a model of plant behavior. Of course, models of plant behavior depend on more than the few variables we typically take from models of climate change. Once we understand the likely plant behavior, such as lower yield of maize with higher temperatures, then we can discuss adaptation. What happens when the farmer irrigates, changes crops, or protects against flooding?

What is climate?

Weather is what one experiences every day. Often weather is summarized into a few numbers: high and low temperature, relative humidity, precipitation, and wind speed and direction. These are the common elements of a weather report. Each day has different weather. Climate is also a familiar construct. When we say Phoenix has a hot and dry climate, we mean that the weather in Phoenix is hot and dry most days. Climate is a statement about what weather one expects to experience.

An example that has only two types of weather is enough to explain climate. The two types of weather are rainy and clear. For instance, rainy days are coldish with a high of 20°C and a low of 12°C. Clear days are sunny with a high of 35°C and a low of 30°C. With just these two types of days, there could be many climates. Some places would have a predominance of rain and we would call that a place with a rainy climate; others have a predominance of sun and those would have a sunny climate. Climate, in this simple example, is just the likelihood of each type of weather. For instance, a 0.2 chance of the rainy weather and a 0.8 chance of the sunny weather are a description of the climate. Climate change would be, for instance, a change to a .3 chance of rainy and a .7 chance of sunny.

To formalize the distinctions among weather and climate, let weather be W, a vector that describes the weather of one day (e.g., high and low temperature, wind speed and direction, precipitation). The probability of seeing weather W is f(W). Climate is summarized by f(W), the likelihood of seeing different types of weather. Climate change is changing f to f', a new probability density function (pdf).

Climate is often crudely summarized as expected weather or even expected temperature – that is, $E_f[W]$. The subscript f means that we should take the expectation with respect to f. Climate change is then $E_{f'}[W] - E_f[W]$, where the first expectation is taken with respect to f'. Generally, this does not line up with the idea that climate change is a change in the pdf. For instance, using just the mean of weather ignores the variance, skewness, and higher moments of f. So, using expected weather as a measure of climate change ignores the possibility that rain will become more concentrated on a few days.

Farmers are concerned with weather. Their planting decisions are influenced first and foremost by the actual weather experienced in a given year. A cold, wet spring puts off the planting of corn. They are also influenced by predictions of future weather. The type of crop one chooses depends on things like the number of frost-free days, the number of days with very high temperatures, the period between rainstorms, and so on. These are all statistics that meaningfully summarize f to farmers. But farmers are definitely interested in f, not just in $E_f[W]$.

What is adaptation and how does it relate to efficiency?

Adaptation means changing practices in response to climate change. It could be adding irrigation or changing crops or adopting low tillage. In general, it could be any change in the vector of inputs and outputs that is made because the climate changed.

Consider a farmer's choice between how much land to use for livestock and how much land to use for row crops. By changing the land allocation, a farmer changes the farm's outputs of livestock and row crops. Allocating more land to livestock is a potential response to a climate change toward hotter and drier weather. A choice of two outputs gives the standard production possibilities frontier, which is the solid curve in Figure 2.1. Both points B and C are on the frontier. Point C has relatively more livestock and point B more crops. A way to get from C to B would be to change the mix of crops and grow less of a hay crop and more of a grain crop. Another way would be to change the type of grain grown from one with a large amount of usable fodder – teff would be an example – to one with little fodder, like a high-yielding, modern short wheat. More extremely, the farm could dedicate land entirely to pasture. Again simplifying the problem, the slope of the straight line in Figure 2.1 is the relative prices of the two outputs and the tangency at B is the best output point for this farm, which maximizes the farm revenue.

Figure 2.1 also has point A, which is a point below the production possibilities frontier. Most developing country farms produce at a point such as A, in the sense that changing varieties, fertilizer, and land preparation would move toward the frontier. In other words, most poor smallholder farmers in sub-Saharan Africa

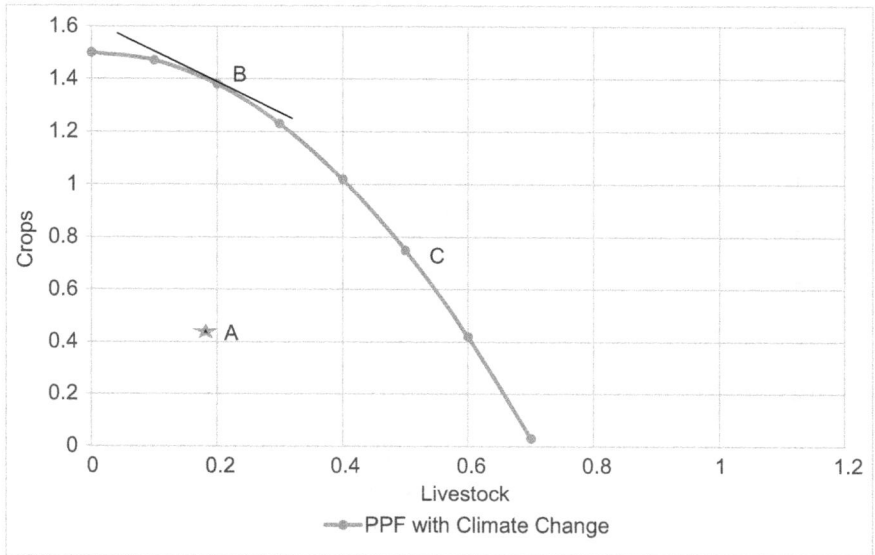

Figure 2.1 Production Possibility Frontier

(SSA) are operating below the frontier. For an example, in a relatively dry row crop region, allocating farm labor toward digging pits in which to plant, which stores more water, moves output up, closer to the frontier. It does so even after accounting for the uses from which the labor was diverted. Agricultural development is about moving from a point such as A to points such as B and C. Improved seeds, subsidized fertilizer, crop insurance, extension service, water harvesting, and contour plowing are all part of the packages of techniques to move a farm to the efficiency frontier.

Global warming changes the location of the frontier. Figure 2.2 shows the original frontier as a solid line and the frontier after global climate change as a dashed line. It is drawn so that livestock is favored and crops are disfavored by climate change. This would correspond to a change to a drier climate. Keeping the price line the same, the tangency is now at point D, giving the farm even more livestock relative to crops. It is no longer possible to produce at point B.

Now let us return to our inefficient farm, producing at point A. Before climate change, it should have adopted techniques to move toward point B; now, after climate change, it should adopt techniques to move to point D. But what of point A itself?

Point A itself shifts inward. Doing exactly what was done in a cooler, wetter climate results in losses in output in a hotter, drier climate. This is shown as point A' on Figure 2.2.

Should the farmer adapt to climate change by trying to get back to point A? Point A was inefficient in the first place, so why adapt by going back there? As Figure 2.2 is drawn, the farmer should aim for point D. All of the farmer's efforts

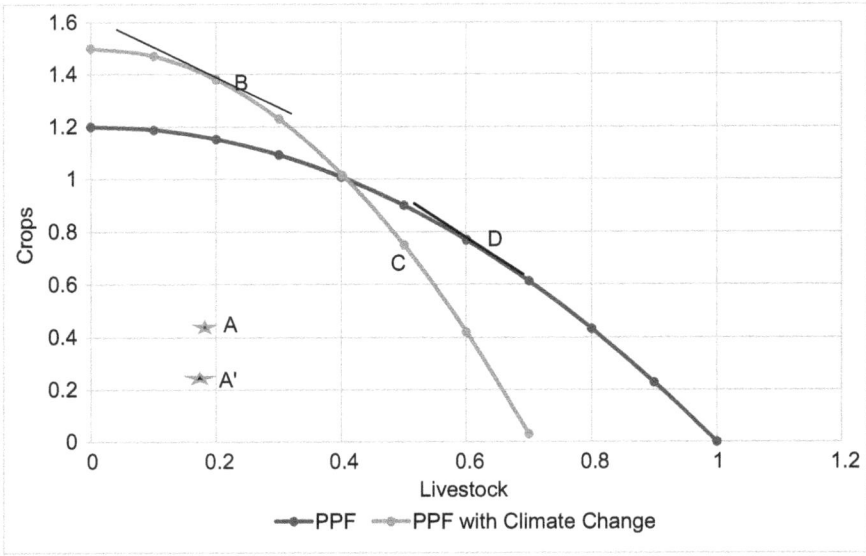

Figure 2.2 Production Possibility Frontier with and without Climate Change

to get toward the frontier and hopefully reach point D are described in the literature as being adaptations to climate change. They are properly called adaptation to climate change in the sense that they go to the frontier after climate change and are caused by the climate change–induced losses of moving from A to A'. However, they are mostly also things needed to move to the frontier before climate change.

Would agricultural extension be a good idea before climate change? Yes. After climate change? Yes. That is the answer for a great many agricultural improvements. So, many of the things that should be done looking forward to climate change should have been done even without climate change. Some things will be different. With enough drying, building bunds to slow down water runoff no longer makes sense. Nor would adopting high-yielding corn if there is no longer water for corn. But, with point A' far enough inside the production possibility frontier, many of the ways of increasing output work the same with and without climate change. Thus, many of the strategies that are advocated for agricultural development are also adaptive to climate change.

Predictions of climate change

Predictions of climate change come from global circulation models (GCMs). GCMs model the distribution of heat in the atmosphere. The beginning point for modeling is the amount of heat that comes from sunlight, which is either is absorbed by the atmosphere, land, and ocean, or radiated back into space.

Global warming happens when less of the incoming radiation is reradiated back into space. To see where that heat goes, GCMs divide the atmosphere into a large number of cubes, and then model how heat is transmitted from one cube to another. They also model how heat moves from the atmosphere to the land and the ocean. In addition to tracking heat, the models track the movement of air (wind) and water vapor (leading to rain). GCMs show a wide range of potential effects for climate change.

There are many GCMs and they are run under many different heat input conditions. They are grouped according to how much additional energy the atmosphere retains, which is the difference between insolation (solar radiation received) and outgoing radiation (called radiative forcing), compared to a baseline before global warming. The more CO_2 equivalent (CO_2e)[1] in the atmosphere, the greater the forcing. The groupings are called representative concentration pathways or RCP. They are grouped by the amount of radiative forcing. For instance, with 2.6 watts per m^2 of radiative forcing, the designation would be RCP2.6. RCP2.6 is associated with 421 ppm of CO_2e in 2100. With less control of CO_2e, there would be more radiative forcing. RCP8.5 leads to CO_2e of greater than 1,000 ppm and is the more extreme scenario. RCP8.5 is the extreme scenario and will result if emissions remain mostly unabated. That scenario is associated with 936 ppm of CO_2 in 2100 – a number that would rise in the years following. Models are run by different groups, all with the same input scenario in terms of radiative forcing. The different models give different outcomes.

Because there are many models and many predictions, it is common to use an average of the models. CMPI5 multi-model mean projections for RCP2.6 show a pretty uniform temperature increase of just under 2°C for all of Africa (Niang et al., 2014). In terms of rainfall, northern SSA is projected to increase precipitation by under 20%, while southern SSA is projected to decrease precipitation by under 20%. The report characterizes these precipitation projections as not being significant because they are small relative to baseline variation and/or less than two-thirds of the models agree on the outcome. However, the models lend some credence to the projection that there will be more variability in rainfall. That is, there will be more extreme rainfall events, leading to flooding, and more extended dry periods, leading to drought.

The models depend on dividing the atmosphere into cells. These cells are quite big (generally over 100 km) and so the models do not provide highly place-specific information. A separate step, downscaling, takes the output of GCMs and provides information at a scale relevant for agriculture – a few km rather than 100 km.

Weather and crops: data sources and their limitations

How do crops respond to a change in climate? To estimate the response to climate change, one needs to know the response of crops to weather. Measuring weather and matching those measurements to the crops on the ground require data. Without knowledge of weather and crops, there is no way to find the change in crop yield caused by a change in weather. Without knowledge of yield change, there is no way to discuss what measures would mitigate the change in yield. Africa is a particularly difficult place to obtain good data on weather and crops.

Data

There are five main sources of data on cropping patterns and crop yields in Africa: satellite, weather stations, administrative data, survey data, and experiment station data. Each of these sources has its limitations, particularly when measurements are required for historic periods.

Satellite data for land cover depends on imagery and interpretation. While there is imagery for the historical period from Landsat and modern imagery from many programs, matching the imagery to crops requires knowing the truth on the ground. Satellite information is the intensity of light of various colors. The color readings for each crop are different, so, with enough plots of land in known crops, statistical methods allow a good match of color signals seen by the satellite and crop type of the ground. The USDA's National Agricultural Statistics Service crop data level for the US Midwest gives a good idea of how well this works. Despite the large effort by multiple US agencies, the crop layer provides good identification for only two major crops: corn and soy. The number of plots on the ground is not sufficient to distinguish pasture from irrigated pasture, from forest, from oats, hay, and so forth. The problem in Africa is far greater. Africa is bigger than the United States and China combined. It also has many more important cropping systems than the Midwest – for instance, millet, sorghum, teff, cassava. Because of the multiplicity of cropping systems and the scale of Africa, there are no accurate satellite crop maps for large regions of Africa.

The potential for satellite data is immense. For instance, at a very fine scale, satellites can read soil temperature and insolation. By finding the change in soil temperature across a morning, an analyst can find out how much soil moisture there is. This permits near real-time drought measurement (Anderson et al., 2011). Further matching of known crops to imagery would allow accurate crop maps, backcasted to the historical period.

The situation with weather stations is that the density of stations is far less than in North America. Local weather is constructed by interpolating between weather stations. Where the stations are close together, as in North America, the interpolation is sufficiently accurate just using information such as elevation. In Africa, using general circulation models provides additional accuracy. However, the accuracy of historical weather across large crop-growing regions in Africa is substantially less than that in North America.

Cropping patterns and yields are typically recorded as administrative data, often at the county level. For North America, for instance, crop coverage data can often be found at the field or even the 30m x 30m pixel level, yet crop yield is found in administrative data only at the county level. Counties can be hundreds of kilometers long and wide. For instance, in this volume, Kabubo-Mariara and Kabara (see Chapter 4, this volume) use Kenyan county-level crop data.

Household survey data yields fine-scale information on cropping patterns and choices for a limited number of households with limited geography. For instance, Komba and Muchaponda (see chapter 7, this volume) examine adaptation in Tanzania with a sample of 556 households.

Finally, there are agricultural experiment stations located across SSA (Lobell et al., 2011) and these stations keep good records on weather and crops, making them ideal for finding the effect of weather on yields. The drawback to the use of experiment stations (or their cooperating farmers) is that these farms are run much closer to optimally than are the small plots of the average farmer. Experiment stations are designed to figure out how to get the most from soil and climate and therefore tend to have better control of fertilization, seeds, weeding, pests, and so on. Put another way, if SSA in general had the same yields as the experiment stations, concern about food would be much less in SSA.

The fundamental lack of weather stations and of crop reporting makes it much more difficult to find an empirical effect of weather on crop yields in Africa. However, the lack of data in Africa is not the only thing making it difficult to predict yield in response to weather. Often, the outputs from the GCMs are not sufficient to predict plant growth because growth depends on much more than temperature and precipitation.

Model of crop growth

The basic model of plant growth is that given sufficient water and nutrients, crop growth is proportional to intercepted solar radiation. As plants grow, they have more leaf area, which intercepts more radiation. Solar radiation is not changed (much) by climate change, but temperature and rainfall are. How do these changes affect plant growth?

The basic model is that, given sufficient water, plants grow. The biggest threat from climate change is to reduce the number of days on which there is sufficient water. Losing sufficient water for plant growth is called drought and the agronomic community has established many indices of drought based upon rainfall (water supply) and temperature (water demand). The most widespread of these indices is the Palmer index.[2] The indices are far from perfect because of how plants get water. Water falls from the sky; some of it is stored by the soil and some runs off; and then the plant retrieves the water from the soil. Deep soils matched to plants that can access deep water (e.g., chickpeas) can go longer between rainstorms than can plants in shallow soils. Similarly, soils with good water-holding capacity (loam and clay) can go longer without rain than sandy soils. Climate change threatens to make rainstorms less frequent. Adaptation to climate change aims to make the soil store more water and release it to the plants when needed.

Temperature is affected by climate change. In North America, which is more heavily studied, higher temperatures cut two ways. First, the North American growing season is limited by cold. Frost ends growing in the fall and cold soil prevents growing too early in the spring. So higher temperatures are helpful on this front. SSA simply isn't cold, at least in most of the places where most of the food is grown, with some exceptions; see, for example, the finding that frost stress discouraged adoption of sustainable intensification practices (Teklewold et al., 2013). So, there is no benefit from a generally warmer climate. On the negative side, warmth, per se, increases evaporation. Warmth increases water demand from soil, yet it does little for plant growth (recall that plant growth is determined by insolation – the incoming solar radiation – not by residual heat). Extreme warmth also kills plants. Even with sufficient water, maize wilts when the temperature is hot enough.

Plant respiration – the process by which plants use the sugars they make to grow – is also sensitive to temperature. Nighttime temperatures are important because respiration cannot happen when the temperature is too low.

Plants are also sensitive to humidity. The inside of plants is wet and the air is usually much drier. This difference in vapor pressure makes it possible for plants to transpire and lose water to the air. The benefit to the plant is that this process pulls nutrients from the plant's roots to its shoots. While GCMs do provide a measure of the vapor pressure deficit, a measure of the driving force that pulls water out of plants, it is so far rare for them to be incorporated in yield equations used for climate change.

Finally, plants can't grow if they are physically knocked over or drowned. Extreme weather in the form of floods or hail is an important influence on yield.

There are two very different traditions for modeling weather-dependent plant growth. Process models closely follow plant science in modeling all of the important processes in a plant, such as transpiration, respiration, size of leaves (leaf area index), and size of roots. Regression models rely on the covariance of weather and yield when other factors have been accounted for. Regression models are much less data-intensive than process models. This is because of the use of fixed effects. In a regression model, the effect of soil is just a dummy variable for place. In a process model, the properties of the soil need to be modeled – how much water can it hold, and so forth. Process models aim to show how the crop will perform given a set of conditions, such as how much fertilizer is applied. Regression models show the yield with the actual amount of fertilizer applied; they implicitly include farmers' behavior as the human factor. The models encountered in this book are regression models.

The evidence on maize and temperature is now voluminous. The seminal paper is by Schlenker and Roberts (2009), again for North America (but see Lobell et al., 2011, for Africa). The authors examine how corn yield varies with growing season temperature and rainfall. Their key insight is that it is only very high heat that causes reductions in yield. The model shows response to temperature in bins – for example, the plants' yield given that there are so many hours between the temperatures of 28°C and 30°C. This gives very different results than a quadratic (or other low-order polynomial) would give because those methods impose a shape on the data that is different from the true response. Indeed, shortfalls in yield are well explained by abnormally high temperature. The results show that every hour over 29°C leads to a noticeable decline in yield. In North America, however, the exceptionally high temperatures come with less rain. So, they proxy for drought. Indeed, rerunning the Schlenker and Roberts (2009) regressions with an interaction term of the form temperature/rainfall shows the significance of even this primitive drought index. We can conclude from this that it is the combination of high temperature and low rainfall that leads to plant failure. In the African case, drought is definitely the phenomenon that causes plant losses and, therefore, needs to be modeled explicitly.

Adaptation

There are a number of major ways to change the agricultural economy in response to a changing climate. Where climate change is adverse to agriculture – for example, where a dry place is becoming drier – adaptation can restore some but not all of the productivity of the system.

Changing crops

How does one adapt to high temperature, per se? By changing crops. Corn is more heat-sensitive than sorghum. So, if heat were really the problem, then moving to a better-suited crop would be the obvious adaptation. The American Midwest has summers that can be both hot and dry. When it is hot and dry, sorghum outyields maize. However, in more normal weather, it is the other way around. So, is planting sorghum adaptive? Yes, it is better than sticking with maize if the climate is hotter and drier. In a study of yields in Kansas (Staggenborg et al., 2008), the relationship was summarized in terms of maize yield. As it gets hotter, maize yield and sorghum yield both decrease, but maize yield decreases at a faster rate. When it is hot and dry enough so that maize yields only 6.4 Mg/ha or less, then (and only then) sorghum yield will be more than corn yield. In good growing conditions, maize reaches its maximal yield. The maximum maize yield in their experiments was 14.6 Mg/ha. The weather that leads to maximal maize yields of 14.6 Mg/ha leads to sorghum yields of only 10.6 Mg/ha. In summary, hotter and drier growing conditions lead to less maize yield, but it makes sense to switch to sorghum only when growing conditions are so poor that maize yield has been cut by more than half. Adaptation, in this case, never restores the high yields of good growing conditions.

Adaptation by changing crops can also be viewed as adaptation by changing time in the ground. Early season maize avoids hot late-season temperatures. Early season millet in Ethiopia does the same. Barley is the short-season crop used in California for rain-fed hot and dry systems. These crops are adapted by using fewer days in the growing season and thus avoiding rainfall failures or extreme heat later in the season. In terms of the basic model of plant growth, since they are not in the ground, they are not receiving insolation, and are not growing. The adaptation is better than non-adaptation, but it is making the best of a bad situation, not restoring the status quo ante.

While response to weather is the most noticed aspect of a plant when one thinks of adaptation to climate change, response to pests can be even more limiting. Many sorghums are susceptible to witch-weed (Striga) and that has become a limiting factor in many parts of Ethiopia (Adugna, 2007). Striga is favored in more marginal land and so climate change to hot and dry favors Striga. Adaptation is to choose resistant sorghums, to hand weed, to fertilize, and to use chemical suppression, if possible.

Water and water management

The goal of soil and water management is to make damp soil available to plants at all times. Deep soils of good water-holding capacity naturally accomplish this when matched with deep-rooted plants. Soil itself is manipulated by adding organic matter (measured as soil carbon) to make it hold more water. However, most adaptation focuses on regulating how quickly water moves across soil.

Making use of rainfall

Slowly moving water percolates into the soil and becomes available for use. Quickly moving water erodes soil, lowering its depth and fertility, and reducing its water-holding capacity. Methods to slow and trap water include digging pits and planting in them, leaving stubble on the land, planting grass buffer strips, and building earthen and stone bunds (walls, terraces.)

On the other side, too much water requires drainage – that is, a way for the excess water to slowly leave a field without waterlogging or eroding it. Climate change is expected both to produce a drier climate on average in SSA (but just barely so) and to produce a climate with more extreme rainfall events. So, adaptation will require extensive work on reshaping fields to both harvest water and withstand flood events.

Summer fallowing concentrates two years of rainfall on one year's crop. In Ethiopia, fallowing during the "short rains" season concentrates the available moisture on a crop grown during the "long rains." In order to fallow, one must suppress weeds. Weeds will grow in the fallow season and will use the moisture that the farmer is trying to conserve; plus, they will create a seed bed of weeds to compete with the main season crop. Weed suppression is traditionally done with plowing. This is the reason why Ethiopian farmers go over their land with the plow many times; they are killing the weeds and concentrating the moisture from both rainy seasons on one crop. The alternative is a no-till or low-till system. The downside of plowing is that it loosens the soil and makes it susceptible to erosion. The downsides of no-till systems are that they use the stubble to enrich the soil and deflect rain drops (instead of feeding the stubble to animals) and they require glyphosate or hand-weeding to kill the weeds. Glyphosate requires accurate application, using a sprayer. It costs money. And some now think that it is a carcinogen. Hand-weeding diverts labor from other activities.

Irrigation

Water management can also be accomplished through irrigation; this is the major method to deal with intra-season climate variability. Large-scale irrigation is not available to most smallholder farmers in Africa; most of them depend on rainfall. However, there are small-scale methods of water management in use on a fair number of smallholder farms. These include borehole wells or diversions from small waterways. In all these cases, adaptation requires conveyance of the water.

Water application methods have different costs and different efficiencies. Efficiency is measured as the ratio of evapo-transpiration (useful water) to total applied water. Water that is not used either percolates deeply into the soil or runs off. Flood irrigation is the cheapest method of putting water on land; however, it makes high points in a field too dry and low points too wet. This reduces irrigation efficiency and yield. In developed countries, laser leveling solves the problem of wet and dry spots. Sprinkler systems are a more expensive way of regulating uniformity of irrigation. Drip irrigation (including underground drip) gives the greatest uniformity and the most control over irrigation. It is also the most expensive. Drip irrigation is currently used in some places in Africa – for instance, in the citrus industry in the Republic of South Africa. The system is very water-efficient, putting water on the plants' roots, not on bare ground, and providing uniform application in time and across the field (Taylor and Zilberman, 2017). This technology is ideally suited to increased variance in rainfall and to a drier climate. However, drip irrigation requires a constant, clean source of water, perhaps a farm pond, and equipment to deliver the water. Equipment, including computerized controllers, requires education. Flooding a field, on the other hand, requires only opening and closing gates. In general, the possibilities for adapting to climate change through irrigation at the farm level depend on the ability of farmers to pay for and manage sophisticated technology. At present, few of these resources are available to smallholder farmers.

Migration, consolidation, education, insurance, and finance

As SSA goes through growth and modernization, there are several processes that will increase agricultural production. These processes, such as increased availability of insurance and credit, are not necessarily a reaction to climate change; however, they do restore some of the productivity lost to a changing climate.

Building roads and land titling would create an economic transformation of the agricultural sector from a predominantly smallholder subsistence system to a moderate-scale commercial system. This transformation would allow farms to specialize. As in North America, the livestock would leave the grain farms and either graze or be fed in feed lots. It would no longer be necessary to grow hay crops on every farm; farms would specialize into corn-bean or wheat systems based on soil and weather. Productivity would rise because the crops would be better suited to their climate and soil. The land that remained in farms would be more intensively cultivated, meaning appropriate use of fertilizer, better seeds, better weed control, and so forth. Agrarian transformation would be adaptive to climate change, even though this is not the reason this transformation is occurring.

Similarly, the expansion of the financial and insurance systems is adaptive to climate change without climate change being the reason for the expansion of these systems. Better financial systems mean higher availability of credit. Better insurance systems mean that the risk of agricultural investment is lessened. Together they can lead to intensification of agriculture.

Finally, increasing education is adaptive to climate change. Modern agricultural methods require reading and knowledge of basic chemistry. So, increased education can lead to adoption of intensive methods that restore some of the losses caused by adverse climate change.

Conclusion

Agricultural adaptation to climate change is the changing of agricultural techniques in response to a changed climate. In this chapter, we began by distinguishing climate from weather, climate being the distribution function of weather and weather being the precipitation, temperature, and so forth that we experience every day. Adaptation is formally defined as a change in production techniques (a change in either inputs or outputs) in response to climate change. For developing countries, the current techniques for production are largely suboptimal, so most of the technique changes now suggested to improve productivity would improve productivity after climate change as well. Climate change is largely in the future, so studying adaptation to climate change is really studying what farmers would do if predictions of climate change came to be true. The studies are a modeling exercise and they depend on general circulation models (GCM) of climate change. The outputs of these GCMs are used as inputs to models of plant growth. If plants (e.g., maize) were unaffected by climate change, there would be no reason to change techniques. Crop models depend on data and data on crop yield is particularly sparse in SSA. The models also depend on underlying agronomic science, such as plant response to humidity, heat, and water. Finally, given the GCMs and the plant models, farmers are assumed to choose new techniques that are adaptive to the after-climate change yield of crops.

The major adaptations envisioned for agriculture are changing agricultural outputs and changing water management. With a hotter and drier climate, a change in crops would be toward those that are more heat-tolerant. Heat tolerance is usually achieved at a loss in yield. For instance, barley is in the ground a shorter amount of time than maize or wheat and is often harvested before the hottest and driest part of the season. Because it uses less of the season, it has less chance for growth and yields less. With a hotter and drier climate, a change in agricultural systems would be toward grazing systems. An alternative (or addition) to changing crops is to change water management. For a hotter and drier climate, trapping more water on the land by using pits would be adaptive. So would modern technology like drip irrigation.

Finally, there are off-farm changes that may be undertaken in response to climate change or adopted even without climate change. Examples include migration, farm consolidation and specialization, and education. These are often included in the adaptation literature because they facilitate change in the agronomic system.

Notes

1 CO_2e is the equivalent in CO_2 of the sum of all greenhouse gasses.
2 The Palmer Drought Severity Index (PDSI) and Crop Moisture Index (CMI) are indices of the relative dryness or wetness affecting water sensitive economies. The data is provided in graphical and tabular formats, for the contiguous United States. www.cpc.ncep.noaa.gov/products/monitoring_and_data/drought.shtml; also www.drought.gov provides the current Palmer drought classification for the United States.

References

Adugna, A. 2007. The Role of Introduced Sorghum and Millets in Ethiopian Agriculture. *Journal of Semi-Arid Tropics Agricultural Research* 3(1). http://ejournal.icrisat.org/mpii/v3i1/news/The%20role%20of%20introduced%20sorghum.pdf

Anderson, M.C., C. Hain, B. Wardlow, A. Pimstein, J.R. Mecikalski and W.P. Kustas. 2011. Evaluations of Drought Indices Based on Thermal Remote Sensing of Evapotranspiration over the Continental United States. *Journal of Climate* 24(8): 2025–2044.

Lobell, D.B., M. Bänzinger, C. Magorokosho and B. Vivek. 2011. Nonlinear Heat Effects on African Maize as Evidenced by Historical Yield Trials. *Nature Climate Change* 1: 42–45.

Niang, I., O.C. Ruppel, M.A. Abdrabo, A. Essel, C. Lennard, J. Padgham and P. Urquhart. 2014. Africa. In *Climate Change 2014: Impacts, Adaptation, and Vulnerability, Part B: Regional Aspects: Contribution of Working Group II to the Fifth Assessment Report of the Intergovernmental Panel on Climate Change*, edited by V.R. Barros, C.B. Field, D.J. Dokken, M.D. Mastrandrea, K.J. Mach, T.E. Bilir, M. Chatterjee, K.L. Ebi, Y.O. Estrada, R.C. Genova, B. Girma, E.S. Kissel, A.N. Levy, S. MacCracken, P.R. Mastrandrea and L.L. White. Cambridge: Cambridge University Press, pp. 1199–1265.

Schlenker, S., and M.J. Roberts. 2009. Nonlinear Temperature Effects Indicate Severe Damages to U.S. Crop Yields under Climate Change. *Proceedings of the National Academy of Sciences of the United States of America* 106(37): 15594–15598.

Staggenborg, S.A., K.C. Dhuyvetter and W.B. Gordon. 2008. Grain Sorghum and Corn Comparisons: Yield, Economic, and Environmental Responses. *Agronomy Journal* 100(6): 1600–1604.

Taylor, R., and D. Zilberman. 2017. Diffusion of Drip Irrigation: The Case of California. *Applied Economic Perspectives and Policy* 39(1): 16–40.

Teklewold, H., M. Kassie and B. Shiferaw. 2013. Adoption of Multiple Sustainable Agricultural Practices in Rural Ethiopia. *Journal of Agricultural Economics* 64(3): 597–623.

3 Mapping vulnerability to climate change of the farming sector in the Nile Basin of Ethiopia

A micro-level perspective*

Zenebe Gebreegziabher, Alemu Mekonnen, Rahel Deribe Bekele, Jonse Bane, and Samuel Abera Zewdie

Introduction

This chapter analyzes vulnerability to climate change of Ethiopian farmers across different agroecological zones by constructing composite vulnerability indices, which integrate both the biophysical conditions of the farming regions and the socioeconomic conditions of the farm households. Findings show that, among the four major agroecological zones in the Nile Basin of Ethiopia, the humid lowlands and drought-prone highland areas are the most vulnerable zones. Findings also show that *enset*-based farming[1] (a local farming system where *enset* is the dominant crop) in moisture-sufficient highland areas has the highest adaptive capacity, while the humid lowland zone is the lowest in terms of adaptive capacity to climate change.

Vulnerability and adaptation options to climate change in Africa and other developing countries have received increasing attention. In particular, Africa is viewed as the most vulnerable region to climate variability and change. Africa is characterized by nature-dependent livelihoods, indicating that the continent is disproportionally hit by adverse effects of climate change. For instance, IPCC (2007) argued that climate change is expected to expose between 75 and 250 million people to water stress by 2020. In addition, by 2020 there will be a significant reduction in arable land and, in some African countries, yields from rain-fed agriculture will decline by as much as 50%, which worsens food insecurity and malnutrition. Vulnerability of a country or region to climate change depends both on its socioeconomic development and on environmental factors (Kelly and Adger, 2000; McCarthy et al., 2001). The extent of vulnerability to risks of climate change is also exacerbated by the interaction of multiple stresses that occur at various levels, limited adaptive capacity, and constrained choice of adaptation options of most households in rural Africa (Boko et al., 2007).

* The authors gratefully acknowledge financial support from the Swedish International Development Cooperation Agency (Sida) through the Environment for Development Initiative at the University of Gothenburg, Sweden, Department of Economics, Environmental Economics Unit.

With a population of over 90 million (UN, 2013), Ethiopia is the second-most populous country in Africa. Ethiopia has witnessed double-digit and broad-based economic growth (averaging about 11% growth in real GDP per year), compared to the regional average of about 5% over the past decade (World Bank, 2014). This economic growth is contributing to poverty reduction and progress toward achieving the Millennium Development Goals. However, the country has been experiencing more frequent and recurrent droughts since the 1980s, with a declining average daily precipitation rate and a rising average temperature (World Bank, 2007). Over the coming decades, frequency of droughts is projected to increase (Oxfam, 2009). This means that Africa's population is highly vulnerable to climate change because the majority (about 80%) are rural dwellers engaged in rain-fed agriculture. The limited economic resources, low literacy level, limited technology and infrastructure, high level of poverty, limited/lack of specialization in cash-yielding activities, and so forth also confound their adaptive capacity and exacerbate their degree of vulnerability to climate change, although these are improving. Consequently, this chapter's examination of different zones, in order to identify the most vulnerable areas of the country, is particularly relevant for the design and targeting of policies aimed at enhancing proactive adaptive capacity, such as planned adaptations.

Some studies have examined households' vulnerability to climate change across regional states and in the Nile Basin of Ethiopia (Deressa, Hassan, Alemu et al., 2008; Deressa, Hassan and Ringler, 2008) and others analyze determinants of households' choice of adaptation options (Deressa, Hassan, Alemu et al., 2008; Deressa, Hassan and Ringler, 2008; Di Falco and Veronesi, 2014; Di Falco et al., 2012; Kabubo-Mariara and Karanja, 2006). Some attempts also have been made to identify the economic aspects, including the economy-wide impact of climate change in Ethiopia using a Ricardian approach and CGE model (Deressa, 2007; Yesuf et al., 2008; Mideksa, 2010; Deressa and Hassan, 2009; Gebreegziabher et al., 2013, 2015). Similarly, some studies have been carried out on vulnerability and adaptation to climate change in other sub-Saharan African (SSA) countries, such as Lesotho (Ziervogel and Calder, 2003), South Africa (Nhemachena and Hassan, 2007; Stringer et al., 2009), Tanzania (Paavola, 2004), and the Sudan (Osman-Elasha et al., 2006). Such studies also include comparison of vulnerability and adaptation to climate change across countries, such as comparisons between Ethiopia and South Africa (Bryan et al., 2009) and between Kenya and Tanzania (Eriksen et al., 2005).

Though these studies provide some useful insights in the area of vulnerability to climate variability and change and possible adaptation options, and reflect the current efforts to understand the relationship between climate change and vulnerability, most of them are at the aggregate level and hence have little policy relevance at the micro level. Except for Deressa, Hassan and Ringler (2008), which attempted to construct an overall vulnerability indicator at regional levels, no study has examined climate change vulnerability by constructing indices at the micro level in Ethiopia. Thus, the aim of this study is to fill this gap through assessing the levels of vulnerability to climate change of different agroecological zones of Ethiopia, taking the Nile Basin region as a case study. The objectives of this study are twofold: first, to construct individual vulnerability indices; and, second, to assess overall vulnerability and compare the extent across different agroecological zones.

Assessing and measuring vulnerability are key to figuring out which places and people are the most vulnerable, as well as the degree of vulnerability and possible adaptation options. More specifically, vulnerability assessment helps to set three policy measures. First, it is used to specify long-term targets for mitigation of climate change; second, to identify vulnerable places and people and to prioritize resource allocation for adaptation; and, finally, to put forward specific adaptation recommendations for specific places and groups (Füssel and Klein, 2006). Thus, this work fits with the ongoing use of vulnerability assessments across policy areas, including famine early warning (USAID, 2007), food insecurity (WFP, 2004), poverty, health, and globalization (O'Brien et al., 2004; UNEP, 2004).

The rest of chapter is organized as follows. The next section reviews climate change hotspots and vulnerability. The following section presents the conceptual framework. The fourth section presents the empirical method (model) employed in the study, while the fifth section presents results and discussion. Finally, the chapter concludes and draws implications.

Climate change hot spots and vulnerability: review of literature

Developing countries in general and Africa in particular are highly vulnerable to climate variability and change. According to IPCC, "Africa is one of the most vulnerable continents to climate change and climate variability" (Boko et al., 2007, pp. 435). Within Africa, countries like Ethiopia, with low levels of economic development, inadequate infrastructure, and lack of institutional capacity, are more vulnerable to climate change (Admassie and Adenew, 2008). Among African dwellers, people living in arid and semiarid, low-lying coastal and water-limited or flood-prone areas are particularly vulnerable to climate change (Watson et al., 1996). Arid and semiarid systems and coastal areas in the East African countries are among the most vulnerable parts of SSA (Thornton et al., 2008). Economic sectors and ecosystem services that are highly vulnerable to the effects of climate change include agriculture, fishery, water, and forests. Mendelsohn et al. (2016) document the negative consequences of severe warming on ecosystems in Africa.

Other scholarly studies on climate change hot spots and vulnerable places and people in SSA include Barrios et al. (2010), Dell et al. (2008) and Vincent (2004). Dell et al. examine the impact of climatic changes on economic activity throughout the world. They find that higher temperatures substantially reduce economic growth in poor countries but have little effect in rich countries. Moreover, higher temperatures appear to reduce growth rates in poor countries, rather than just the level of output. In addition, higher temperatures have wide-ranging effects in poor nations, reducing agricultural output, industrial output, and aggregate investment, and increasing political instability. Barrios et al. investigate the role of rainfall trends in explaining the poor growth performance of sub-Saharan African nations as compared to other developing countries. They find that rainfall has been a significant determinant for the poor economic growth in Africa, but not for other developing countries. They argue that had it not been for rainfall decline, today's gap in African GDP per capita relative to the rest of the developing world could have narrowed by a magnitude of around 15% to 40%. Vincent created an index to empirically assess relative levels of social vulnerability to climate change–induced variations in water availability and to allow cross-country comparison in Africa. Results shows that current vulnerability to climate change–induced changes in water availability puts Niger, Sierra Leone, Burundi, Madagascar, Burkina Faso, and Uganda as the most vulnerable countries in Africa, while Djibouti, Mauritius, Algeria, Tunisia, South Africa, and Libya are the least vulnerable. In eastern Kenya, people living in the dry-lands are highly vulnerable to climate extremes, such as drought.

Adaptation to climate variability is exacerbated by limited access to natural resources and infrastructure (Owuor et al., 2005). Thus, poor and marginalized groups are more vulnerable to climate change and their adaptation options are constrained by social setting and access to resources. Thornton et al. (2008) argue that such households have limited adaptive capacity and hence are highly vulnerable to the adverse effects of climate change. Gbetibouo (2009) emphasizes that household size, farming experience, wealth, access to credit and water, and off-farm activity diversification affect adaptive capacity to climate change in the Limpopo River Basin of South Africa. Climate change vulnerabilities are severe in food-insecure parts of Zambia (Thurlow et al., 2009). In Lesotho, marginalized social groups in terms of resources, infrastructure, and information access are more vulnerable to climate change (Ziervogel and Calder, 2003). Household-level evidence from Kenya and Tanzania reveals that individuals who are engaged in many activities at low intensity are more vulnerable than individuals specialized in one favored activity. In these two nations, household coping options are constrained by access to resources and infrastructure (Eriksen et al., 2005). Using a livelihood vulnerability index, Hahn et al. (2009) find that various regions in Mozambique have different levels of vulnerability and hence propose different adaptation options.

According to Gbetibouo and Ringler (2009), vulnerability to climate change and variability is intrinsically linked with social and economic development. Provinces with high levels of infrastructure development, high literacy rates, and low shares of agriculture in total GDP are relatively low on the vulnerability index. In contrast, the highly vulnerable regions are characterized by densely populated rural areas, large numbers of small-scale farmers, high dependency on rain-fed agriculture, and serious land degradation. In addition, households with limited fixed assets, such as livestock, and households that depend on rain-fed agriculture are more vulnerable to climate change (Shewmake, 2008). Deressa, Hassan, Alemu, et al. (2008) also find that the degree of vulnerability varies across regional states of the country, with relatively less developed arid and semiarid regions (Afar and Somali) being highly vulnerable to climate change. Drought-prone regions, such as Oromia in Ethiopia's eastern and lowland areas and Tigrai, are also highly vulnerable to climate change. Deressa, Hassan and Ringler (2008) measure the vulnerability of farmers to climatic extremes, such as droughts, floods, and hailstorms, by employing the "vulnerability as expected poverty" approach in the Nile Basin of Ethiopia. Their results show that farmers' vulnerability is highly sensitive to the minimum daily requirement for survival (or poverty line) in the area. The results further indicate that farmers in *kola* agroecological zones (which are warm and semiarid) are the most vulnerable to extreme climatic events.

In climate change literature, two components of vulnerability are identified: internal and external dimensions. The internal dimension refers to frailty, insecurity, and capacity to anticipate, cope with, resist, and recover from adverse effects of shocks. The external dimension involves exposure to risks and shocks. However, in its Second Assessment Report, IPCC defined vulnerability as the degree to which climate change may cause damage to a system and noted that the level of vulnerability depends on the sensitivity of the system to shocks and its ability to adapt to negative impacts of climate change (Watson et al., 1996). At the time of the Second Assessment Report, the report changed its focus concerning vulnerability from internal and external dimensions to sensitivity and adaptive capacity. It is argued that vulnerability to climate change is severe where a system is highly sensitive to climate change and adaptability is the lowest. The IPCC Third Assessment Report (TAR) reconciles both sides by adding a third component to vulnerability, defining it as

> The degree to which a system is susceptible to, or unable to cope with, adverse effects of climate change, including climate variability and extremes. Vulnerability is a function of the character, magnitude, and rate of climate variation to which a system is exposed, its sensitivity, and its adaptive capacity.
>
> (McCarthy et al., 2001)

According to this definition, vulnerability includes an external dimension that is represented by the exposure of a system to climate variations, as well as a more complex internal dimension comprising its sensitivity and adaptive capacity to these stressors (Füssel and Klein, 2006). The IPCC Fourth Assessment Report (AR4), which reports recent advances in climate change, contains a vulnerability definition consistent with that of the Third Assessment Report (IPCC, 2007). Under this framework, a highly vulnerable system would be one that is very sensitive to modest changes in climate, where the sensitivity includes the potential for substantial harmful effects, and for which the ability to adapt is severely constrained. Hence, vulnerability is a function of three components: exposure, sensitivity, and adaptive capacity, which are influenced by a range of physical, environmental, and socioeconomic and political factors. These factors are outlined for the case of Ethiopia in Figures 3A1, 3A2, and 3A3.

A common thread in the climate change impacts and vulnerability literature is that countries, regions, economic sectors, and social groups differ in their degree of vulnerability to climate change (Bohle et al., 1994). Similarly, in vulnerability analysis, poverty and inequality are key factors, as inequality and marginalization are important determinants of the degree of vulnerability to climate change (Ribot, 1996). Thus, it is paramount to focus on the poor people in developing countries rather than emphasizing poor countries themselves.

Conceptual framework

Different international organizations define vulnerability in various ways. Definitions of vulnerability by the Intergovernmental Panel on Climate Change (IPCC) are related specifically to climate change. Vulnerability to climate variability and change can be defined in several ways as pre-adaptation and post-adaptation (end point) vulnerability, outcome vulnerability, and conceptual vulnerability. According to the post-adaptation definition, vulnerability refers to the residual impacts of climate change after adaptation options have been taken into account. That is, "the level of vulnerability is determined by the adverse consequences that remain after the process of adaptation has taken place" (Kelly and Adger, 2000, p. 327). This approach specifically focuses on the physical dimensions of vulnerability. The pre-adaptation (or starting point) definition of vulnerability has its origin in the literature on food security and famine (Bohle et al., 1994) and vulnerability to natural hazards (Blaikie et al., 1994). According to this approach, limited access to resources and political instability and exposure are the main causes of vulnerability. However, in this study we adopt the definition of the IPCC Third Assessment Report (IPCC, 2001), which defines vulnerability as a function of three dimensions: sensitivity, exposure, and adaptive capacity.

Figure 3.1 provides the conceptual framework for analyzing vulnerability. As the figure illustrates, our study context could be viewed as exposed to both gradual climate change (top left corner of Figure 3.1), mainly involving change in temperature and precipitation, and climate extremes (top right corner of Figure 3.1), such as drought (von Braun and Webb, 1995) and flood (World Bank, 2008). Exposure in turn affects sensitivity – that is, exposure to higher frequencies and intensities of climate risk seriously affect economic and social outcomes, such as crop yield, income, and health. Exposure is also related to adaptive capacity. Specifically, higher adaptive capacity either reduces the potential damage from or provides resilience against exposure to higher climate risk.

This conceptual framework for vulnerability also suggests that sensitivity and adaptive capacity are interlinked. That is, given some fixed level of exposure, adaptive capacity influences the level of sensitivity. Lower adaptive capacity results in higher sensitivity and vice versa. Hence, sensitivity and adaptive capacity together with exposure add up to overall (total) vulnerability. The conceptual framework also captures socioeconomic vulnerability, which mainly deals with variations within a society (Adger, 1999), and biophysical vulnerability, which emphasizes the adverse effects of environmental factors on human and natural systems (Füssel and Klein, 2006). We employ the integrated approach, which tries

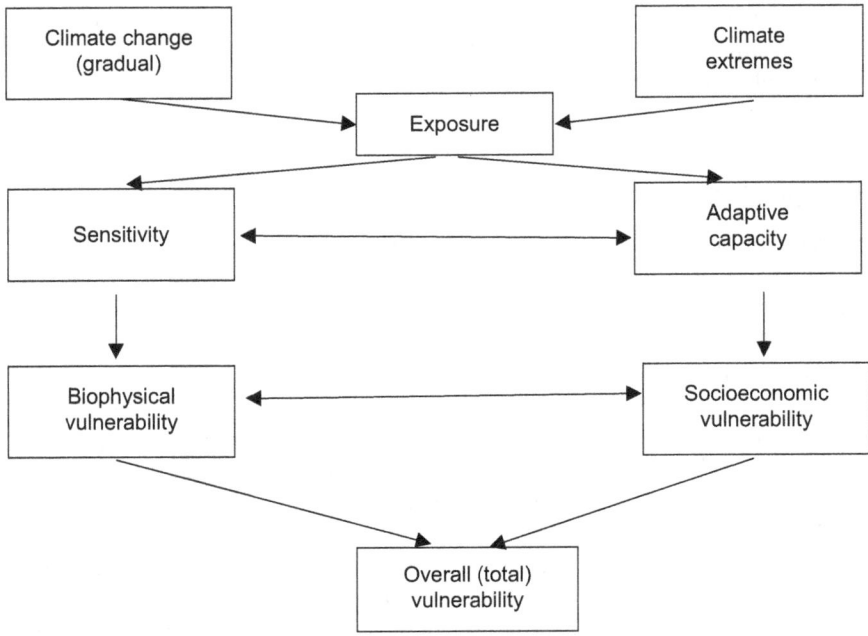

Figure 3.1 Conceptual Framework for Vulnerability Analysis

Source: Adapted from Deressa, Hassan and Ringler (2008).

to integrate both biophysical and socioeconomic factors in analyzing vulnerability to climate change (Füssel, 2007).

Adaptive capacity

The IPCC (2001) describes "adaptive capacity" as the potential or ability of a system, region, or community to adjust to the effects or impacts of climate change (including climate variability and extremes).[2] Adaptive capacity is a context-specific concept and can vary from country to country, from community to community, among social groups and individuals, and over time (IPCC, 2001; Smit and Wandel, 2006). Besides, according to McCarthy et al. (2001), adaptive capacity is considered "a function of wealth, technology, education, information, skills, and infrastructure, access to resources, and stability and management capabilities." Therefore, analyzing vulnerability must involve identifying not only the threat but also the "resilience,"[3] or the potential responsiveness of the system and its ability to exploit opportunities and resist or recover from the negative effects of a changing environment. The means of resilience could be the assets and entitlements that an individual, a household, or a community can mobilize and manage in the face of hardship. It should also be noted that there are close linkages between vulnerability and livelihoods, and thus building resilience is a question of expanding and sustaining these assets (Moser, 1998). Put differently, vulnerability is envisaged to be closely linked to asset ownership. The more assets people have, the less vulnerable they are; conversely, the greater the erosion of people's assets, the greater their vulnerability (insecurity).

Following Gbetibouo and Ringler (2009), in this study, we postulate that adaptive capacity is explained as dependent upon four livelihood assets:

1   Social capital represents social networks as captured by number of *ider* and *iqub* members and number of *debbo* and *jigge* participants in close relationship within the *got*. (A got is a group of villages; ider is a social network which operates as a funeral association; iqub is a rotating saving and credit network; and debbo and jigge refer to social networks where people support each other during peak seasons of agricultural activities.) Social capital is envisaged to have three roles. It can be a means of transferring information about new technology. It also facilitates smoother financial transactions among farmers. In addition, it enables farmers to overcome collective action dilemmas. Therefore, it is hypothesized that social capital positively influences adaptation to climate change.

2   Human capital is represented in the study by the literacy rate among household heads, as well as age and attendance at training. This is envisaged to reduce vulnerability by increasing households' capabilities and access to information and, thus, enhancing households' abilities to cope with climate extremes and variability.

3   Financial capital includes farm holding size, number of livestock (proxied by tropical livestock units or TLU), and whether the primary residence is constructed with a metal roof. These are considered indicators of a household's wealth. Households with fewer TLU, smaller farm size, and no metal roof are assumed to be less economically sound and this makes them more susceptible to climate change.

4   Physical capital includes variables such as distance to and major source of domestic water, access to extension services and new technologies, and distance to the nearest input and output markets. Better access to infrastructure and markets is believed to reduce transaction costs and strengthen the links between input and product markets.

5   The effect of variables such as marital status and gender of the household head is assumed to be indeterminate.

Exposure

O'Brien et al. (2004) relate exposure to the degree of climate stress affecting a unit of analysis – that is, the magnitude and frequency of extreme events to which a particular area or unit of analysis is exposed. This could be in the form of either long-term changes in climate conditions or changes in climate variability (IPCC, 2001). In our study, variables such as decadal change in temperature and precipitation are regarded as factors that affect exposure to climate change. We take the base year as 1994/5. It is also assumed that large decadal changes in both temperature and precipitation lead to higher exposure.

Sensitivity

Sensitivity explains the human and environmental conditions that can either worsen the hazard or trigger an impact (Gbetibouo and Ringler, 2009). In this study, we look at the following factors that may have impact on the sensitivity of farmers in the study area:

> Climate extremes: In different parts of the study area, the main constraints of agriculture are drought, floods, or hailstorms.
> Population density: It is assumed that districts with high population density are more sensitive to climate change, because more people are exposed to climate extremes and variations.

Empirical method and data

Calculating the vulnerability indices

There are two broad approaches to empirically calculating vulnerability: (1) econometric and (2) indicator methods. The former expresses vulnerability as expected poverty, low expected utility, and uninsured exposure to risk (Hoddinott and Quisumbing, 2003). The latter tries to assess vulnerability by integrating indicators to form a composite index, which can be at a local level (Adger, 1999; Hahn et al., 2009), national level (O'Brien et al., 2004), or global level (Brooks et al., 2005). The basic challenge in constructing indices is the lack of standard ways of assigning weight to each indicator. The two most common weighting methods used to combine indicators are equal and unequal weighting schemes. The former method assigns equal weight to each indicator. The latter method assigns different weights to various indicators using expert opinion (Vincent, 2007), complex fuzzy logic (Eakin and Bojorquez-Tapia, 2008), or a principal component analysis (Easter, 1999).

In this study, we use an integrated approach to construct a composite vulnerability index based on principal component analysis (PCA) weighting schemes. An integrated approach is chosen because it incorporates both socioeconomic and environmental factors in assessing vulnerability. Specifically, composite indices have long been used in a wide variety of disciplines to measure complex, multidimensional concepts that cannot be observed or measured directly (Füssel, 2007). Their power lies in their ability to synthesize a vast amount of diverse information into a simple and usable form.

Overall vulnerability is calculated as the net effect of adaptive capacity, sensitivity, and exposure. Following Moss et al. (2001), we assigned a negative value to sensitivity and exposure, because their impact is assumed to be negative.

Similarly, factors which are listed under adaptive capacity are assigned positive signs, on the assumption that people with higher adaptive capacity are less sensitive to damages from climate extremes and variations, keeping the level of exposure constant. Thus, we specify overall vulnerability as:

$$\text{Vulnerability} = (\text{adaptive capacity}) - (\text{sensitivity} + \text{exposure}) \tag{1}$$

Note that in Equation (1) a higher net value indicates lesser vulnerability and vice versa.

The next step is to attach weights to indicators. The literature suggests three different methods for attaching weights: (1) expert judgment (Brooks et al., 2005; Moss et al., 2001); (2) arbitrary choice of equal weight (Lucas and Hilderink, 2004; O'Brien et al., 2004; Patnaik and Narayanan, 2005); and (3) statistical methods, such as principal component analysis (PCA) or factor analysis (Cutter et al., 2003; Thornton et al., 2006). We use the third method. That is, we use statistical methods (PCA) to generate the weights and we use STATA software to run PCA. Given a set of variables, PCA is a technique for extracting those few orthogonal linear combinations of variables that most successfully capture the common information from a broad set of variables. The linear index of all the variables that captures the largest amount of information common to all the variables is defined as the first principal component of a set of variables (Filmer and Pritchett, 2001).

Let us assume that there is a set of Z-variables (β^*_{1j} to β^*_{zj}) which represent the Z-indicators (attributes) of each agroecological zone. PCA starts by specifying each indicator, normalized by its mean and standard deviation.[4] The selected variables are expressed as linear combinations of a set of underlying components for each agroecological zone j.

$$\begin{aligned} \beta_{1j} &= x_{11}W_{1j} + x_{12}W_{2j} + \ldots + x_{1z}W_{zj} \text{ where } j=1 \ldots J \\ \beta_{z1j} &= x_{z1}W_{1j} + x_{z2}W_{2j} + \ldots + x_{zz}W_{zj} \end{aligned} \tag{2}$$

where the W's are the components and the x's are the coefficients on each component for each variable that does not vary across agroecological zones. The solution to the problem is indeterminate because only the left side of each line is observed. PCA overcomes this problem through finding the linear combination of the variables with maximum variance (i.e., the first principal component $W1j$) and then finding a linear combination of the variables orthogonal to the first and with maximum remaining variance and so on. Technically, the procedure solves the equations $(\mathbf{R}-\lambda\mathbf{I})v_n = 0$ for λ_n and vn, where $\mathbf{R}$ is the matrix of correlations between the scaled variables (the β's) and v_n is the vector of coefficients on the nth component for each variable. Solving these equations gives the characteristic roots of $\mathbf{R}$, λ_n, which are known as eigen values, and their associated eigen vectors, v_n. The final set of estimates is produced by scaling each v_n so that their squares sum up to the total variance; this is another restriction imposed to achieve determinacy of the problem.

By inverting the system, which is implied by Equation (2), the scoring factors from the model are recovered. This yields a set of estimates for each of the Z-principal components:

$$W_{1j}=a_{11}\beta_{1j}+a_{12}\beta_{2j}+\ldots+a_{1z}\beta_{zj}\,j=1\ldots J$$
$$W_{zj}=a_{z1}\beta_{1y}+a_{z2}\beta_{2j}+\ldots+a_{zz}\beta_{zj} \tag{3}$$

where the a's are the factor scores. Therefore, the first principal component, expressed in terms of the variables, is an index for each agroecological zone based on the following expression:

$$W_{1j}=a_{11}(\beta^{*}_{1j}\text{-}\beta^{*}_{1})/(s^{*}_{1})+\ldots+a_{1z}(\beta^{*}_{zj}\text{-}\beta^{*}_{z})/(s^{*}_{z}) \tag{4}$$

Though our scale of analysis is agroecological zone, we also provide the results of district-level analysis in the appendices.

Variables and hypothesis

As mentioned earlier, we base our definition of vulnerability on IPCC (2001), where a region's vulnerability to climate change and variability is described by three elements: exposure, sensitivity, and adaptive capacity. In addition, based on an extensive review of previous studies, we identify the indicators of vulnerability components selected, and we present the research hypothesis. In particular, we draw on Aandahl and O'Brien (2001), Moss et al. (2001), Cutter et al. (2000, 2003), TERI (2003), O'Brien et al. (2004), Lucas and Hilderink (2004), Brenkert and Malone (2005), Brooks et al. (2005), Patnaik and Narayanan (2005), Thornton et al. (2006), Deressa, Hassan and Ringler (2008), Gbetibouo and Ringler (2009), and Hahn et al. (2009). Accordingly, the following indicator variables are considered and corresponding hypotheses (expected signs) are put forward (see Table 3.1).

Data and study area description

We use a survey of 1,000 farm households conducted during the 2004/05 fiscal year in the Nile Basin of Ethiopia. The survey was conducted by the International Food Policy Research Institute (IFPRI) in collaboration with the Ethiopian Development Research Institute (EDRI). The sampled districts for the survey were selected to represent the different attributes of the basin, including typologies of the regions' agroecological zones, average annual rainfall, rainfall variability, and vulnerability (food aid–dependent population). Peasant associations (administrative units lower than districts) were also purposefully selected to include households that irrigate their farms. One peasant association was selected from each of 20 sampled districts, for a total of 20 sampled peasant associations. Once the peasant associations were chosen, 50 farming households were randomly selected from each peasant association (peasant associations have more

Table 3.1 Components of Indicators of Vulnerability and Expected Signs

Determinants of vulnerability	Component indicator	Expected sign
Adaptive capacity	Gender of the hh head	
	Age of the hh head	
	Marital status of the hh head	(+)
	No. of years of education of the hh head	(+)
	Primary residence with metal roof matter	(+)
	Major source of domestic water	(+)
	Distance of water source from home (domestic) in km	(+)
	Have you attended training on crop production?	(+)
	Have you attended training on livestock activities?	(+)
	Do you get advice from extension workers on crop production?	(+)
	Do you get advice from extension workers on livestock activities?	(+)
	Distance to nearest market place for selling products (Hrs.)	(+)
	Distance to nearest market place for obtaining inputs (Hrs.)	(+)
	Total area of farmland	(+)
	Highly fertile or not	(+)
	Total value of insecticide used	(+)
	Total value of herbicide used	(+)
	Total value of fungicide used	(+)
	Tropical Livestock Units	(+)
	Number of IDIR members in close relationship within the GOT	(+)
	Number of IQUB members in close relationship within the GOT	(+)
	Number of DEBBO participants in close relationship within the GOT	(+)
	Number of JIGGE participants in close relationship within the GOT	(+)
Exposure	Change in temperature	(-)
	Change in precipitation	(-)
Sensitivity	Type of shock – drought	(-)
	Type of shock – flood	(-)
	Type of shock – hailstorm	(-)
	Population density	(-)

than one village) for a total of 1,000 interviewed households (Deressa, Hassan, Alemu et al., 2008) (see Table 3.2).

Data on precipitation and temperature for 2004/04 and 1994/05 was also obtained from the Ethiopian Metrological Agency (EMA). The analysis is based on four agroecological zones – namely, moisture-reliable humid lowlands,

Table 3.2 Surveyed Districts and Peasant Associations

Regional state	Zone	District	Agroecological zone	Peasant association	Number of households
Tigray	East Tigrai	Hawzein	Drought-prone highland	Selam	50
		Atsbi Wonberta	Drought-prone highland	Felge Woine	50
	South Tigrai	Endomehoni	Drought-prone highland	Mehan	50
Amhara	North Gondar	Debark	Humid lowlands	Mekara	50
		Chilga	Humid lowlands	Teber Sekro	50
		Wogera	Humid lowlands	Sak Debir	50
	South Gondar	Libo Kemkem	Humid lowlands	Angor	50
	East Gojam	Bichena	Humid lowlands	Aratband Bichena	50
	West Gojam	Quarit	Humid lowlands	Gebez	50
Oromiya	West Wellega	Gimbi	Humid lowlands	Were Sayo	50
		Haru	Humid lowlands	Genti Abo	50
	East Shoa	Bereh Aleltu	Cereal-based moisture-sufficient highlands	Welgewo	50
	East Shoa	Hidabu Abote	Cereal-based moisture-sufficient highlands	Sira Marase	50
	East Wellega	Limu	Humid lowlands	Areb Gebya	50
		Nunu Kumba	Humid lowlands	Bachu	50
	Jimma	Kersa	Cereal-based moisture-sufficient highlands	Merewa	50
Benishangul-Gumuz	Metekel	Wonbera	Drought-prone highland	Addis Alem	50
	Asosa	Bambasi	Humid lowlands	Sonka	50
	Kamashi	Sirba Abay	Drought-prone highland	Koncho	50
SNNP	Zone 1	Gesha Daka	Enset-based moisture-sufficient highlands	Kicho	50
Total					1000

moisture-sufficient highlands (cereal-based), moisture-sufficient highlands (enset-based), and drought-prone highlands. In addition, district-level results are attached in the appendices.

The Nile Basin of Ethiopia is the area of focus in this study. The Nile Basin of Ethiopia covers a total area of about 358,889 km², which is equivalent to 34% of Ethiopia's total area, and contains about 40% of the country's population. Portions of six different regional states of Ethiopia are contained within the basin. Specifically, the basin encompasses 38% of the total land area of Amhara, 24% of Oromia, 15% of Benishangul-Gumuz, 11% of Tigrai, 7% of Gambella, and 5% of Southern Nations Nationalities and Peoples (SNNP). The basin contains three major rivers: the Abbay River, which originates from the central highlands; the Tekeze River, which originates from the northwestern part of the country; and the Baro-Akobo River, which originates from the southwestern part of the country. The total annual surface runoff of the three rivers is estimated at 80.83 billion cubic meters per year, which amounts to nearly 74% of the total runoff from Ethiopia's 12 river basins (Deressa, 2007).

Results and discussion

Descriptive statistics

Our summary statistics in Table 3.3 show that not much difference is observed across agroecological zones in household variables such as gender, age, marital status, and education. In all the zones, distance to the nearest water source is greater

Table 3.3 Summary Statistics (Mean) of Normalized Values of the Original Data by Their Respective Means and Standard Deviations by Agroecology

Description	AEZ1		AEZ2		AEZ3		AEZ4	
	Mean	SD	Mean	SD	Mean	SD	Mean	SD
Gender of the hh head is male	0.912	0.283	0.840	0.368	1.000	0.000	0.892	0.312
Age of the hh head	44.370	13.681	47.767	15.142	43.340	12.103	44.759	13.468
Marital status of the hh head (married)	0.901	0.298	0.827	0.380	0.940	0.240	0.815	0.389
No. of years of education of the hh head	1.732	2.827	1.653	2.788	1.860	2.828	1.614	2.630
Primary residence with metal roof	0.589	0.493	0.308	0.463	0.061	0.242	0.333	0.472
Major source of domestic water	0.395	0.489	0.387	0.489	0.380	0.490	0.500	0.501
Distance of water source from home (domestic) in km	0.855	1.624	5.302	32.044	2.732	14.107	3.469	27.048

(Continued)

Table 3.3 (Continued)

Description	AEZ1		AEZ2		AEZ3		AEZ4	
	Mean	SD	Mean	SD	Mean	SD	Mean	SD
Have you attended training on crop production?	0.493	0.500	0.067	0.250	0.060	0.240	0.256	0.437
Have you attended training on livestock activities?	0.460	0.499	0.040	0.197	0.000	0.000	0.232	0.423
Do you get advice from extension workers on crop production?	0.691	0.463	0.140	0.348	0.080	0.274	0.544	0.499
Do you get advice from extension workers on livestock activities?	0.605	0.489	0.113	0.318	0.020	0.141	0.500	0.501
Distance of nearest market place for selling products (hrs.)	5.498	4.381	5.393	3.683	7.207	2.405	5.815	5.244
Distance of nearest market place for obtaining inputs (hrs.)	5.615	4.088	5.526	3.724	7.467	2.468	5.632	4.587
Total area of farmland	2.174	1.676	2.240	1.360	2.923	3.817	1.742	1.363
Highly fertile or not	0.451	0.498	0.487	0.501	0.360	0.485	0.516	0.501
Total value of insecticide used	3.842	21.157	0.882	7.766	0.000	0.000	2.272	16.791
Total value of herbicide used	13.353	193.228	17.242	33.195	118.640	226.962	3.599	14.634
Total value of fungicide used	0.131	2.287	0.373	4.572	3.040	17.474	0.100	1.581
Tropical Livestock Units	4.516	3.036	5.379	3.873	6.859	3.515	4.512	3.401
Number of IDIR members in close relationship within the GOT	16.689	39.622	16.827	28.836	16.400	19.008	22.616	48.368
Number of IQUB members in close relationship within the GOT	0.575	2.859	0.687	4.055	3.660	11.705	2.512	6.812
Number of DEBBO participants in close relationship within the GOT	6.936	14.737	7.647	9.203	15.460	16.066	7.040	10.903
Number of JIGGE participants in close relationship within the GOT	1.224	5.147	2.640	6.029	0.220	1.418	0.760	1.969

than 0.5 km, indicating that water supply is difficult to access in those regions of the country. The lowest use of training and extension services is observed in enset-based moisture-sufficient highlands. Humid lowland moisture-reliable areas show the highest frequency of extreme events, such as drought, flood, and hailstorms. On the other hand, enset-based moisture-sufficient highland areas did not experience any extreme climate events over the prior five years. Farmers living in enset-based moisture-sufficient highland areas are wealthier than those in the other regions in terms of livestock (TLU). Similarly, those farmers have the highest social interaction of all the regions. Drought-prone highlanders used the least technology – that is, insecticides, pesticides, and fungicides. The highest incremental increase in temperature from 1994/95 to 2004/05 is found in drought-prone areas of the Nile Basin region. Of all the agroecological zones, the most densely populated area is the cereal-based moisture-sufficient highland.

Empirical results

In what follows, we first discuss the results in relation to each of the three dimensions of vulnerability and then turn to overall vulnerability.

Adaptive capacity

Figure 3.2 presents the adaptive capacity index, which suggests large differences across agroecological zones of the Nile Basin of Ethiopia. Our findings also show that, among the four major agroecological zones in the basin, the enset-based moisture-sufficient highland areas have the highest adaptive capacity, with an

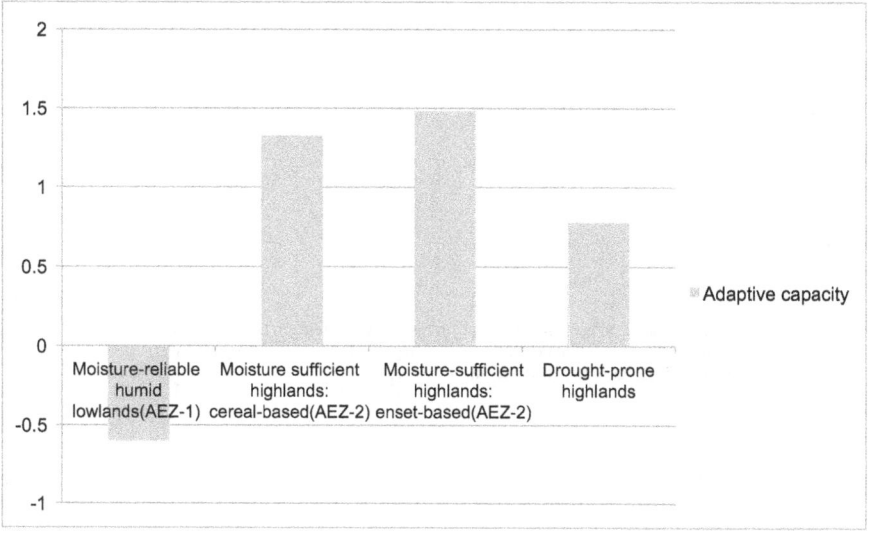

Figure 3.2 Adaptive Capacity across Agroecological Zones of the Nile Basin of Ethiopia

index of 1.48. This is perhaps due to better access to infrastructure, asset accumulation, and social networks. By contrast, the humid lowland zone is the lowest in terms of adaptive capacity to climate change, with a value of −0.6.

Exposure

The overall exposure of the farming sector across the four agroecological zones of the Nile Basin of Ethiopia is presented in Figure 3.3. Drought-prone highland areas are the most exposed to climate change (0.6). The area is characterized by high temperature that is increasing over time. The least exposed agroecological zone is the enset-based moisture-sufficient highlands (0.47), followed by moisture-reliable humid lowland areas (−0.16), even though, in both zones, the amount of precipitation has been declining over time.

Sensitivity

Regarding sensitivity, even though population density is highest in the moisture-sufficient highland areas, those areas tend to be the least sensitive zones in the study area (Figure 3.4). As in the exposure index, drought-prone highland areas are the most sensitive zone to climate change. The area is characterized by higher frequency of drought, flooding, and hailstorms, in addition to high temperature that is increasing over time.

Overall vulnerability

Following the foregoing results, Figure 3.5 shows potential impacts (exposure and sensitivity) and coping capacity (adaptive capacity) across different

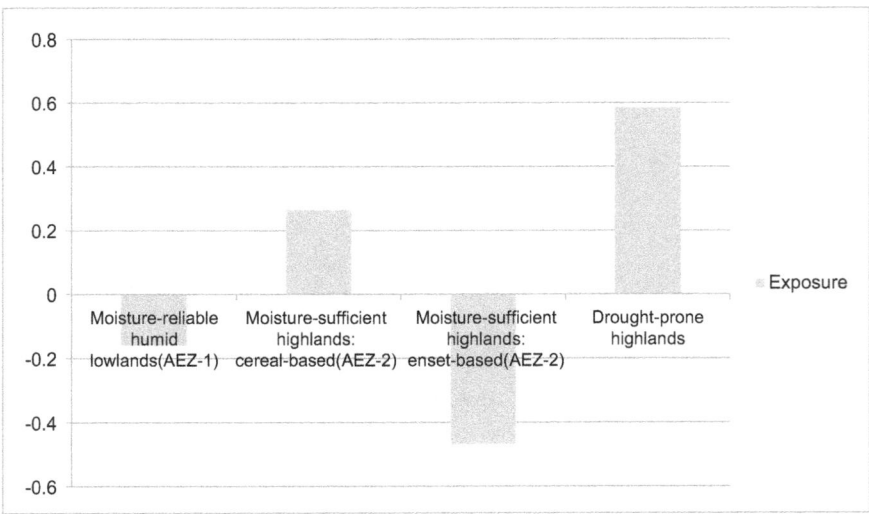

Figure 3.3 Exposure across Agroecological Zones of the Nile Basin of Ethiopia

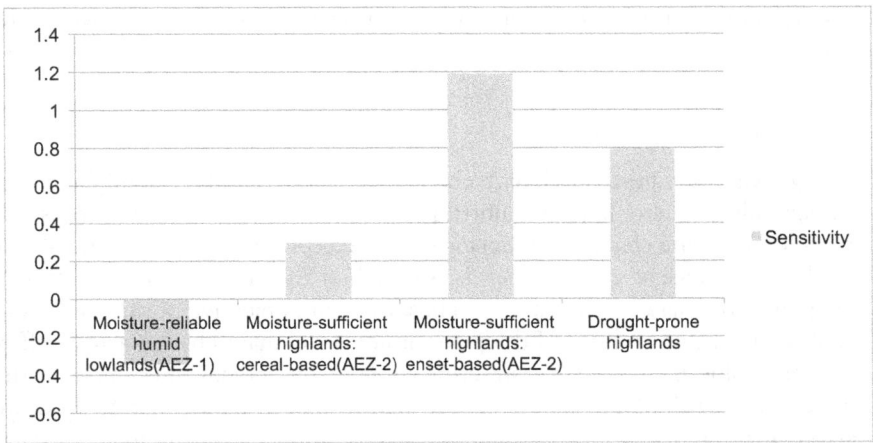

Figure 3.4 Sensitivity across Agroecological Zones of the Nile Basin of Ethiopia

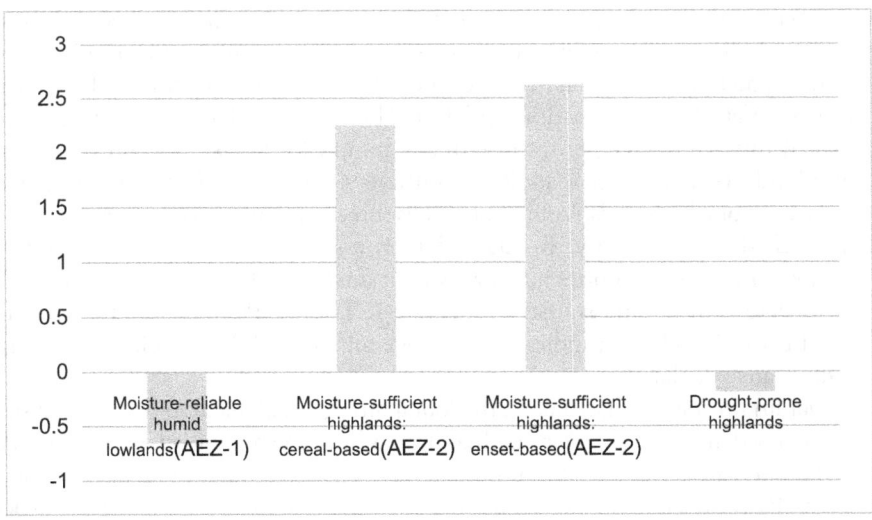

Figure 3.5 Overall Vulnerability across Agroecological Zones of the Nile Basin of Ethiopia

agroecological zones of the Nile Basin. The results of the overall vulnerability index also identified the humid lowlands (−0.68) and drought-prone highland areas (−0.16) of the Nile Basin of Ethiopia as the most vulnerable zones. This is because of their lower adaptive capacity and higher exposure and sensitivity to climate change. The less vulnerable zones are the enset-based moisture-sufficient highland zone (2.56) and the cereal-based moisture-sufficient highland zone

(2.25).The aggregated and disaggregated results of the overall vulnerability indices are consistent with the results at the district level, which are shown in the appendix.

Conclusions and implications

The growing literature on climate change and agriculture has highlighted that the agricultural sector in poor countries in tropical and subtropical regions is vulnerable to climate change. Ethiopia is not an exception. Assessing vulnerability to climate change at various levels is essential for the planning and targeting of proactive and planned adaptations. Thus, the main objective of this study is to analyze the vulnerability of Ethiopian farmers to climate change across different agroecological zones by constructing vulnerability indices and comparing the indices, taking the Nile Basin as a case study. Using the indicator method, we assess vulnerability by integrating indicators to form a composite index. We use IPCC's (2001) definition of vulnerability. Our vulnerability indicator approach is integrated, in that the selected indicators represent both the biophysical conditions of the farming regions and the socioeconomic conditions of the farm households.

We use a survey of 1,000 farm households conducted during the 2004/05 fiscal year in the Nile Basin of Ethiopia. The survey was administered by the International Food Policy Research Institute (IFPRI) in collaboration with the Ethiopian Development Research Institute (EDRI). The analysis is based on four agroecological zones – namely, moisture-reliable humid lowlands, moisture-sufficient highlands (cereal-based), moisture-sufficient highlands (enset-based), and drought-prone highlands. Vulnerability has three components: exposure, sensitivity, and adaptive capacity. To quantitatively assess overall vulnerability and its components, we run a principal component analysis (PCA) with 30 indicators, using data analysis and statistical software (STATA). PCA is used to generate weights for the different indicators, and overall vulnerability and its component indices are calculated.

Our findings show that, among the four main agroecological zones in the basin, enset-based moisture-sufficient highland areas have the highest adaptive capacity, perhaps due to better access to infrastructure, asset accumulation, and social networks. By contrast, humid lowland zones are the lowest in terms of adaptive capacity to climate change. Regarding exposure and sensitivity, even though population density is highest in the moisture-sufficient highland areas and the precipitation amount is declining over time, those areas tend to be the least exposed and sensitive zones in the study area. Drought-prone highland areas are the most exposed and sensitive zones to climate change. The area is characterized by higher frequency of drought and flood, in addition to greater temperature increases over time. The results of the overall vulnerability index also identified the humid lowlands and drought-prone highland areas of the Nile Basin of the country as the most vulnerable zones. This is because of their lower adaptive capacity and higher exposure and sensitivity to climate change. The less vulnerable zones are the enset-based moisture-sufficient highland zone and the cereal-based moisture-sufficient highland zone.

Generally, our agroecological analysis provides useful insights and captures local specific variations that often have been overlooked in earlier studies. The findings imply that climate change should be placed within the broader context of development strategy and rural poverty reduction. Because there are large spatial differences of vulnerability across agroecological zones, the development and rural poverty reduction policies and strategies should be tailored to location-specific circumstances. In particular, great effort should be exerted in expansion of education, training, and extension to marginalized areas. Moreover, public investment in technology and infrastructure is essential to enhance adaptive capacity. Investment in different water, land, and forest conservation programs should be implemented at both community and household levels, in order to lessen sensitivity and exposure to climate risk of rural households in the Nile Basin of Ethiopia.

Notes

1 *Enset* (i.e., Enset ventricosum) is an edible crop (plant). It is the edible species of a separate genus of the banana family, thus named "false banana," but the enset fruit is not edible. Enset is a genus of monocarpic flowering plants native to tropical regions of Africa and Asia. It is also common in southern parts of Ethiopia. The plant is cut before flowering, the pseudostem (false stem) and leaf midribs are scraped, the pulp is fermented for 10–15 days and finally steam-baked flat-bread is prepared out of it and used as food (Shank and Ertiro, 1996). An enset-based farming system is where enset is the dominant crop but other crops, such as cereals (maize or sorghum) and tuber crops (potatoes and sweet potatoes), are also grown (Brandt et al., 1997). The cereal-based system is different from the enset-based system in that cereals, such as teff, wheat, and barley, are the dominant crops grown.
2 For example, communities may rely on informal nonmonetary arrangements and social networks along with livelihood diversification and financial remittances through extended family networks to cope with shocks such as drought and storm damage (Adger, 2001; Barnett, 2001; Sutherland et al., 2005; Yilma et al., 2012).
3 Though it can be envisaged that they are interrelated, it is important to note that the terms "vulnerability" and "resilience" are not one and the same. "Vulnerability" is essentially the pre-event, inherent characteristics of a system (social or natural) that create the potential for harm, whereas "resilience" is the ability of a social system to respond to and recover from disasters (Colburn and Seara, 2011).
4 For example, $(\beta^*_{1j} - \beta^*_1)/s^*_1$, where β^*_1 is the mean of β^*_{1j} across agroecological zones and s^*_1 is its standard deviation.

References

Aandahl, G., and K. O'Brien. 2001. Vulnerability to Climate Changes and Economic Changes in India's Agriculture. Paper Presented at the Biannual Conference of the Nordic Association for South Asian Studies. "Waters of Hope: The Role of Water in South Asian Development." September 20–22, Voss, Norway.

Adger, W.N. 1999. Social Vulnerability to Climate Change and Extremes in Coastal Vietnam. *World Development* 27(2): 249–269.

Adger, W.N. 2001. Scales of Governance and Environmental Justice for Adaptation and Mitigation of Climate Change. *Journal of International Development* 13(7): 921–931.

Admassie, A., and B. Adenew. 2008. *Stakeholders' Perceptions of Climate Change and Adaptation Strategies in Ethiopia, EEA Research Report*. Addis Ababa: Ethiopian Economic Association. www.eeaecon.org/Research%20Materials.htm

Barnett, J. 2001. Adapting to Climate Change in Pacific Island Countries: The Problem of Uncertainty. *World Development* 29: 977–993.

Barrios, S., L. Bertinelli and E. Strobl. 2010. Trends in Rainfall and Economic Growth in Africa: A Neglected Cause of the African Growth Tragedy. *The Review of Economics and Statistics* 92(2): 350–366.

Blaikie, P., T. Cannon, I. Davis and B. Wisner. 1994. *At Risk: Natural Hazards, People's Vulnerability and Disasters*. London: Routledge.

Bohle, H.G., T.E. Downing and M.J. Watts. 1994. Climate Change and Social Vulnerability: The Sociology and Geography of Food Insecurity. *Global Environmental Change* 4(1): 37–48.

Boko, M., I. Niang, A. Nyong, C. Vogel, A. Githeko, M. Medany, B. Osman-Elasha, R. Tabo and P. Yanda. 2007. 'Africa' in Climate Change 2007: Impacts, Adaptation and Vulnerability. In *Contribution of Working Group II to the Fourth Assessment Report of the Intergovernmental Panel on Climate Change (IPCC)*, edited by M.L. Parry, O.F. Canziani, J.P. Palutikof, P.J. van der Linden and C.E. Hanson. Cambridge, UK: Cambridge University Press, pp. 433–467.

Brandt, S.A., A. Spring, C. Hiebsch, J.T. McCabe, E. Tabogie, M. Diro, G. Wolde-Michael, G. Yntiso, M. Shigeta and S. Tesfaye. 1997. *The 'Tree against Hunger': Enset-Based Agricultural Systems in Ethiopia*. Washington, DC: American Association for the Advancement of Science.

Brenkert, A.L., and E.L. Malone. 2005. Modeling Vulnerability and Resilience to Climate Change: A Case Study of India and Indian States. *Climatic Change* 72: 57–102.

Brooks, N., W.N. Adger and P.M. Kelly. 2005. The Determinants of Vulnerability and Adaptive Capacity at the National Level and the Implications for Adaptation. *Global Environmental Change* 15: 151–162.

Bryan, E., T.T. Deressa, G.A. Gbetibouo and C. Ringler. 2009. Adaptation to Climate Change in Ethiopia and South Africa: Options and Constraints. *Environmental Science and Policy* 12: 413–426.

Colburn, L., and T. Seara. 2011. Resilience, Vulnerability, Adaptive Capacity, and Social Capital. 2nd National Social Indicators Workshop, NOAA Fisheries Service. September 27–29, Silver Spring, Maryland, U.S.

Cutter, S.L., B.J. Boruff and W.L. Shirley. 2003. Social Vulnerability to Environmental Hazards. *Social Science Quarterly* 84(2): 242–261.

Cutter, S.L., J.T. Mitchell and M.S. Scott. 2000. Revealing the Vulnerability of People and Places: A Case Study of Georgetown County, South Carolina. *Annals of the Association of American Geographers* 90(4): 713–737.

Dell, M., B.F. Jones and B.A. Olken. 2008. Climate Change and Economic Growth: Evidence from the Last Half Century. National Bureau of Economic Research Working Paper No. 14132, Washington, DC: NBER.

Deressa, T.T. 2007. Measuring the Economic Impact of Climate Change on Ethiopian Agriculture: Ricardian Approach. World Bank Policy Research Working Paper No. 4342. Washington, DC: World Bank.

Deressa, T.T., and R.M. Hassan. 2009. Economic Impact of Climate Change on Crop Production in Ethiopia: Evidence from Cross-Section Measures. *Journal of African Economies* 18(4): 529–554.

Deressa, T.T., R.M. Hassan, T. Alemu, M. Yesuf and C. Ringler. 2008. Analyzing the Determinants of Farmers' Choice of Adaptation Methods and Perceptions of Climate Change in the Nile Basin of Ethiopia. IFPRI Discussion Paper No. 00798. Washington, DC: International Food Policy Research Institute.

Deressa, T.T., R.M. Hassan and C. Ringler. 2008. Measuring Ethiopian Farmers' Vulnerability to Climate Change across Regional States. IFPRI Discussion Paper No. 00806. Washington, DC: International Food Policy Research Institute.

Di Falco, S., G. Köhlin and M. Yesuf. 2012. Strategies to Adapt to Climate Change and Farm Productivity in the Nile Basin of Ethiopia. *Climate Change Economics* 3(2): 1–18.

Di Falco, S., and M. Veronesi. 2014. Managing Environmental Risk in the Presence of Climate Change: The Role of Adaptation in the Nile Basin of Ethiopia. *Environmental and Resource Economics* 57(4): 553–577.

Eakin, H., and L.A. Bojorquez-Tapia. 2008. Insights into the Composition of Household Vulnerability from Multicriteria Decision Analysis. *Global Environmental Change* 18(2008): 112–127.

Easter, C. 1999. Small States Development: A Commonwealth Vulnerability Index. *The Round Table* 351(1): 403–422.

Eriksen, S.H., K. Brown and P.M. Kelly. 2005. The Dynamics of Vulnerability: Locating Coping Strategies in Kenya and Tanzania. *Geography Journal* 171: 287–305.

Filmer, D., and L.H. Pritchett. 2001. Estimating Wealth Effects without Expenditure Data – or Tears: An Application to Educational Enrolments in States of India. *Demography* 38(1): 115–131.

Füssel, H.M. 2007. Vulnerability: A Generally Applicable Conceptual Framework for Climate Change Research. *Global Environmental Change* 17(2): 155–167.

Füssel, H.M., and R.J.T. Klein. 2006. Climate Change Vulnerability Assessments: An Evolution of Conceptual Thinking. *Climatic Change* (75): 301–329.

Gbetibouo, G.A. 2009. Understanding Farmers' Perceptions and Adaptations to Climate Change and Variability: The Case of Limpopo Basin, South Africa. IFPRI Discussion Paper No. 00849. Washington, DC: International Food Policy Research Institute Washington.

Gbetibouo, G.A., and C. Ringler. 2009. Mapping South African Farming Sector Vulnerability to Climate Change and Variability: A Subnational Level Assessment. IFPRI Discussion Paper No. 00885. Washington, DC: International Food Policy Research Institute.

Gebreegziabher, Z., A. Mekonnen, R. Deribe, S. Abera and M.M. Kassahun. 2013. Crop-Livestock Inter-Linkages and Climate Change Implications for Ethiopia's Agriculture: A Ricardian Approach. Environment for Development Initiative Discussion Paper No. 13–14. Gothenburg, Sweden: Environment for Development and Washington, DC: Resources for the Future.

Gebreegziabher, Z., J. Stage, A. Mekonnen and A. Alemu. 2015. Climate Change and the Ethiopian Economy: A Computable General Equilibrium Analysis. *Environment and Development Economics* 21(2): 205–225.

Hahn, M., A. Reiderer and S. Foster. 2009. The Livelihood Vulnerability Index: A Pragmatic Approach to Assessing Risks from Climate Variability and Change: A Case Study in Mozambique. *Global Environmental Change* 19: 74–88.

Hoddinott, J., and A. Quisumbing. 2003. Methods for Microeconometric Risk and Vulnerability Assessments. Social Protection Discussion Paper Series No. 0324. Social Protection Unit, Human Development Network. Washington, DC: World Bank.

IPCC (Intergovernmental Panel on Climate Change). 2001. Climate Change 2001: Impacts, Adaptation, and Vulnerability. Contribution of Working Group II to the Third

Assessment Report of the Intergovernmental Panel on Climate Change. Cambridge, UK: Cambridge University Press.

IPCC (Intergovernmental Panel on Climate Change). 2007. Climate Change 2007: Impacts, Adaptation and Vulnerability. In *Contribution of Working Group II to the Fourth Assessment Report of the Intergovernmental Panel on Climate Change*, edited by M.L. Parry, O.F. Canziani, J.P. Palutikof, P.J. van der Linden and C.E. Hanson. Cambridge, UK: Cambridge University Press, p. 976.

Kabubo-Mariara, J., and F.K. Karanja. 2006. The Economic Impact of Climate Change on Kenyan Crop Agriculture: A Ricardian Approach. CEEPA Discussion Paper No. 12. University of Pretoria, South Africa, Centre for Environmental Economics and Policy in Africa.

Kelly, P.M., and W.N. Adger. 2000. Theory and Practice in Assessing Vulnerability to Climate Change and Facilitating Adaptation. *Climatic Change* 47: 325–352.

Lucas, P.L., and H.B.M. Hilderink. 2004. The Vulnerability Concept and Its Application to Food Security. Report 550015004/2004. Netherlands: National Institute for Public Health and the Environment (RIVM).

McCarthy, J., O.F. Canziani, N.A. Leary, D.J. Dokken and C. White (Eds.). 2001. Climate Change 2001: Impacts, Adaptation, and Vulnerability. Contribution of Working Group II to the Third Assessment Report of the Intergovernmental Panel on Climate Change. Cambridge, UK: Cambridge University Press. https://www.ipcc.ch/publications_and_data/ar4/wg2/en/frontmattersg.html

Mendelsohn, R., I.C. Prentice, O. Schmitz, B. Stocker, R. Buchkowski and B. Dawson. 2016. The Ecosystem Impacts of Severe Warming. *American Economic Review* 106(5): 612–614.

Mideksa, T.K. 2010. Economic and Distributional Impacts of Climate Change: The Case of Ethiopia. *Global Environmental Change* 20(2): 278–286.

Moser, C.O.N. 1998. The Asset Vulnerability Framework: Reassessing Urban Poverty Reduction Strategies. *World Development* 26(1): 1–19.

Moss, R.H., A.L. Brenkert and E.L. Malone. 2001. Vulnerability to Climate Change: A Quantitative Approach. Research Report Prepared for the U.S. Department of Energy.

Nhemachena, C., and R. Hassan. 2007. Micro-Level Analysis of Farmers' Adaptation to Climate Change in Southern Africa. IFPRI Discussion Paper No. 00714. Washington, DC: International Food Policy Research Institute.

O'Brien, K., R. Leichenko, U. Kelkar, H. Venema, G. Aandahl, H. Tompkins, A. Javed, S. Bhadwal, S. Barg, L. Nygaard and J. West. 2004. Mapping Vulnerability to Multiple Stressors: Climate Change and Globalization in India. *Global Environmental Change* 14: 303–313.

Osman-Elasha, B., N. Goutbi, E. Spanger-Siegfried, W. Dougherty, S. Hanafi, S. Zakieldeen, A. Sanjak, H. Abdel and H.M. Elhassan. 2006. Adaptation Practices and Policies to Increase Human Resilience against Climate Variability and Change: Lessons from the Arid Regions of Sudan. Working Paper No. 42. Washington, DC: Assessments of Impacts and Adaptations to Climate Change.

Owuor, B., S. Eriksen and W. Mauta. 2005. Adapting to Climate Change in a Dryland Mountain Environment in Kenya. *Mountain Research and Development* 25(4): 310–315.

Oxfam. 2009. Oxfam Annual Report 2009–10 for the Stitching Oxfam International Trustees Report. www.oxfam.org/en/about/accountability

Paavola, J. 2004. Livelihoods, Vulnerability and Adaptation to Climate Change in the Morogoro Region, Tanzania. Working Paper EDM No. 04–12. Centre for Social and Economic Research on the Global Environment. Norwick, UK: University of East Anglia.

Patnaik, U., and K. Narayanan. 2005. Vulnerability and Climate Change: An Analysis of the Eastern Coastal Districts of India. Paper Presented at the "Human Security and Climate Change International Workshop." June 21–23, Asker, Norway.

Ribot, J.C. 1996. Climate Variability, Climate Change and Vulnerability: Moving Forward by Looking Back. In *Climate Variability, Climate Change and Social Vulnerability in the Semi-Arid Tropics*, edited by J.C. Ribot, A.R. Magalhães and S.S. Panagides. Cambridge, UK: Cambridge University Press.

Shank, R., and C. Ertiro. 1996. *A Linear Model for Predicting Enset Plant Yield and Assessment of Kocho Production in Ethiopia*. Addis Ababa: UNDP and WFP.

Shewmake, S. 2008. Vulnerability and the Impact of Climate Change in South Africa's Limpopo River Basin. IFPRI Discussion Paper No. 804. Washington, DC: International Food Policy Research Institute, Washington.

Smit, B., and J. Wandel. 2006. Adaptation, Adaptive Capacity and Vulnerability. *Global Environmental Change* 16: 282–292.

Stringer, L.C., J.C. Dyer, M.S. Reed, A.J. Dougill, C. Twyman and D. Mkwambisi. 2009. Adaptations to Climate Change, Drought and Desertification: Local Insights to Enhance Policy in Southern Africa. *Environmental Science and Policy* 12(7): 748–765.

Sutherland, K., B. Smit, V. Wulf and T. Nakalevu. 2005. Vulnerability to Climate Change and Adaptive Capacity in Samoa: The Case of Saoluafata Village. *Tiempo* 54: 11–15.

TERI (The Energy Research Institute). 2003. *Coping with Global Change: Vulnerability and Adaptation in Indian Agriculture*. Delhi, India: The Energy and Resource Institute.

Thornton, P.K., J. Jones, T. Owiyo, R.L. Kruska, M. Herrero, V. Orindi, S. Bhadwal, P. Kristjanson, A. Notenbaert, N. Bekele and A. Omolo. 2008. Climate Change and Poverty in Africa: Mapping Hotspots of Vulnerability. *African Journal of Agricultural and Resource Economics* 2(1): 24–44.

Thornton, P.K., P.G. Jones, T. Owiyo, R.L. Kruska, M. Herrero, P. Kristjanson, A. Notenbaert, N. Bekele and A. Omolo with contributions from V. Orindi, B. Otiende, A. Ochieng, S. Bhadwal, K. Anantram, S. Nair, V. Kumar and U. Kulkar. 2006. Mapping Climate Vulnerability and Poverty in Africa. Report to the Department for International Development, London, UK. International Livestock Research Institute, Nairobi, Kenya.

Thurlow, J., T. Zhu and X. Diao. 2009. The Impact of Climate Variability and Change on Economic Growth and Poverty in Zambia. IFPRI Discussion Paper No. 00890. Washington, DC: International Food Policy Research Institute.

UN. 2013. *World Population Prospects: The Revision 2012*, Volume 1. Comprehensive Tables (United Nations Publications ST/ESA/SER.A/336). Herndon, VA, US.

UNEP. 2004. Poverty-Biodiversity Mapping Applications. Presented at IUCN World Conservation Congress. November 17–25, Bangkok, Thailand.

USAID. 2007. Famine Early Warning Systems Network (FEWS-NET). www.fews.net/

Vincent, K. 2004. Creating an Index of Social Vulnerability to Climate Change for Africa. Technical Report No. 56. Norwich, UK: Tyndall Centre for Climate Change Research, University of East Anglia.

Vincent, K. 2007. Uncertainty in Adaptive Capacity and the Importance of Scale. *Global Environmental Change* 17(1): 12–24.

Von Braun, J., and P. Webb. 1995. *Famine and Food Security in Ethiopia: Lessons for Africa*. Washington, DC: IFPRI.

Watson, R.T., M.C. Zinyowera and R.H. Moss (Eds.). 1996. *Climate Change 1995: Impacts, Adaptations and Mitigation of Climate Change: Scientific-Technical Analyses, Contribution of Working Group II to the Second Assessment Report of the Intergovernmental Panel on Climate Change.* Cambridge, UK: Cambridge University Press.

WFP (World Food Programme). 2004. Annual Report. www.wfp.org/sites/default/files/2004_wfp_annual_report.pdf

World Bank. 2007. *World Development Report: Development and the Next Generation.* Washington, DC: World Bank.

World Bank. 2008. Ethiopia: A Country Study on the Economic Impacts of Climate Change. Environment and Natural Resource Management Report No. 46946-ET. Washington, DC: Sustainable Development Department, Africa Region, World Bank.

World Bank. 2014. Ethiopia Overview. www.worldbank.org/en/country/ethiopia/overview (accessed July 22, 2014).

Yesuf, M., S. Di Falco, C. Ringler and G. Köhlin. 2008. Impact of Climate Change and Adaptation to Climate Change on Food Production in Low-Income Countries: Household Survey Data Evidence from the Nile Basin of Ethiopia. IFPRI Discussion Paper No. 00828. Washington, DC: International Food Policy Research Institute.

Yilma, Z., A. Mebratie, R. Sparrow, D. Abebaw, M. Dekker, G. Alemu and A.S. Bedi. 2012. Coping with Shocks in Rural Ethiopia. *Journal of Development Studies* 50(7): 1009–1024.

Ziervogel, G., and R. Calder. 2003. Climate Variability and Rural Livelihoods: Assessing the Impact of Seasonal Climate Forecasts in Lesotho. *Area* 35(4): 403–417.

Appendix

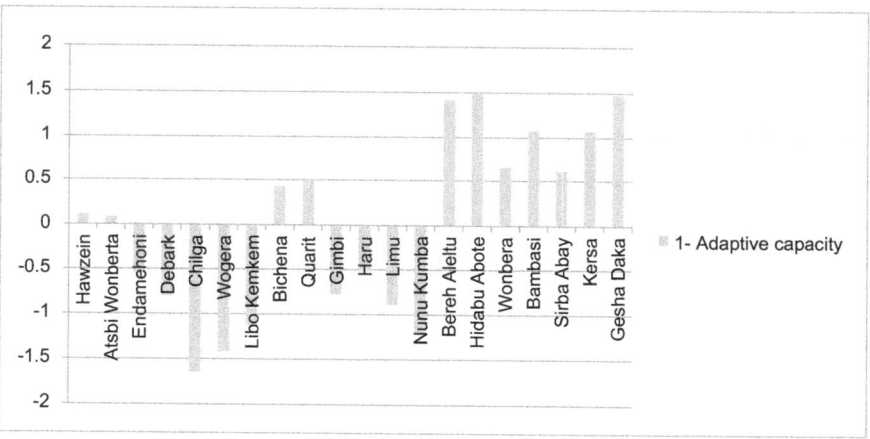

Figure 3A1 Result of Vulnerability Indices across District Level of the Nile Basin of Ethiopia

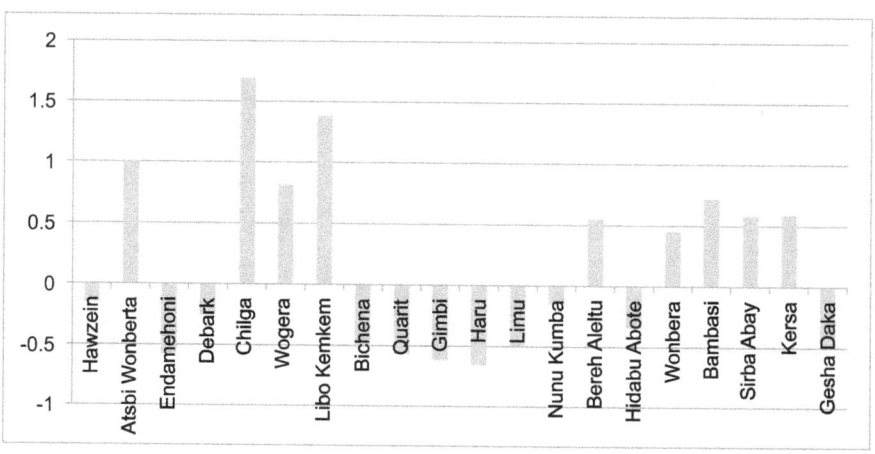

Figure 3A2 Exposure

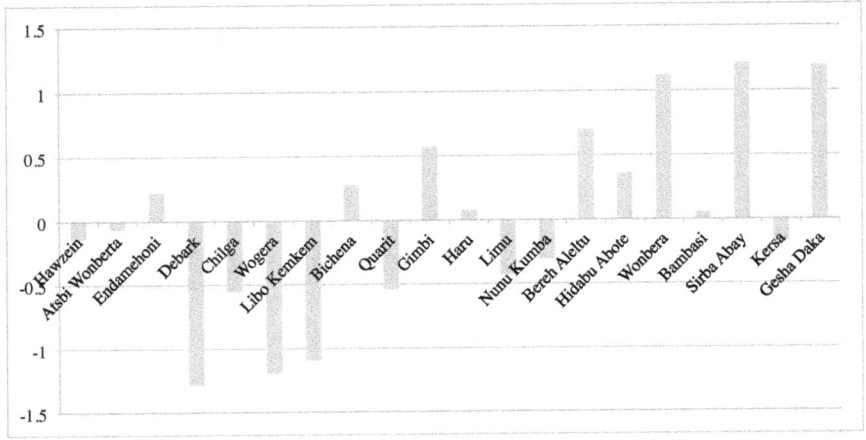

Figure 3A3 Sensitivity

4 Climate change and food security in Kenya*

Jane Kabubo-Mariara and Millicent Kabara

Introduction

Climate change has intensified the risk of catastrophic natural disasters all over the world. Residents of developing countries are particularly vulnerable to these catastrophic risks for three reasons: first, they rely primarily on natural resource–dependent income sources for their livelihoods; second, they have few resources with which to adapt to the anticipated change in climatic patterns; and, third, lack of planning and poor management at the central level impedes or delays recovery from climate-related shocks, and in some cases even leads to increased economic and social damage.

The World Food Programme (2011) notes that climatic change threatens to significantly increase the number of people at risk of hunger and undernutrition. Predictions are that more powerful and more frequent droughts and storms will wreak greater devastation. Rising sea levels will ruin fertile farmland. Changing rainfall patterns will deplete harvests. Increasingly scarce resources will exacerbate social tensions and may spark conflict. Millions more people will be at risk of hunger and undernutrition. Most of them will be in the world's poorest countries, where hunger, undernutrition and food insecurity are already widespread. Sub-Saharan Africa is the region worst affected by hunger and undernutrition. In some countries, yields from rain-fed agriculture could fall by 50% by 2020 (World Food Programme, 2011).

According to the Intergovernmental Panel on Climate Change (IPCC, 2001), climate change will lead to increases in the frequency and intensity of natural disasters and extreme weather events, such as droughts, floods and hurricanes; rising sea levels and the contamination/salinization of water supplies and agricultural lands; changes in rainfall patterns, with an expected reduction in agricultural productivity in already fragile areas, especially in sub-Saharan Africa, and declining water quality and availability in arid and semiarid regions. Diminishing water

* The authors gratefully acknowledge financial support from the Swedish International Development Cooperation Agency (Sida) through the Environment for Development Initiative at the University of Gothenburg, Sweden, Department of Economics, Environmental Economics Unit.

availability and quality, together with rising water demand, have created immense challenges in poor and vulnerable communities. The effects of these changes on hunger and undernutrition have been felt across the world, with a disproportionate impact on vulnerable communities in less developed countries – those with the least resources and capacities to adapt and respond.

Climate change is expected to affect food security in several respects: increased vulnerability to climate change due to dependence on rain-fed agriculture; high levels of poverty; and low levels of human and physical capital as well as generally poor infrastructure. IPCC predicts that, by 2050, crop yields in sub-Saharan Africa will have declined by 14% (rice), 22% (wheat) and 5% (maize), pushing the vast number of already poor people, who depend on agriculture for their livelihoods, deeper into poverty and vulnerability. It also predicts decreased food availability by 500 calories less (a 21% decline) per person in 2050 and a further increase in the number of malnourished children by over 10 million – a total of 52 million in 2050 in sub-Saharan Africa alone.

Against this background, this chapter analyzes the effect of climate variables on food security in Kenya, a low-middle income country characterized by low and declining crop productivity. The country can be divided into seven agro-climate zones (Sombroek et al., 1982). These zones differ in terms of moisture index, rainfall, vegetation and farming systems. High-potential areas (Zones I, II and III) have a moisture index greater than 50% but account for only 12% of Kenya's land area. These are located above an altitude of 1200 m and have mean annual temperatures of below 18°C. These areas are mainly suitable for livestock farming (mostly cattle and sheep), cash crops (coffee, tea and pyrethrum) and key food crops (maize, beans and wheat). Medium-potential zones favor farming systems similar to the high-potential areas, but barley, cotton, cassava, coconut and cashew nuts also are cultivated. Areas with moisture indexes of less than 50% are semi-humid to arid regions (Zones IV, V, VI and VII) and account for about 80% of the land area. Most arid and semiarid areas lie below 1260m, have relatively high temperatures and are less suited for arable agriculture. Sorghum, millet, livestock and wildlife are the main farming systems (Table 4A1).[1]

Given this agroecological setting, agricultural production is undermined by unpredictable weather and climate variations, especially in the less arable zones. Climatic variations affect crop and livestock systems both directly and indirectly and could have severe socioeconomic impacts, such as shortages of food, water, energy and other essential basic commodities, as well as long-term food insecurity (Kabubo-Mariara, 2008a). There is a growing body of literature on the impact of climate change in Africa. Most studies have concentrated on the impact of climate change on crop and livestock productivity, while other studies have assessed adaptation to climate change (see, e.g., Kabara and Kabubo-Mariara, 2011; Herrero et al., 2010; Kabubo-Mariara, 2008b, 2009; Deressa et al., 2009; Dinar et al., 2008; Hassan and Nhemachena, 2008; Kabubo-Mariara and Karanja, 2007; Deressa et al., 2005; Gbetibouo and Hassan, 2005; Turpie et al., 2002). To design policy measures for averting catastrophes such as those experienced in the Horn of Africa in 2011, there is a need for research on the impact of climate change on food security. There is, however, a dearth of literature on the relationship between climate change and food security in Kenya. This study addresses the gap. The general objective of this study is to analyze the linkages between climate change and food security in Kenya, and to formulate policy options for mitigating the effect of climate change on food security in Kenya.

The rest of the chapter is organized as follows. The next section discusses the data. The section that follows presents the framework for analysis and methods. The fourth section presents the results and discussion. The last section concludes with policy implications.

Data

This study uses county-level data for the period 1975 to 2012. Crop data was sourced mainly from the Ministry of Agriculture,[2] supplemented by data from the International Livestock Research Institute (ILRI), which provided more refined data at the division level. The data sets for four climate variables (precipitation, temperature, runoff, and total cloud cover) were obtained from the European Centre for Medium-Range Weather Forecasts (ECMWF) reanalysis-interim (ERA-Interim) model, which archives the data in both daily and twice daily ten-day and monthly formats. The resolution of the data is 1.5 degree by 1.5 degree (approx. 150 km).[3] ECMWF is the most comprehensive model in terms of archiving most of the common and uncommon weather parameters. Precipitation and runoff data are measured in millimeters, while temperature data is measured in degrees Celsius. Cloud cover or precipitable moisture/cloudiness, which indicates the amount of water vapor in a vertical column of the atmosphere (Revuelta et al., 1985), is measured as a percentage of the sky covered by clouds. Population data was sourced from the Kenya National Bureau of Statistics. The soil database used in this study is a combination of the Kenya Soil Survey (KSS) soil map of 1982 (Sombroek et al., 1982), revised in 1997, and the FAO et al. (2012) classification. The agroecological zone layer was sourced from the Food and Agriculture Organization (1998). Base maps were generated from the Kenya census of 1999. The layers were sourced from the ILRI GIS database, which was generated in collaboration with the Kenya National Bureau of Statistics (KNBS).

Framework for analysis and methods

This study adopts the framework on climate change and food security developed by FAO (2008). The framework illustrates how adaptive adjustments to food system activities will be needed all along the food chain to cope with the impacts of climate change. Climate change affects food security outcomes for the four components of food security – food availability, food accessibility, food utilization and food system stability – in various direct and indirect ways (FAO, 2008; Schmidhuber and Tubiello, 2007). The transmission mechanism from climate change to food insecurity is complex. Climate change variables influence biophysical factors and how they are managed through agricultural practices and land use for food production. They also influence physical and human capital, which indirectly affect the economic and sociopolitical factors that govern food access and utilization. The framework shows that the impact of climate change on food security occurs through the following mechanisms: the CO_2 fertilization effect of increased greenhouse gas concentrations in the atmosphere; increasing mean, maximum and minimum temperatures; gradual changes in precipitation; increase in the frequency, duration and intensity of dry spells and droughts; changes in the timing, duration, intensity and geographic location of rain and snowfall; increase in the frequency and intensity of storms and floods; and greater seasonal weather variability and changes in the start/end of growing seasons. These climate change variables have been shown in the literature to impact agricultural productivity and global food supply. Following this framework, the study assesses the impact of climate change on the food availability dimension of food security. We investigate the impact of climate variability and change on major food crops (maize, beans, sorghum and millet).

This chapter uses a modified form of the Massetti and Mendelsohn (2011) approach to estimate the effects of climate change on food insecurity. This is a modified Ricardian approach applied to panel data – an extension of the Deschenes and Greenstone (2007) approach. The basic starting point is the Ricardian approach, which assumes that the value of farm land per acre is affected by climate change (Mendelsohn et al., 1994). Deschenes and Greenstone (2007) extended the Ricardian model and used county-level panel data to estimate the effect of weather on agricultural profits, conditional on county and state-by-year fixed effects. Their approach differs from the hedonic approach in a few key ways. First, under an additive separability assumption, its estimated parameters are purged of the influence of all unobserved time-invariant factors. Second, land values cannot be used as the dependent variable once the county fixed effects are included because land values reflect long-run averages of weather, rather than annual deviations from these averages, and there is no time variation in such variables. Third, the approach can be used to approximate the effect of climate change on agricultural land values, though land value is not the independent variable. Deschenes and Greenstone (2007) caution that there are two issues that could undermine the validity of using annual variation in weather to infer the impacts of climate change. First, short-run variations in weather may lead to temporary changes in prices that obscure the true long-run impact of climate change. Second, farmers cannot undertake the full range of adaptations in response to a single year's weather realization. The authors, however, show evidence to discount the possibility that the results obtained were affected by either of these two concerns.

Deschenes and Greenstone (2007) start with the hedonic cross-sectional model (Mendelsohn et al., 1994) and estimate the following equation:

$$Y_{ct} = \alpha_c + \gamma_t + X'_{ct}\beta_t + Z'_c\varphi_t + W'_c\theta_t + \varepsilon_{ct} \tag{1}$$

where Y_{ct} is the value of agricultural land per acre in county c in year t, measured by county-level agricultural profits. α_c and γ_t are the county fixed effects and year indicators respectively – γ_t controls for annual differences in the dependent variable that are common across counties. X_{ct} is a vector of observable time-varying determinants of farmland values. Z_c is a vector of time-invariant control variables. W_c' is a vector of climate variables (annual realizations of weather). ε_{ct} is the stochastic error term, which has two components: a permanent, county-specific component, α_c, and an idiosyncratic shock, u_{ct}. β, γ and θ are all time-variant coefficients. θ captures the true effect of climate on farmland values. Equation (1) is further extended to include interactions of irrigated and non-irrigated farms and climate variables. Deschenes and Greenstone (2007) make appropriate assumptions and adjustments[4] to correct for econometric problems arising in the estimation of Equation (1), including omitted variable bias.

Deschenes and Greenstone (2007) estimate the foregoing model using repeated cross sections. Massetti and Mendelsohn (2011), however, argue that Equation (1) is mis-specified because, in an ideal panel data model, the coefficients of the time-invariant variables should not change over time. They propose the following modified form of Equation (1), with β, γ and θ as time-invariant coefficients:

$$Y_{ct} = X_{ct}'\beta + Z_c'\varphi + W_c'\theta_t + \varepsilon_{ct} \tag{2}$$

We estimate two variants of Equation (2) in this study. First, we use crop productivity as a proxy for food security, as has been done in the literature (Gregory et al., 2005; Parry et al., 2004; Xiong et al., 2007). The dependent variable in this study is, however, crop yields (production per acre) rather than land values or profits. Second, we estimate another variant where the dependent variable is the probability of being food-insecure.[5] The explanatory variables include climate-related variables (temperature and precipitation), runoff, population, agroecological zones and soil variables. Fixed effects and random effects models are estimated for both crop productivity and food insecurity, with the random effects model testing the effect of time-invariant variables (Mundlak, 1978). Sensitivity analysis is carried out through simulations based on the Special Report on Emissions Scenarios (SRES) and atmospheric oceanic global circulation models (AOGCMs).

Results and discussion

Descriptive statistics

The summary statistics are presented in Table 4.1. The data shows that, on average, yields ranged from 1.81 to 3.13 tons per hectare. Millet reported the highest mean yields, while the lowest was from beans, which also recorded the highest variability. Average total yields were estimated at 2.15 tons per hectare.

For climate/weather-related variables, we use two different definitions of seasons to estimate a series of models for the main food crops. First, there are two main cropping seasons in Kenya: the extended long rains season, which runs from March to August, and the short-rains season, which runs from September to February. The long and short rains refer to the extended wet and dry conditions,

Table 4.1 Summary Statistics

Variable	Mean	Std. dev.
Maize yields	2.21	3.98
Bean yields	1.81	22.54
Millet yields	3.13	10.45
Sorghum yields	1.92	3.41
Spring precipitation (Sept–Nov)	10.54	12.62
Fall precipitation (Dec–Feb)	7.58	9.45
Summer precipitation (Mar–May)	15.82	19.29
Winter precipitation (June–Aug)	7.63	10.29
Short-rains (Sept–Feb) precipitation	9.06	10.28
Long-rains (Mar–Aug) precipitation	11.72	13.89
Spring temperature (Sept–Nov)	17.95	3.37
Fall temperature (Dec–Feb)	18.12	3.61
Summer temperature (Mar–May)	18.96	3.50
Winter temperature (June–Aug)	17.22	3.46
Short-rains temperature	18.03	3.47
Long-rains temperature	18.09	3.46
Summer runoff (mm)	0.20	0.40
Winter runoff	0.12	0.23
Spring runoff	0.13	0.26
Fall runoff	0.07	0.15
Short-rains runoff	0.16	0.31
Long-rains runoff	0.10	0.19
Population density	269.06	577.57
Agroecological zone	2.49	1.47
Soil drainage	7.13	2.17
Soil depth	4.62	2.67
Silt soil	18.03	14.47

respectively. In Kenya, long rains fall between March and May and short rains between October and December. The extended rain seasons are, however, longer, covering the whole cropping season. Long-rains crops planted in early March are harvested in August. Farms are then prepared and planted in September and the crops harvested in February. In this study, the long rains season is therefore defined as March to August and the short-rains season as September to February. We further define seasons to correspond to three-month spells, or short seasons, where each of the two rain seasons is broken down into two seasons: March to May (summer), June to August (winter), September to November (spring) and December to February (fall).[6] The same definition of seasons is applied to runoff.

The data shows little variation in long- versus short-rains temperatures, which are on average about 18°C. There is more variation in the spell-length seasonal temperatures, with the highest recorded for summer and lowest for winter, with a range of almost 2°C. Precipitation shows more significant variation between seasons. Average runoff varies from 0.07 to 0.2 cubic meters. Summer experiences the highest runoff and variability, compared to fall, which exhibits the lowest runoff and least variability. Long-rains runoff was much lower than short-rains runoff. The average population density stood at 269.06 persons per square kilometer. Agroecological zones are based on the length of the growing period (LGP) – that is, the period in which the soil has enough moisture to sustain plant growth. As mentioned earlier, the country is subdivided into seven major agroecological zones, ranging from humid to very arid, with the regions varying in rainfall, moisture availability and temperature highs and lows. Regarding soils, the only information available at the county level is on three types of soil (clay, silt and sand, with clay soils being the dominant type, followed by sand), soil depth and soil drainage. The distribution of type of soil, depth and drainage varies by agroecological setting. Agroecological zone and soil data are time-invariant.

Effect of climate variables on food security

Crop productivity analysis

Regression results

Table 4.2 presents fixed effects regression results using the short and long rain season definitions. The results show that both short- and long-rains precipitation exhibit an inverted U-shaped relationship with most crops. The coefficients are significant for maize (for both long and short rains) and for sorghum (only for long rains), but are insignificant for beans and millet.[7] Long-rains precipitation has larger impacts than short-rains precipitation. The results suggest that high rainfall is crucial for increased crop productivity and thus food security, but excessive rainfall is harmful. This is because flooding and waterlogging destroy crops at the formative periods, while heavy rains during the harvest season lead to rotting of mature crops. Short-rains temperature exhibits a U-shaped relationship with maize and sorghum yields, while long-rains temperature exhibits a U-shaped relationship, which is significant for maize only. The results support literature that has found nonlinear effects of temperature and precipitation on agricultural production (Mendelsohn et al., 1994, 2003; Kurukulasuriya and Mendelsohn, 2008; Kabubo-Mariara and Karanja, 2007).[8] Short-rains runoff is associated with higher yields, while long-rains runoff is associated with lower yields. Earlier studies found a nonlinear relationship between hydrological factors and crop revenue (Kabubo-Mariara and Karanja, 2007), but we uncovered no significant effect when we introduced the quadratic term.

We further investigate the impact of population density on crop yields.[9] The literature suggests that population density is a proxy for agricultural adaptation options (Kurukulasuriya and Mendelsohn, 2008). Population density could also capture availability of farm labor. The positive significant effect on maize and

Table 4.2 Fixed Effect Estimates of Weather Variability on Crop Yields, Short- and Long-Rains Seasons

Variables	Maize	Sorghum	Bean	Millet
Short-rains (Sept–Feb) precipitation	0.2170***	0.0478	0.0004	−0.0231
	[0.065]	[0.062]	[0.002]	[0.256]
Short-rains precipitation squared	−0.0056***	−0.001	0.0000	−0.0002
	[0.001]	[0.001]	[0.000]	[0.005]
Long-rains (Mar–Aug) precipitation	0.2974***	0.1279**	0.0011	−0.1748
	[0.052]	[0.050]	[0.001]	[0.195]
Long-rains precipitation squared	0.0002	−0.0003	0.0000	0.0028
	[0.001]	[0.001]	[0.000]	[0.003]
Short-rains temperature	−7.8370***	−4.2380**	−0.0333	3.1717
	[1.795]	[1.751]	[0.056]	[9.457]
Short-rains temperature squared	0.1609***	0.0851*	0.0009	−0.1009
	[0.049]	[0.047]	[0.002]	[0.284]
Long-rains temperature	3.8528**	0.3461	−0.0143	−3.6579
	[1.601]	[1.589]	[0.050]	[7.994]
Long-rains temperature squared	−0.0721*	−0.0035	0.0004	0.0953
	[0.043]	[0.042]	[0.001]	[0.236]
Short-rains runoff	4.2094***	0.7331	0.0171	−0.6717
	[1.304]	[1.302]	[0.038]	[4.595]
Long-rains runoff	−5.5259***	−1.4086	0.0215	0.7192
	[2.074]	[2.040]	[0.060]	[7.300]
Population density	0.0016***	0.0011	0.0001**	−0.0041
	[0.001]	[0.001]	[0.000]	[0.010]
Constant	38.7435***	40.3589***	0.8992**	14.9148
	[13.637]	[13.754]	[0.407]	[65.367]
Observations	1,362	1,105	1,462	696
R-squared	0.198	0.055	0.01	0.003
Number of code	47	44	43	24

Note: Standard errors in brackets; *** $p<0.01$,

** $p<0.05$,

* $p<0.1$.

beans could be interpreted as suggesting that adaptation to climate change and availability of family labor are associated with increased yields and thus food security. Similar conclusions were also drawn by Nhemachena et al. (2010), who found that factor endowment (land, labor, capital, technology) contributes to higher net revenue. However, sorghum and millet are found to be unresponsive to population density.

Table 4.3 shows the effect of short seasonal weather variability on crop yields. The results once again suggest that beans and millet are largely unresponsive to weather variability. We find that both spring and winter precipitation exhibit an

Table 4.3 Fixed Effect Estimates of Seasonal Weather Variability on Crop Yields

Variables	Maize	Sorghum	Bean	Millet
Spring precipitation (Sept–Nov)	0.1865***	0.0253	0.0009	−0.0231
	[0.045]	[0.045]	[0.001]	[0.179]
Spring precipitation squared	−0.0046***	−0.0016**	0.0000	0.0000
	[0.001]	[0.001]	[0.000]	[0.003]
Fall precipitation (Dec–Feb)	−0.0406	−0.0124	−0.0008	−0.1105
	[0.046]	[0.045]	[0.001]	[0.250]
Fall precipitation squared	0.0009	0.0012	0.0000	0.0015
	[0.001]	[0.001]	[0.000]	[0.007]
Summer precipitation (Mar–May)	−0.0815**	−0.0584*	0.0006	−0.3171**
	[0.035]	[0.034]	[0.001]	[0.136]
Summer precipitation squared	0.0019***	0.0010**	0.0000	0.0026*
	[0.000]	[0.000]	[0.000]	[0.002]
Winter precipitation (June–Aug)	0.6052***	0.3758***	0.0009	0.238
	[0.054]	[0.052]	[0.002]	[0.204]
Winter precipitation squared	−0.0064***	−0.0051***	0.0000	−0.0022
	[0.001]	[0.001]	[0.000]	[0.004]
Spring temperature (Sept–Nov)	0.5032	1.0839	0.0888	5.6035
	[1.669]	[1.666]	[0.054]	[8.968]
Spring temperature squared	−0.0146	−0.031	−0.0024	−0.1751
	[0.046]	[0.045]	[0.002]	[0.272]
Fall temperature (Dec–Feb)	−4.3586***	−2.6975**	−0.1034**	3.7938
	[1.348]	[1.355]	[0.045]	[7.403]
Fall temperature squared	0.0889**	0.0562	0.0028**	−0.0901
	[0.036]	[0.035]	[0.001]	[0.216]
Summer temperature (Mar–May)	−3.2914**	−3.6622**	0.0207	−18.0565**
	[1.584]	[1.601]	[0.053]	[7.964]
Summer temperature squared	0.0473	0.0853**	−0.0003	0.4017*
	[0.039]	[0.039]	[0.001]	[0.215]
Winter temperature (June–Aug)	3.2444**	1.7625	−0.0532	7.0621
	[1.423]	[1.411]	[0.045]	[6.356]
Winter temperature squared	−0.04	−0.0374	0.0013	−0.1404
	[0.040]	[0.038]	[0.001]	[0.191]
Summer runoff	1.8362***	0.7027	0.0214	0.2446
	[0.666]	[0.677]	[0.020]	[2.547]
Fall runoff	−4.3032**	−2.7322	0.02	−1.3764
	[1.726]	[1.569]	[0.052]	[8.003]
Population density	0.0017***	0.0012	0.0001**	−0.003
	[0.001]	[0.001]	[0.000]	[0.010]
Constant	43.2779***	39.4000***	0.8652**	40.3216
	[13.613]	[14.048]	[0.427]	[68.552]
Observations	1,362	1,105	1,462	696
R-squared	0.28	0.111	0.015	0.024
Number of code	47	44	43	24

Note: Standard errors in brackets; *** p<0.01,

** p<0.05,

* p<0.1.

inverted U-shaped (hill) relationship with maize and sorghum yields, which supports the results for short and long-rains seasons' precipitation. Summer precipitation exhibits a U-shaped relationship with maize and sorghum yields, but the effect of fall precipitation is insignificant for all crops. This implies that high summer precipitation is not necessarily beneficial for crops, as this is the formative period for crop growth. Previous studies have shown that high precipitation is not always beneficial for farmers (Kabubo-Mariara, 2009) and could therefore also be associated with food insecurity. Turning to temperature, we find that fall temperature exhibits a U-shaped relationship with maize and beans yields, while similar relationships are found between summer temperature and all crop yields except beans. Winter temperatures exhibit a hill-shaped relationship with most crops, but this is significant only for maize. High temperatures during the planting period slow down or destroy crop growth, while moderately high winter temperatures are crucial for crop maturity (Kabubo-Mariara and Karanja, 2007). The results for the short spell-length seasonal climate variables support the results for the short and long rain season variables and illustrate the nonlinear relationship between climate variables and crop productivity.

We further investigate the effect of summer and fall runoff on crop yields. We find that summer runoff has a positive impact on crop yields, but the effect is significant only for maize. Fall runoff exhibits a negative impact, but the coefficient is significant only for maize and sorghum. As with the first model, the impact of population density is positive and significant for maize and beans yields, suggesting that population density is associated with higher productivity and thus better food security.

In Table 4.4, we introduce time-invariant variables into the model, following Mundlak (1978). The variables include agroecological zones and soil characteristics. First, we find that the results of all the time-variant variables are consistent

Table 4.4 Random Effects Estimates of Seasonal Weather Variability on Crop Yields

Variables	Maize	Sorghum	Beans	Millet
Spring precipitation (Sept–Nov)	0.1364***	−0.0114	−0.0008	−0.0023
	[0.043]	[0.041]	[0.001]	[0.172]
Spring precipitation squared	−0.0041***	−0.0011	0.0000	−0.0003
	[0.001]	[0.001]	[0.000]	[0.003]
Fall precipitation (Dec–Feb)	−0.0406	0.0212	−0.0004	−0.087
	[0.043]	[0.042]	[0.001]	[0.237]
Fall precipitation squared	0.0018**	0.0013	0.0000	0.0007
	[0.001]	[0.001]	[0.000]	[0.006]
Summer precipitation (Mar–May)	−0.1738***	−0.0840***	0.0004	−0.3072**
	[0.032]	[0.031]	[0.001]	[0.129]
Summer precipitation squared	0.0023***	0.0011***	0.0000	0.0026*
	[0.000]	[0.000]	[0.000]	[0.001]

Variables	Maize	Sorghum	Beans	Millet
Winter precipitation (June–Aug)	0.2380***	0.2256***	0.0049***	0.2287
	[0.038]	[0.038]	[0.001]	[0.186]
Winter precipitation squared	0.0003	−0.0023***	−0.0001***	−0.0022
	[0.001]	[0.001]	[0.000]	[0.004]
Spring temperature (Sept–Nov)	0.1819	1.6345	0.0419	5.3659
	[1.562]	[1.533]	[0.052]	[8.730]
Spring temperature squared	−0.0009	−0.0457	−0.001	−0.1697
	[0.043]	[0.042]	[0.001]	[0.265]
Fall temperature (Dec–Feb)	−0.9203	−1.095	0.024	3.3393
	[1.069]	[1.076]	[0.039]	[7.159]
Fall temperature squared	0.0198	0.027	−0.001	−0.0724
	[0.028]	[0.028]	[0.001]	[0.211]
Summer temperature (Mar–May)	−3.6934**	−3.6612**	0.0201	−17.3613**
	[1.484]	[1.460]	[0.051]	[7.727]
Summer temperature squared	0.0602	0.0908**	−0.0008	0.3846*
	[0.037]	[0.036]	[0.001]	[0.209]
Winter temperature (June–Aug)	3.2681**	2.6516**	−0.0920**	6.9111
	[1.280]	[1.267]	[0.043]	[6.289]
Winter temperature squared	−0.0592*	−0.0674**	0.0029**	−0.1419
	[0.035]	[0.034]	[0.001]	[0.190]
Summer runoff	0.0363	0.1497	0.0375*	0.1732
	[0.640]	[0.631]	[0.020]	[2.494]
Fall runoff	−3.8253**	−2.5774	−0.026	−1.5876
	[1.657]	[1.640]	[0.051]	[7.761]
Population density	0.0004	0.0005	0.00001**	−0.0052
	[0.000]	[0.000]	[0.000]	[0.009]
Agroecological zone	0.5970***	0.4734**	−0.0073	−1.1662
	[0.154]	[0.206]	[0.007]	[1.492]
Soil drainage	0.2122***	0.0067	0.0034	0.061
	[0.076]	[0.082]	[0.004]	[1.374]
Soil depth	0.0962	−0.009	−0.0033	−0.408
	[0.061]	[0.080]	[0.003]	[0.691]
Silt soil	−0.0201*	−0.0141	0.0014	−0.0112
	[0.011]	[0.012]	[0.001]	[0.185]
Constant	15.0954**	7.7277	0.5485**	45.7872
	[6.432]	[6.879]	[0.262]	[54.145]
Observations	1,362	1,105	1,462	696
Number of code	47	44	43	24

Note: Standard errors in brackets; *** p<0.01,

** p<0.05,

* p<0.1.

with the fixed effects estimates presented in Table 4.2. The results for the new variables show that maize is the only crop that is responsive to time-invariant variables. Specifically, maize productivity responds positively to favorable agro-ecological zones, soil drainage and depth, but performs poorly on silt soils. Sorghum productivity responds positively to favorable agroecological zone, but the effect of other time-invariant factors is insignificant. We also uncover no significant effect of the time-invariant factors on beans and millet. The results support literature that has shown that favorable development domain dimensions and population density are associated with higher crop productivity (Kabubo-Mariara, 2012) and thus better food security.

Simulating the effect of climate change on crop productivity

To estimate possible effects of future climate change on crop productivity, the study used a set of climate change scenarios (atmosphere-ocean global circulation models; AOGCM) predicted by the Intergovernmental Panel on Climate Change (IPCC, 2001). We used estimated model coefficients and corresponding variable means to examine how changes in climate are likely to affect future productivity of Kenyan crops. Predicted changes in temperature and precipitation are used to adjust benchmark values of crop yields and the impact evaluated. The simulations in this study are based on the A2 and B2 Special Report on Emissions Scenarios (SRES), as the two have been integrated by many AOGCMS because of the assumptions on which each is based.[10] Ten scenarios have been derived for Kenya by using five different models in conjunction with two different emission scenarios: A2 and B2 (Strzepek and McCluskey, 2006). The five models are: CGCM (coupled general circulation model), CSIRO (the Commonwealth Scientific and Industrial Research Organization model), ECHAM (the European Centre Hamburg model), HADCM (the Hadley Centre coupled model) and PCM (the parallel climate model). Based on these models, Strzepek and McCluskey (2006) predicted various temperature and precipitation changes for Kenya for various years up to 2100. This chapter uses the predictions for 2050 and 2100 to simulate the likely effect of future climate change on crop yields (Appendix Table 4A2). The numbers present the predicted decadal average changes in annual climate variables for 2050 and 2100, relative to the year 2000.[11]

   To evaluate the effect of climate change of crop yields, we first added the predicted change in temperature from each AOGCM to the mean temperature values for each county, and then evaluated the impact on crop yields. We further adjusted the mean precipitation by the predicted percentage to get the new precipitation levels. We then compared the predicted crop yield to the baseline yield and computed the percentage change. The results (Table 4.5) show that, except for HADCM3, which predicts a modest gain in maize yield in 2100, farms are expected to suffer declining crop yields due to climate change. Specifically, for the year's crop yield in 2050, the largest damage from a combination of increased temperatures and precipitation is predicted from the CSIRO2, HADCM3 and CGCM2 for both A2 and B2 SRES. The largest damage is estimated at a 69%

Table 4.5 Predicted Damage to Maize Crop Yields from Different AOGCM Climate Scenarios

Scenario	A2				B2			
Model	Year	Predicted yield	Net loss	% damage	Year	Predicted yield	Net loss	% damage
HADCM3								
	2050	0.88	−1.43	−62	2050	0.81	−1.50	−65
	2100	2.44	0.13	6	2100	1.10	−1.21	−52
PCM								
	2050	1.18	−1.13	−49	2050	1.14	−1.17	−51
	2100	0.95	−1.36	−59	2100	0.84	−1.47	−64
ECHAM								
	2050	1.19	−1.12	−48	2050	1.29	−1.02	−44
	2100	1.98	−0.33	−14	2100	1.37	−0.94	−41
CSIRO2								
	2050	0.89	−1.42	−62	2050	0.71	−1.60	−69
	2100	2.11	−0.20	−9	2100	0.90	−1.41	−61
CGCM2								
	2050	0.89	−1.42	−61	2050	0.92	−1.39	−60
	2100	1.50	−0.81	−35	2100	0.72	−1.59	−69

Note: Base yield = 2.31 tons per hectare.

decline in yields. The lowest potential damage is from the ECHAM B2 scenario, at 44%. For the year 2100, the greatest damages are predicted to come from the B2 SRES scenario and the CGCM2, PCM and CSIRO2 models. A2 SRES predicts rather modest losses for 2100. Kenya is therefore likely to suffer severe food insecurity by the year 2100 unless farmers mitigate and undertake adaptation measures against climate change. The simulation results support other results that have shown that global warming will damage crop production, alter crop choices and diversification, reduce livestock productivity and influence livestock adaptation options. The results also reveal differential effects predicted by different AOGCMs, as in the literature (Kabubo-Mariara, 2009, 2008a, 2008b; Kabubo-Mariara and Karanja, 2007).

Food security analysis

Regression results

We use subjective definitions of food availability to capture food security. The data available for this study is for crop productivity measured by yields. Productivity directly translates into food availability in that the lower the productivity, the lower the amount that will be available for consumption. In this chapter, we define households as food-insecure if they fall short of a relative poverty line. The poverty literature suggests that relative poverty lines can be set at various percentiles of the welfare measure: 25th percentile, 40th percentile and 60th percentile. In this case, our measure of food availability is yield and so we construct poverty thresholds based on the three percentiles of yields. A county would have experienced food insecurity if yields fell short of the 25th, 40th or 60th percentile of the yields in any one period.

Using this 60th percentile poverty threshold, the highest incidence of food insecurity is associated with maize (52%), followed by beans and sorghum (39% each). Millet is lowest at 27%. The same trend is observed for the other two percentile definitions of poverty, with maize recording an incidence of 23% and 35% for the 25th and 40th percentiles, respectively; beans and sorghum at 16% and 26% for the 25th and 40th percentiles, respectively; and millet only 11% and 18% for the two percentiles, respectively. The incidence of food insecurity taking into account all crops is basically equivalent to that of maize. This is probably because maize is the main food crop and so anyone insecure in maize is likely to be food-insecure.

We estimate panel regression models for the probability of being food-insecure at the three poverty thresholds for two variants of weather variability – the short- and long-rains definition, and also seasonal spells definitions. The discussion of the results is based on the 60th percentile poverty line.[12] The results for the short- vs. long-rains seasons' definition of weather variability are presented in Table 4.6. The results suggest that short-rains and long-rains precipitation exhibit U-shaped relationships with food insecurity for all crops, suggesting a nonlinear relationship between precipitation and food insecurity. The impact of short-rains precipitation is largest for millet crops, followed by maize. The largest impact of long-rains precipitation is on maize insecurity. The impact of the quadratic term is significant only for maize.

The short-rains temperatures exhibit inverted U-shaped relationships with food insecurity, again highlighting a nonlinear relationship between temperature and food insecurity, but we find no significant effect of long-rains temperature, although it has an inverted U-shaped relationship with all food crop insecurity. The short-rains effect is largest for millet and maize, while the long-rains effect is insignificant for all food crops. The results support other studies which have found a nonlinear relationship between climate variables and farm values (Mendelsohn et al. 1994, 2003; Kabubo-Mariara and Karanja, 2007; Kurukulasuriya and Mendelsohn, 2008; Deressa et al., 2005). The results also support earlier studies that have found that global warming is likely to have adverse effects on farm productivity in Africa (Massetti and Mendelsohn, 2011; Kabara and Kabubo-Mariara, 2011; Mohamed et al., 2002; Molua, 2002, 2008; Nhemachena et al., 2010).

We do not uncover any significant effect of short- and long-rains runoff on food insecurity. Population density is, however, associated with lower food insecurity. This is because areas of high population density are associated with higher crop

Table 4.6 Random Effects Estimates of Short- and Long-Rains Seasons Weather Variability on Food Security – Poverty Line Set at 60% Percentile of Yields

Variables	Maize	Sorghum	Bean	Millet
Short-rains (Sept–Feb) precipitation	−0.0866***	−0.0408**	−0.0840***	−0.1177***
	[0.024]	[0.021]	[0.022]	[0.031]
Short-rains precipitation squared	0.0011**	0.0002	0.0010*	0.0012*
	[0.001]	[0.000]	[0.001]	[0.001]
Long-rains (Mar–Aug) precipitation	−0.0821***	−0.0303*	−0.0366**	−0.0186
	[0.021]	[0.017]	[0.019]	[0.025]
Long-rains precipitation squared	0.0007*	0.0003	0.0003	0.0003
	[0.000]	[0.000]	[0.000]	[0.000]
Short-rains temperature	2.2624***	1.3781**	1.1282*	3.7882***
	[0.633]	[0.545]	[0.684]	[1.138]
Short-rains temperature squared	−0.0554***	−0.0321**	−0.0273	−0.1092***
	[0.017]	[0.014]	[0.019]	[0.033]
Long-rains temperature	−0.5985	−0.0237	−0.4578	−0.7621
	[0.552]	[0.515]	[0.656]	[0.942]
Long-rains temperature squared	0.0069	0.0031	0.0084	0.0207
	[0.014]	[0.013]	[0.018]	[0.027]
Short-rains runoff	−0.0941	0.2334	−0.725	0.5319
	[0.573]	[0.483]	[0.564]	[0.616]
Long-rains runoff	0.4206	−0.5356	0.4589	−0.7515
	[0.922]	[0.782]	[0.914]	[0.992]
Population density	−0.0013***	−0.0001	−0.0007***	−0.0065***
	[0.000]	[0.000]	[0.000]	[0.001]
Agroecological zone	0.0797	−0.0839	−0.3186*	−0.3451
	[0.284]	[0.127]	[0.163]	[0.375]
Soil drainage	−0.0856	0.0400	0.102	0.3183*
	[0.163]	[0.068]	[0.093]	[0.187]
Soil depth	−0.0077	0.0322	−0.099	0.0179
	[0.136]	[0.056]	[0.076]	[0.186]
Silt soil	0.017	0.0002	−0.0357*	−0.0278
	[0.024]	[0.010]	[0.019]	[0.054]
Constant	−11.8841**	−13.2540***	−3.9975	−25.9010***
	[5.041]	[3.887]	[4.293]	[7.955]
lnsig2u	1.5830***	−0.3012	0.2669	1.7535***
	[0.440]	[0.295]	[0.312]	[0.327]
Observations	1,598	1,598	1,598	1,598
Number of code	47	47	47	47

Note: Standard errors in brackets; *** $p<0.01$,

** $p<0.05$,
* $p<0.1$.

productivity. The effects are, however, quite marginal. Favorable agroecological zones are associated with reduced food insecurity, but the effect is significant only for beans. Good soil drainage and silt soil are also associated with lower millet and bean crop insecurity. The effect of these time-invariant factors could be explained by the expected positive correlation between food crop production and favorable development domains (Kabubo-Mariara, 2012).

Table 4.7 presents results for the effect of seasonal weather variability on food insecurity. The results suggest that food insecurity responds to seasonal precipitation and the latter generally exhibits a U-shaped relationship with the probability of being food-insecure. Spring precipitation is significant for maize, beans and millet. Beans and millet insecurity respond significantly to fall precipitation, while only millet responds to summer precipitation. Insecurity for all crops responds significantly to winter precipitation. Very high summer precipitation will damage crops at the early cropping season up to some threshold. Turning to temperatures, the results suggest that spring and fall temperatures exhibit an inverted U-shaped relationship with food insecurity. Insecurity in maize and sorghum responds significantly to spring temperature, while only sorghum insecurity responds significantly to fall temperatures. Summer runoff exhibits a significant negative effect on beans insecurity, but we uncover no significant effect of fall runoff on food insecurity. Population density is associated with lower food insecurity and the effect is significant for all crops except sorghum. Favorable agroecological zones reduce food insecurity, but the effect is significant only for beans.

Despite using different approaches, the results support earlier studies on the impact of climate change on food security (Arndt et al., 2011; Schmidhuber and Tubiello, 2007). Xiong et al. (2007) found that, in the absence of adaptation, climate change is likely to adversely affect rice, wheat and maize production in China. Similarly, Yates and Strzepek (1998) found that despite increased water

Table 4.7 Random Effects Estimates of Seasonal Weather Variability on Food Insecurity – Poverty Line Set at 60% Percentile of Yields

Variables	Maize	Sorghum	Bean	Millet
Spring precipitation (Sept–Nov)	−0.0655***	−0.0138	−0.0342**	−0.0695***
	[0.016]	[0.015]	[0.015]	[0.022]
Spring precipitation squared	0.0009***	0.0000	0.0003	0.0003
	[0.000]	[0.000]	[0.000]	[0.000]
Fall precipitation (Dec–Feb)	0.0052	−0.0057	−0.0451**	−0.0452
	[0.022]	[0.020]	[0.018]	[0.030]
Fall precipitation squared	−0.0001	−0.0002	0.0009**	0.0018**
	[0.001]	[0.001]	[0.000]	[0.001]
Summer precipitation (Mar–May)	−0.014	0.0057	0.0055	0.0290*
	[0.014]	[0.012]	[0.012]	[0.016]
Summer precipitation squared	−0.0001	0.0000	−0.0001	0.0000
	[0.000]	[0.000]	[0.000]	[0.000]

Variables	Maize	Sorghum	Bean	Millet
Winter precipitation (June–Aug)	−0.1239***	−0.0827***	−0.0816***	−0.1206***
	[0.022]	[0.019]	[0.020]	[0.030]
Winter precipitation squared	0.0019***	0.0011***	0.0012***	0.0017***
	[0.000]	[0.000]	[0.000]	[0.001]
Spring temperature (Sept–Nov)	1.6415***	−0.5138	0.9402	3.7155***
	[0.632]	[0.576]	[0.716]	[1.085]
Spring temperature squared	−0.0528***	0.0085	−0.0332	−0.1112***
	[0.017]	[0.015]	[0.021]	[0.032]
Fall temperature (Dec–Feb)	0.7444	1.4264***	0.4237	0.6085
	[0.525]	[0.453]	[0.565]	[0.903]
Fall temperature squared	−0.0096	−0.0320***	−0.0054	−0.0205
	[0.014]	[0.012]	[0.016]	[0.026]
Summer temperature (Mar–May)	−0.9411	0.6344	0.3242	1.1303
	[0.625]	[0.560]	[0.686]	[1.025]
Summer temperature squared	0.0300**	−0.0076	−0.0015	−0.0174
	[0.015]	[0.013]	[0.018]	[0.027]
Winter temperature (June–Aug)	0.2922	0.0188	−0.7021	−1.9908***
	[0.539]	[0.497]	[0.571]	[0.765]
Winter temperature squared	−0.0188	−0.0083	0.0131	0.0505**
	[0.015]	[0.013]	[0.016]	[0.022]
Summer runoff	−0.0709	−0.1761	−0.8972***	0.5384
	[0.313]	[0.268]	[0.342]	[0.383]
Fall runoff	−0.2089	0.0092	1.2754	−1.7497
	[0.857]	[0.740]	[0.906]	[1.170]
Population density	−0.0012***	−0.0001	−0.0007***	−0.0065***
	[0.000]	[0.000]	[0.000]	[0.001]
Agroecological zone	−0.1931	−0.2048	−0.4360**	−0.3047
	[0.253]	[0.141]	[0.177]	[0.383]
Soil drainage	−0.1673	−0.0172	0.0463	0.2764
	[0.145]	[0.075]	[0.101]	[0.180]
Soil depth	−0.0578	0.0083	−0.12	0.0681
	[0.120]	[0.062]	[0.082]	[0.176]
Silt soil	0.024	0.0072	−0.0336	−0.021
	[0.022]	[0.011]	[0.021]	[0.037]
Constant	−11.4608**	−14.7019***	−6.7598	−31.5097***
	[5.192]	[4.296]	[4.739]	[8.609]
lnsig2u	1.3098***	−0.1049	0.4347	2.0331***
	[0.439]	[0.305]	[0.324]	[0.330]
Observations	1,598	1,598	1,598	1,598
Number of code	47	47	47	47

Note: Standard errors in brackets; *** p<0.01,

** p<0.05,
* p<0.1.

availability in Egypt, production was still likely to be adversely affected by climate change. Other studies have found differential effects of climate change on food insecurity. Arndt et al. (2011) found that the impact of climate change is likely to vary by climate scenario, sector and region. Parry et al. (2004) found that climate change is likely to lead to declining crop yields and to increase the disparities in cereal yields between developed and developing countries. Gregory et al. (2005) also found that the impact of climate change on food security varies between regions and between different societal groups within a region. Kabubo-Mariara (2009) found that, in the long term, climate change is likely to lead to increased poverty, vulnerability and loss of livelihoods.

Simulating the effect of climate change on food insecurity

Table 4.8 presents the predicted changes in food crop insecurity based on the A2 and A3 SRES and the five AOGCMs for the years 2050 and 2100. The results show that the largest predicted increases in food insecurity will be from the A2, relative to B2, SRES. B2 actually predicts very minor changes in food insecurity for all crops. This supports findings by Parry et al. (2005), who found that the impact of climate change on the risk of hunger under the B2 SRES is characterized by much lower levels of risk than under A2. The greatest increases in food insecurity are predicted to be for maize, Kenya's staple food crop, estimated at 21% for the year 2100 by the HADCM3 model. This suggests that, unless measures are adopted to adapt to or mitigate against climate change effects, about 73% (base of

Table 4.8 Predicted Changes in Food Insecurity from Different AOGCM Climate Scenarios

Scenario		A2				B2			
Model	Year	Maize	Sorghum	Beans	Millet	Maize	Sorghum	Beans	Millet
HADCM3									
	2050	4.04	0.15	0.55	0.58	3.96	0.17	0.53	0.58
	2100	21.15	5.08	2.83	2.67	11.34	2.06	1.52	1.50
PCM									
	2050	1.68	−0.22	0.23	0.28	1.8	−0.21	0.24	0.20
	2100	8.56	1.28	1.15	1.15	4.44	0.26	0.60	0.63
ECHAM									
	2050	2.79	−0.05	0.39	0.39	2.91	−0.02	0.41	0.39
	2100	15.21	3.19	2.05	1.90	7.7	1.04	1.05	0.97
CSIRO2									
	2050	3.63	0.09	0.49	0.53	3.82	0.12	0.51	0.58
	2100	18.90	4.36	2.53	2.40	11.09	2.00	1.48	1.50
CGCM2									
	2050	2.80	−0.07	0.38	0.43	2.25	−0.16	0.30	0.36
	2100	15.36	3.27	2.05	1.99	6.43	0.73	0.86	0.90

52% + 21%) of all Kenyans may be maize-insecure by 2100. The lowest predicted increase in maize insecurity by the year 2100 is from the PCM model, which predicts an 8.56% increase in insecurity, suggesting that 60% of all Kenyans are likely to be maize-insecure by 2100. The predictions for sorghum are more promising, probably because it is more drought-resistant than maize. The results suggest that sorghum crop insecurity may actually witness modest improvements by the year 2050, as predicted by the PCM, ECHAM and CGCM2 models, but modest increases in insecurity are predicted by all AOGCMS for the two SRES for the year 2100. The models also predict rather modest increases in beans and millet insecurity. The results support the simulation results for crop productivity.

Conclusion

Climate change affects food security in several respects: increased vulnerability to climate change due to dependence on rain-fed agriculture; high levels of poverty; low levels of human and physical capital; and poor infrastructure. Severe droughts in Kenya have continued to interrupt rainfall patterns, leaving behind serious consequences, such as harvest failure, deteriorating pasture conditions, decreased water availability and livestock losses. To design policy measures for averting famine catastrophes in parts of Kenya, there is a need for research on the impact of climate change on food security. This study sought to address this gap. Fixed and random effects regressions for food crop security were estimated. The study further simulated the expected impact of future climate change on food insecurity based on the Special Report on Emissions Scenarios (SRES) and atmospheric oceanic global circulation models (AOGCMs). The study is based on county-level panel data for yields of four major crops (maize, beans, sorghum and millet) and daily climate data spanning over three decades.

The results show that climate variability affects food security irrespective of how food security is defined. First, food security is proxied by food crop yields. The results show that rainfall during short seasonal spells, as well as during long vs. short rains, exhibits an inverted U-shaped relationship with most food crops; the effects are most pronounced for maize and sorghum. Beans and millet are found to be largely unresponsive to climate variability and also to time-invariant factors. We find that long-rains precipitation has larger impacts than short-rains precipitation. The results suggest that high rainfall is therefore crucial for increased crop productivity and thus food security, but that excessive rainfall is harmful. Short-rains and fall and summer temperature exhibit a U-shaped relationship with yields for most crops, while long-rains temperature exhibits an inverted U-shaped relationship. Winter temperatures, however, exhibit a hill-shaped relationship with most crops. The differential impacts of temperature are explained by the fact that high temperatures during the planting period slow down or destroy crop growth, while moderately high winter temperatures are crucial for crop maturity. Simulations of the effects of future climate change scenarios on crop productivity show that climate change will adversely affect food security, with up to 69% decline in yields by the year 2100. Population density has a positive effect on crop yields, suggesting that population density is associated with higher productivity. Maize productivity responds positively to favorable agroecological zones, soil drainage and depth, but performs poorly on silt soils. Sorghum productivity responds positively to favorable agroecological zone.

The results for food insecurity show that climate variables have a nonlinear relationship with food insecurity. Increased seasonal precipitation is associated with reduced food insecurity, but excessive precipitation will increase insecurity due to damage to crops. Temperature generally exhibits an inverted U-shaped relationship with food insecurity, suggesting that increased temperatures will increase food crop insecurity. Maize and millet may, however, benefit from increased summer and winter temperatures. Population density and favorable agroecological zones are associated with lower food insecurity. Good soil drainage and silt soil are also associated with lower millet and bean crop insecurity. High river runoff is associated with lower bean crop insecurity. The simulated effects of different climate change scenarios on food insecurity suggest that adverse climate change will increase food insecurity in Kenya. The largest increases in food insecurity are predicted for the A2 SRES, relative to B2. Climate change is likely to have the largest effects on maize insecurity, which is likely to increase by between 8.56% and 21% by the year 2100, other factors constant. The results further suggest that sorghum, a relatively more drought-resistant crop than maize, may actually witness modest improvements in terms of food insecurity by the year 2050, followed by modest increases in insecurity by the year 2100. The models also predict rather modest increases in beans and millet insecurity.

The results of this study point at the need for policies that safeguard agriculture against the adverse effects of climate change in order to alleviate food insecurity in Kenya. Food insecurity is found to be responsive to climate variability and change. Different food crops respond differently to climate change variables. It is therefore important that climate change mitigation is given much more priority in policy planning and also implementation. Mitigation against global warming can take two main forms: reduction of human emission of greenhouse gases and increasing the capacity of carbon sinks through reforestation. Though Kenya makes a relatively small contribution to greenhouse gas emission, a bigger role in mitigation can be played by encouraging reforestation throughout the country, especially in the more arid areas, where drought-resistant trees and crops could be introduced. This is supported by Thorlakson and Neufeldet (2012), who found that agroforestry increased farm productivity.

The government can also play a bigger role by ensuring policy-driven adaptation to climate change, especially in more vulnerable counties. Though farmers may be aware of the effects of climate change, they may not always have full information about adaptation options available to them. Continuous climate change monitoring, intensified early warning systems and dissemination of relevant information to farmers are crucial. Even where farmers have information, they may not have the means to adopt suggested adaptation options. Yet, previous studies suggest that adaptation strategies are important for very vulnerable farm households who have the least capacity to produce food (Di Falco and Veronesi,

2011). For instance, one of the key strategies to reduce vulnerability is by encouraging irrigation, especially in drier and more marginal areas. This is, however, constrained by water availability, and, even where water is available, by lack of finances for required irrigation infrastructure. Lessons from FAO agricultural projects in Turkana County in the recent past serve as a good illustration of how irrigation interventions can serve as a solution in vulnerable areas. Lessons can also be learned from practices from relatively dry countries, such as Israel, where agriculture thrives, and the United Arab Emirates, where trees and other vegetation are grown using recycled water. There is also a need to encourage intensive rain water harvesting, particularly in drier areas, to supplement any available water. In the dry counties, a lot of water could be harvested during heavy rains (when severe floods hit such areas) and stored for use during the dry seasons.

There are two key caveats to the results in this chapter. First, the study does not take into account household-level characteristics and other factors that could affect the response of food insecurity to climate change. Future research should incorporate the impact of household-level factors, farm characteristics, and other development domains in order to test the effect of climate change controlling for other factors. Second, in this study, simulations predict the impact of climate change over time, *ceteris paribus*. In the very long run, it is likely that other factors (e.g., prices of inputs and outputs and technology) may change and thus affect crop farming and food security. Farmers may also adopt appropriate adaptation measures to counter the effect of climate change on agriculture. Future research should address these concerns.

Appendix

Table 4A1 Characteristics of Agro-climate Zones and Farming Systems in Kenya

Zone	Moisture index (%)	Climate classification	Average annual rainfall (mm)	Average annual potential evaporation (mm)	Vegetation	Farming system
I	>80	Humid	1100–2700	1200–2000	Moist forest	Dairy, sheep, coffee, tea, maize, sugarcane
II	65–80	Subhumid	1000–1600	1300–2100	Moist and dry forest	Maize, pyrethrum, wheat, coffee, sugarcane
III	50–65	Semi-humid	800–1400	1450–2200	Dry forest and moist woodland	Wheat, maize, barley coffee, cotton, coconut, cassava
IV	40–50	Semi-humid to semiarid	600–1100	1550–2200	Dry woodland and bush land	Ranching, cattle sheep, barley, sunflower, maize, cotton, cashew nuts, cassava
V	25–40	Semiarid	450–900	1650–2300	Bush land	Ranching, livestock, sorghum, millet
VI	15–25	Arid	300–550	1900–2400	Bush land and scrubland	Ranching
VII	<15	Very arid	150–350	2100–2500	Desert scrub	Nomadism and shifting grazing

Source: Kabubo-Mariara and Karanja (2007).

Table 4A2 Predicted Decadal Average Changes in Precipitation and Temperature: 2050–2100

Precipitation (Percentage change)

	CGCM2		CSIRO2		ECHAM		HADCM3		PCM	
Year	2050	2100	2050	2100	2050	2100	2050	2100	2050	2100
A2-scenarios	106	116	109	123	113	134	110	124	106	115
B2-scenarios	104	109	105	109	116	129	108	115	106	110

Temperature (increases °C)

Year	2050	2100	2050	2100	2050	2100	2050	2100	2050	2100
A2-scenarios	3.0	7.4	3.4	8.2	2.8	7.2	3.6	8.7	2.2	5.4
B2-scenarios	2.7	4.7	3.6	6.3	2.8	4.9	3.6	6.3	2.3	3.8

Source: Strzepek and McCluskey (2006) and associated district-level database for Kenya.

Notes

1 This section borrows from Kabubo-Mariara and Karanja (2007).
2 Source: Ministry of Agriculture (2009). Kenya Agricultural Sector Data Compendium, Vol. 2 – Crop Production.
3 Data with finer resolutions was available, but the problem was the time span. For instance, we have daily precipitation satellite data retrieved from the Climate Prediction Centre (CPC) of the National Oceanic and Atmospheric Administration (NOAA) with a resolution of 0.1 degree by 0.1 degree (about 10 km). We also have precipitation from the UK Met Office with a resolution of 50 km by 50 km. This data was used mostly for exploratory spatial analysis to compare climate variability, agroecological zones and population density, among other variables, and crop yields. A collection of daily parameters from the ECMWF model can be found at http://data-portal.ecmwf.int/data/d/interim_daily/.
4 The assumptions and adjustments relate to: (1) lack of correlation between W_c and X_{ct}; (2) possible spatial correlation of the error terms; and (3) weighting of Equation (1) to take into account county variations in cropland and revenue.
5 Food insecurity is based on the FGT index. The FGT poverty measures can be defined as $P_\alpha = 1/n\Sigma[(z - y_i)/z]^\alpha$; $\alpha \geq 0$, for $y < z$, where z is the poverty line and y_i is a measure of the economic welfare of household i (say, consumption expenditure), ranked as $y_1 \leq y_2 \cdots y_q \leq z \leq y_{q+1} \ldots \leq y_n$. The household equivalents of the headcount index, poverty gap index and squared poverty gap index are obtained when $\alpha = 0$, 1 or 2, respectively (Foster et al., 1984). In this chapter, the probability that a county is food-insecure at time i is based on $P_\alpha = 0$. Relative rather than absolute definitions of poverty are, however, used (see "Results and Discussion").
6 The rainy seasons vary depending on the region. For instance, the long rains in the Western, Nyanza and Rift Valley highlands are longer and extend from February to September. Long rains in the Central and Eastern areas are much shorter, spanning from March to June, and coincide very closely with long rains in pastoral areas and cropping in the South-eastern and Coastal areas (source: www.fews.net/Pages/timelineview.aspx?gb=ke&tln=en&l=en, accessed July 18, 2013). The definition of summer, fall, spring and winter is also variable depending on location and so the definition adopted here is indicative (Kabubo-Mariara, 2009).
7 Beans and millet are found to be largely unresponsive to weather variability and, for this reason, the discussion is based on results for maize and sorghum.

8 These findings, however, contrast with Roberts and Schlenker (2013), who argue that the relationship between temperature and ethanol production is nothing like quadratic and that only very high temperatures are harmful.

9 It is possible, however, that good crop yields could lead to high population density on average, but this has not been explored in this chapter. Future research may want to look into this.

10 The scenarios related to future greenhouse gases and aerosols emissions have been developed based on certain assumptions of population and economic growth, land use, technological change and energy availability (Houghton et al., 2001). See Kabubo-Mariara (2009) for a discussion of the SRES and simulations.

11 When we compare the data that we have for various years, there is little variation between climate variables from the year 2000 to 2012, so the base of 2000 used by Strzepek and McCluskey (2006) is still valid for our predictions. The advantage of using these scenarios is that decadal predictions are available for each county.

12 The results for the 25th and 40th percentiles are consistent in terms of signs of coefficients, though levels of significance differ. They are not presented to save space, but are available from the authors.

References

Arndt, C., W. Farmer, K. Strzepek and J. Thurlow. 2011. Climate Change, Agriculture and Food Security in Tanzania. UNU-WIDER Working Paper No. 2011/52. Helsinki, Finland: United Nations University – World Institute for Development Economics Research.

Deressa, T., R. Hassan and D. Poonyth. 2005. Measuring the Economic Impact of Climate Change on South Africa's Sugarcane Growing Regions. *Agrekon* 44(4): 524–542.

Deressa, T., R. Hassan, C. Ringler, T. Alemu and M. Yesuf. 2009. Determinants of Farmers' Choice of Adaptation Methods to Climate Change in the Nile Basin of Ethiopia. *Global Environmental Change* 19(2): 248–255.

Deschenes, O., and M. Greenstone. 2007. The Economic Impacts of Climate Change: Evidence from Agricultural Output and Random Fluctuations in Weather. *The American Economic Review* 97(1): 354–385.

Di Falco, S., and M. Veronesi. 2011. On Adaptation to Climate Change and Risk Exposure in the Nile Basin of Ethiopia. IED Working Paper No. 15. Zurich: ETH/IED.

Dinar, A., R. Hassan, R. Mendelsohn and J. Benhin. 2008. *Climate Change and Agriculture in Africa: Impacts Assessment and Adaptation Strategies*. London, UK: Earthscan.

FAO. 1998. Prediction of Cattle Density, Cultivation Levels and Farming Systems in Kenya. Consultancy Report Prepared by Environmental Research Group Oxford Ltd. and TALA Research Group. Rome, Italy: Department of Zoology, University of Oxford, for the Animal Health Division of FAO.

FAO. 2008. *Climate Change and Food Security: A Framework Document*. Rome, Italy: United Nations Food and Agriculture Organization.

FAO, IIASA, ISRIC, ISSCAS and JRC. 2012. *Harmonized World Soil Database Version 1.2*.

Foster, J., J. Greer and E. Thorbecke. 1984. A Class of Decomposable Poverty Measures. *Econometrica* 52: 761–766.

Gbetibouo, G.A., and R.M. Hassan. 2005. Measuring the Economic Impact of Climate Change on Major South African Field Crops: A Ricardian Approach. *Global and Planetary Change* 47(2–4): 143–152.

Gregory, P.J., J.S.I. Ingram and M. Brklacich. 2005. Climate Change and Food Security. *Philosophical Transactions of the Royal Society B* 360: 2139–2148.

Hassan, R., and C. Nhemachena. 2008. Determinants of African Farmers' Strategies for Adapting to Climate Change: Multinomial Choice Analysis. *African Journal of Agricultural Resource Economics* 2(1): 83–104.

Herrero, M., C. Ringler, J. van de Steeg, P. Thornton, T. Zhu, E. Bryan, A. Omolo, J. Koo and A. Notenbaert. 2010. *Kenya: Climate Variability and Climate Change and Their Impacts on the Agricultural Sector.* Washington, DC: IFPRI.

Houghton, J.T., Y. Ding, D.J. Griggs, M. Noguer, P.J. van der Linden, X. Dai, K. Maskell and C.A. Johnson. 2001. Climate Change: The Scientific Basis. Contribution of Working Group I to the Third Assessment Report of the Intergovernmental Panel on Climate Change. New York: Cambridge University Press.

Intergovernmental Panel on Climate Change (IPCC). 2001. Climate Change 2001: Impacts, Adaptation Vulnerability. Contribution of Working Group II to the Third Assessment Report of the Intergovernmental Panel on Climate Change. Geneva: UNEP/WMO.

Kabara, M., and J. Kabubo-Mariara. 2011. Global Warming in the 21st Century: The Impact on Agricultural Production in Kenya. In *Global Warming in the 21st Century*, edited by J.M. Cossia. New York: Nova Science, Chapter 8, pp. 199–214.

Kabubo-Mariara, J. 2008a. Climate Change Adaptation and Livestock Activity Choices in Kenya: An Economic Analysis. *Natural Resources Forum* 32: 132–142.

Kabubo-Mariara, J. 2008b. Crop Selection and Adaptation to Climate Change in Kenya. *Environmental Research Journal* 2(2–3): 177–196.

Kabubo-Mariara, J. 2009. Global Warming and Livestock Husbandry in Kenya: Impacts and Adaptations. *Ecological Economics* 68: 1915–1924.

Kabubo-Mariara, J. 2012. Institutional Isolation, Soil Conservation and Crop Productivity: Evidence from Machakos and Mbeere Districts in Kenya. *African Journal of Social Sciences* 2(3): 1–26.

Kabubo-Mariara, J., and F. Karanja. 2007. The Economic Impact of Climate Change on Kenyan Crop Agriculture: A Ricardian Approach. *Global and Planetary Change* 57: 319–330.

Kurukulasuriya, P., and R. Mendelsohn. 2008. A Ricardian Analysis of the Impact of Climate Change on African Cropland. *African Journal of Agricultural and Resource Economics* 2(1): 1–23.

Massetti, E., and R. Mendelsohn. 2011. Estimating Ricardian Models with Panel Data. *Climate Change Economics* 2(4): 301–319.

Mendelsohn, R., A. Basist, P. Kurukulasuriya and A. Dinar. 2003. Climate and Rural Income. New Haven, CT: Yale University, School of Forestry and Environmental Studies.

Mendelsohn, R., W. Nordhaus and D. Shaw. 1994. The Impact of Global Warming on Agriculture: A Ricardian Analysis. *American Economic Review* 84: 753–771.

Ministry of Agriculture. 2009. *Kenya Agricultural Sector Data Compendium*, Volume 2. *Crop Production.* Nairobi, Kenya.

Mohamed, A.B., N. van Duivenbooden and S. Abdoussallam. 2002. Impact of Climate Change on Agricultural Production in the Sahel – Part 1: Methodological Approach and Case Study for Millet in Niger. *Climatic Change* 54(3): 327–348.

Molua, E.L. 2002. Climate Variability, Vulnerability and Effectiveness of Farm-Level Adaptation Options: The Challenges and Implications for Food Security in Southwestern Cameroon. *Environment and Development Economics* 7: 529–545.

Molua, E.L. 2008. Turning up the Heat on African Agriculture: The Impact of Climate Change on Cameroon's Agriculture. *African Journal of Agricultural and Resources Economics(AFJARE)* 2(1): 45–64.

Mundlak, Y. 1978. On the Pooling of Time Series and Cross Section Data. *Econometrica* 46: 69–85.

Nhemachena, C., R. Hassan and P. Kurukulasuriya. 2010. Measuring the Economic Impact of Climate Change on African Agricultural Production Systems. *Climate Change Economics* 1(1): 33–55.

Parry, M.L., C. Rosenzweig, A. Iglesias, M. Livermore and G. Fischer. 2004. Effects of Climate Change on Global Food Production under SRES Emissions and Socio-Economic Scenarios. *Global Environmental Change* 14: 53–67.

Parry, M.L., C. Rosenzweig and M. Livermore. 2005. Climate Change, Global Food Supply and Risk of Hunger. *Philosophical Transactions of the Royal Society B* 360: 2125–2138.

Revuelta, A., C. Rodriguez, J. Mateos and J. Garmendia. 1985. A Model for the Estimation of Precipitable Water. *Tellus* 378: 210–215.

Roberts, M.J., and W. Schlenker. 2013. Identifying Supply and Demand Elasticities of Agricultural Commodities: Implications for the US Ethanol Mandate. *American Economic Review* 103(6): 2265–2295.

Schmidhuber, J., and F.N. Tubiello. 2007. Global Food Security under Climate Change. *PNAS* 104(50): 19703–19708.

Sombroek, W.G., H.M.H. Braun and B.J.A. Van de Pouw. 1982. The Exploratory Soil Map and Agro-Climatic Zone Map of Kenya. Report No. E1. Nairobi: Kenya Soil Survey.

Strzepek, K., and A. McCluskey. 2006. District-Level Hydroclimatic Time Series and Scenario Analysis to Assess the Impacts of Climate Change on Regional Water Resources and Agriculture in Africa. CEEPA Discussion Paper No. 13. Centre for Environmental Economics and Policy in Africa, University of Pretoria, South Africa.

Thorlakson, T., and H. Neufeldet. 2012. Reducing Subsistence Farmers' Vulnerability to Climate Change: Evaluating the Potential Contributions of Agro-Forestry in Western Kenya. *Agriculture and Food Security* 1: 15. doi:10.1186/7010-1-15.

Turpie, J., H. Winkler, R. Spalding, R. Fecher and G. Midgley. 2002. Economic Impacts of Climate Change in South Africa: A Preliminary Analysis of Unmitigated Damage Costs. Cape Town: University of Cape Town.

World Food Programme. 2011. www.wfp.org/ (accessed November 12, 2011).

Xiong, W., E. Lin, H. Ju and Y. Xu. 2007. Climate Change and Critical Thresholds in China's Food Security. *Climatic Change* 81: 205–221.

Yates, D.N., and K.M. Strzepek. 1998. An Assessment of Integrated Climate Change Impacts on the Agricultural Economy of Egypt. *Climatic Change* 38: 261–287.

Part II

On-farm practices related to food crop productivity

5 Adaptation to climate change in sub-Saharan agriculture

Assessing the evidence and rethinking the drivers*

Salvatore Di Falco

Introduction

Climate change is a fundamental threat to agricultural productivity, food security and development prospects in sub-Saharan Africa. In this part of the world, production conditions can be particularly challenging. Millions of small-scale subsistence farmers, generally with less than one hectare of land, produce food crops while facing a combination of low land productivity, missing markets, low technology adoption and harsh weather conditions (e.g., high average temperature, erratic rainfall). These result in very low yields and food insecurity (Di Falco and Chavas, 2009). Given the reliance on rainfall and the limited opportunities for economic diversification, sub-Saharan Africa's development prospects have been closely associated with climate. Climate change is projected to further reduce food security (Rosenzweig and Parry, 1994; Parry et al., 2005; Cline, 2007; Lobell et al., 2008; McIntyre et al., 2009; Schlenker and Lobell, 2010). As indicated in the fourth Intergovernmental Panel on Climate Change (IPCC), at lower latitudes, crop productivity is expected to decrease "for even small local temperature increases (1–2° C)" (IPCC, 2007). In many African countries, access to food will be severely affected; "yields from rain fed agriculture could be reduced by up to 50% by 2020" (IPCC, 2007, p. 10). Scientists tell us that future warming is not preventable. As a matter of fact, even if agreements to limit emissions are achieved and implemented, farmers will still face a warmer production environment and agriculture will be more vulnerable.

The agricultural sector is more crucial where economic development is most needed. Christiaensen et al. (2011) showed the important contribution of agriculture to poverty reduction among the poorest and most vulnerable populations. Diao et al. (2010), using an economy-wide model, stressed the key role that agriculture still plays in Africa. This is because of the job opportunities provided to the poorest compared to industrial growth. They conclude that there is little

* This chapter was originally published in 2014, substantially in its current form, in the *European Review of Agricultural Economics*, 41(3): 405–430, as "Adaptation to Climate Change in Sub-Saharan Agriculture: Assessing the Evidence and Rethinking the Drivers." It has been revised and updated. The authors, editors and publisher thank the journal for permission to use the previous work.

evidence showing that "African countries can bypass a broad-based agricultural revolution to successfully launch their economic transformations" (Diao et al., 2010, p. 8). Climate change poses a serious risk of reversing progress toward achieving the Millennium Development Goals. Climate change is both a development and an environmental challenge. Given these premises, there is no question that achieving successful adaptation processes in agriculture is of paramount importance. This entails understanding the barriers to and drivers of adaptation and the identification of the implications in terms of welfare.

This study's contribution is twofold. First, it provides a brief review of the evolution of the literature on climate change economics in agriculture over the last 20 years. The focus is mostly on micro studies in east Africa. I highlight some of the main findings and underscore how the literature on the impact of climate change on agriculture has developed separately from the adaptation literature. I argue that adaptation and impact need to be modelled jointly. Therefore, I present an econometric procedure that allows analyzing adaptation and its effect on a given outcome (e.g., productivity, food security, revenues). Second, it provides some preliminary evidence on the behavioural dimensions of climate change. I examine the causal effects of climate change on risk preferences. How do climatic events affect some behavioural parameters? Are people who are exposed to more or less rainfall, for instance, more likely to be risk-averse? I provide empirical evidence on the role of climatic factors in determining farmers' risk aversion.

The chapter proceeds as follows. The next section provides some background on climate change in sub-Saharan Africa. In the following section, a structural model that can be used for estimating both adaptation and its implications is presented. The fourth section provides the empirical analysis of the role of climate in risk aversion and rate of time preferences. The fifth section offers some reflections on future research. The final section concludes the chapter.

Agriculture and climate change in sub-Saharan Africa

Sub-Saharan Africa (SSA) is the part of Africa that extends below the Sahara Desert. Its surface is equal to an area of $2.4*10^9$ ha. It is rich in natural resources (although not uniformly distributed) and presents different climatic and geological features (IAASTD, 2009).

Sub-Saharan Africa's economy is mostly driven by the agriculture sector, in spite of the issue of land degradation and desertification and poor land management systems (UNEP, 2002; IAASTD, 2009). Over 60% of the population rely on agriculture for their livelihood but only 8% of the land is suitable for staple production (IAASTD, 2009). Characterized by crop production and traditional livestock, it is the sector where the majority of the labour force is employed and contributes to a considerable part of the GDP, about 40% (Barrios et al., 2008). However, SSA countries are different from one to another. For instance, Ethiopia and Rwanda are land-locked countries, highly populated; agriculture is the main economic activity and about four-fifths of inhabitants live in rural areas. Nevertheless, Rwanda, with high altitude, has geological features that represent a hurdle to its agricultural development, compared to the coastal countries, such as Ghana and Kenya, which present settled agro-processing and industrial sectors. By contrast, Uganda is a land-locked country but, together with Ghana, shows agriculture growth and a stable GDP (Diao et al., 2010). Yet, in Tanzania, over 70% of the population rely on subsistence rain-fed agriculture. It accounts for 50% of GNP and 66% of export earnings (Mary and Majule, 2009). In Kenya, agriculture contributes to about 24% of GDP and employment, with about 70% of households living in rural areas. The main output is crop production, which depends on soils, hydrological and climate characteristics. The majority of the land is classified as arid or semi-arid, and therefore dedicated for extensive livestock production, while only 12% of the land is positively used for farming or intense livestock production (Kabubo-Mariara and Karanja, 2007). In Ethiopia, 85% of the population bases its livelihood on agriculture, which accounts for more than 40% of national GDP and 90% of exports, and provides basic needs and income for more than 90% of the poor (Diao et al., 2010, p. 5). Cereals are the main product and source of Ethiopians' daily calorie intake (62%). About 70% of land is used for cereal production, which is chiefly concentrated in the western regions (Diao et al., 2010). Generally, in SSA it is possible to distinguish four different kinds of farming systems:

- The maize-mixed system, which is based primarily on maize, cotton, cattle and goats.
- The cereal/root crop-mixed system, which is based on maize, sorghum, millet, cassava, yams and cattle.
- The irrigated system, based on maize, sorghum, millet, cassava, yams and cattle.
- The tree crop–based system, anchored in cocoa, coffee, oil palm and rubber, mixed with yams and maize (IAASTD, 2009).

They are mainly subsistence-oriented and the use of technology is almost absent. The farms are small-scaled and the average size decreased from 1.5 hectares in 1970 to 0.5 hectares in 1990 (IAASTD, 2009) in part due to the "exhaustion of land frontiers" (IAASTD, 2009).

The diverse topography that characterizes SSA is reflected in farm households' operating areas.[1] In Kenya, for example, there are seven agro-climate zones identified following the moisture index.[2] When the index is greater than 50%, it is favourable to crop production but those areas (Zones I, II, III) account for 18% of the land. Generally, they are situated above 1200 m of altitude, with mean annual temperature lower than 18 degrees. On the other hand, the majority of the country, almost 80%, presents a moisture index less than 50%, annual mean temperature between 22°C and 40°C, and mean annual rainfall less than 1100 mm, and is situated below 1260 m. As a result, they are semi-humid to arid regions (Zones IV, V, VI and VII) and, hence, low-potential farming zones (Kabubo-Mariara and Karanja, 2007). According to Hurni's studies (1998), Ethiopia is divided into five major agro-ecological zones that are dissimilar in altitude and soil and consequently in food production. Those are *Bereha* (hot, arid lowlands), *Kolla* (warm, semi-arid lowlands), *Woina Dega* (temperate, cool sub-humid highlands), *Dega* (cool, humid highlands) and *Wurch* (cold highlands). Zones are largely based on altitude and climate, which dictate the crop types grown in each zone. The highland zones contain much of Ethiopia's crop production, while the lowland zones are dominated by pastoral production systems. Precipitation and temperature changes will not be consistent across these agro-ecological zones, each of which supports differing livelihoods. For example, the lowlands are more reliant on livestock grazing (Hurni, 1998). Yet, although Ethiopia has 3.5 million hectares of irrigable land, only 160,000 hectares (accounting for 3% of the land available) are irrigated. There are two main agriculture seasons that correspond to the rainy seasons *Belg* and *Kirmet*. *Meher* is the name of the agriculture season during *Kirmet*. It is extended between June and September, the period that registers the highest quantity of output, 90%–95% of the national production.

There is clearly a dependence of agriculture on the effects of weather and climate. In fact, the occurrence of drought during the growing season, for instance, will generate a decrease of production and, consequently, a negative influence on food security (Giorgis et al., 2006). Climate variability in Ethiopia is well documented and closely linked with the country's economic growth (Diao and Pratt, 2007; World Bank, 2010).

Climate extremes are not a novelty to Ethiopia but studies underline that drought has occurred more often during the last few decades, particularly in the lowlands (Lautze et al., 2003). A study undertaken by the national meteorological service (published in 2007) highlights that annual minimum temperature has been increasing by about 0.37 degrees Celsius every 10 years over the past 55 years. Rainfall has been more erratic, with some areas becoming drier and others relatively wetter. These findings point out that climatic variations have already happened. Further climate change can exacerbate this very difficult situation. Most climate models converge in forecasting gloomy scenarios of increased temperatures for most of Ethiopia (Dinar et al., 2008).

SSA is the most vulnerable region in the world to climate change, although its contribution to global GHG emissions is only 2%–3% (IAASTD, 2009). Its vulnerability depends on poverty and the limited capacity to adapt (IPCC, 2007). Adaptation measures in agriculture might include water conservation and irrigation (Mendelsohn and Dinar, 2003), crop species switching, improved seed varieties, improved on-farm technology, climate and weather forecasting, application of fertilizers and soil nutrients (which is the lowest in the world), access to extension services, consideration of topographical heterogeneity, and investments in research and development (Di Falco and Veronesi, 2013).

The economics of climate change: impact and adaptation as two separate issues

Until recently, most of the literature on the economics of climate change developed two independent streams. One focused on the study of the impact of climate change on agriculture;[3] the other, on the estimation of the barriers to and drivers of adaptation. In this section, I provide a review of these two areas of research. Let me begin with the former. I highlight micro studies undertaken in sub-Saharan Africa.

Impact of climate change on agriculture

Since the pioneering work by Mendelsohn et al. (1994), the so-called Ricardian approach represents the workhorse within the economic analysis of the impact of climate change. It has been very frequently used to estimate the impact of climatic variables on agriculture. In its traditional application, it is a cross-section analysis that measures the long-term impact of climate and other variables on agricultural performance (e.g., land value and farm revenues). This approach has the advantage that it considers both the consequences of climate on productivity and how farmers adapt to it. Therefore, it overcomes the main critique to the use of production functions in the context of impact of climate change. Notably, Mendelsohn et al. (1994) claimed that the production function approach consistently overestimates production damage by omitting the variety of adaptations that farmers customarily make in response to changing economic or environmental conditions.[4] It is the "dumb-farm scenario" (Mendelsohn et al., 1994). In particular, the authors highlighted the role of adaptation actions in which new activities displace activities no longer (or less) profitable due to changes in climate variables.

Kurukulasuriya et al. (2006) used a Ricardian model to study how climate variables influence farmers' revenues in 11 African countries: Burkina Faso, Cameron, Egypt, Ethiopia, Ghana, Kenya, Niger, Senegal, South Africa, Zambia and Zimbabwe. In this case, the total net farm revenues resulted from the incomes generated by dryland crops that are rain-fed, irrigated crops and livestock. Separate regressions were estimated in order to highlight the response of the three activities. Using data from a survey of about 9,000 farmers, the study confirms that African crop production is sensitive to climate and the hotter and drier regions are likely to be affected most. Increases in temperature and decreases in precipitations have a negative impact on the net revenue. Similar but to a smaller extent will be the effect on the revenues generated by irrigated land because this is less vulnerable to warming. To the contrary, an increase of rainfall brings an overall beneficial effect.

Revenues show a quadratic relationship with both temperature and precipitation. As a result, the marginal impact of climate change varies according to the temperature and rainfall levels presented across different farms. In fact, although Africa is overall hot and dry, the impacts of climate change are not uniform across the country. For instance, dry land throughout sub-Saharan Africa is at risk; cropland with irrigation support located around the Nile or highlands of Kenya has some benefits from warming; drier countries, such as Egypt, Niger or Senegal, have benefits for livestock from rainfall increase (Kurukulasuriya et al., 2006). With an increase of 1°C, the African net revenues in dryland and livestock showed a decrement of about US $27 and $379 per hectare, respectively, while the irrigated areas experienced a rise (an average of $30 per hectare). This effect is due to the fact that crops benefit from irrigation when rainfall shortages occur and also because they are generally located in cool areas. Similarly, the marginal effects of precipitation vary among the activities. A 1 millimetre/month increase in rainfall generates an aggregate enhancement of net revenues of $67 per farm.

Also, the elasticities confirm that the sensitivity of crop and livestock to temperature is greater than the effect of precipitation. The figures support this interpretation. The temperature elasticities are −1.9, 0.5 and −5.4, compared to the elasticities of precipitation, which are equal to 0.1, 0.4 and 0.8 for dryland, irrigated crops and livestock respectively (Kurukulasuriya et al., 2006).

Deressa and Hassan (2010) provided an application of this method in Ethiopia.[5] In their paper, net crop revenue per hectare was regressed on climate, household and soil variables. The results showed that these variables have a significant impact on the net crop revenue per hectare. The seasonal marginal impact analysis indicated that marginally increasing temperature during summer and winter would significantly reduce crop net revenue per hectare, whereas marginally increasing precipitation during spring would considerably increase net crop revenue per hectare.

Another example of the application of the Ricardian model is offered by Kabubo-Mariara and Karanja (2007) to measure the impact of climate on net crop revenues. Using a sample of 816 households, they underlined the negative influence of climate change on productivity.

A slightly different approach was offered by Benhin (2008), who utilized a revised Ricardian approach to assess the impact of climate change on crop production in South Africa. Including hydrological variables (river flow and water resources), which are particularly affected by climate change, he extended the earlier study proposed by Deressa and Hassan (2010). The mean annual estimates indicated that if the temperature rises by 1%, the net crop revenue will increase by about $80, whereas rainfall decreases of 1 millimetre/month cause revenues to fall by $2. However, the impacts are different according to the season. The study also predicts a 90% decrease of crop net revenue by 2100 (Benhin, 2008).

The Ricardian model has been criticized from different points of view.[6] It does not take into account transaction costs – for instance, the costs generated by the decision to change production abruptly. Secondly, it can suffer from an omitted variable problem because it does not take into account time-independent, location-specific factors, such as soil quality and unobservable skills of farmers (Barnwal and Kotani, 2010). Another drawback is that it does not consider variables that are invariant with respect to the space – for example, carbon fertilization effect (Cline, 1996). This is a weak point in climate change studies conducted at a local level, especially in low-income countries, where each meteorological station controls a large portion of the territory (Di Falco et al., 2011). Yet, assuming constant prices leads to errors in measurement of losses and benefits (Cline, 1996; Kurukulasuriya et al., 2006).

Further, despite its successful application in over 27 countries (Mendelsohn and Dinar, 2009), apart from a few exceptions, it normally refers to a single year. Deschenes and Greenstone (2007) argued that, in order to obtain stable results over time on climate change impacts, we should refer to intertemporal variation of the weather instead of cross sections. Nevertheless, Massetti and Mendelsohn (2011) claimed that weather changes are not useful to explain climate change effects because farmers do not have a chance to adapt in the short run. Using weather data may provide a biased evaluation of the long-term effect of climate change (Massetti and Mendelsohn, 2011).

Estimating drivers of adaptation

Sub-Saharan Africa is the area most vulnerable to climate change because warming will be greater than the predicted worldwide average, and agriculture, mainly rain-fed and managed on a small scale, represents the main source of subsistence for African rural communities. Hence, climate change is a threat and adaptation has primary importance in reducing vulnerability and, thus, ensuring livelihood while achieving food and water security and biodiversity (Bryan et al., 2013).

In order to design specific and effective adaptation policies, it is crucial to identify what factors influence farmers' adaptation to climate variations and how to measure them (Bryan et al., 2009; Below et al., 2012).

In this sub-section, we will provide a review of the evidence on the barriers to and drivers of adaptation capacity. Most of the literature has focused on the micro level, where the process of adaptation is independently implemented by farmers or private firms in the field. This is what is called "autonomous adaptation," distinguished from the "planned adaptation" decided on by the government. During the last few decades, the Ricardian method has been the main tool in forecasting autonomous adaptation to climate change, providing useful information to policymakers in order to develop and promote well-targeted policies (Stage, 2010). A number of micro-econometric studies are focused on adaptation and agricultural productivity (Kurukulasuriya et al., 2006; Seo and Mendelsohn, 2008; Di Falco et al., 2011; Di Falco et al., 2012; Bryan et al., 2013) and others on the determinants of using adaptation methods (Maddison, 2007; Hassan and Nhemachena, 2008; Bryan et al., 2009; Deressa et al., 2009). It is crucial to understand how the social, economic, institutional and ecological contexts mediate the climate impacts and influence the adaptation response (Kato et al., 2009).

According to Maddison (2007), there are two important components of adaptation: perception and adoption of strategies. Adaptation can, thus, be thought of as a two-step process. The first step requires that the farmers perceive a change in the climatic conditions. In the second step, farmers implement a set of strategies to deal with the different conditions (Maddison, 2007).

Based on Heckman's probit model, the analysis conducted by Maddison (2007) on a sample of selected households in 11 African countries reveals that experience increases the likelihood of perception of climate change but education seems to be the main determinant in using at least one adaptation strategy. In addition, agriculturists who have easier access to the market where they sell their products and have access to free extension advice show higher willingness to adapt. Changing the crop variety (particularly when temperature varies) and changing dates of planting (following rainfall variations) are, overall, the most common methods of adaptation (Maddison, 2007).

Bryan et al. (2009) studied the adaptation strategies adopted by farmers in Ethiopia and South Africa and the drivers that contributed to choice adaptation. Based on a sample of 1,800 farm households, it emerges that farmers generally use the following methods of adaptation: use of different crops or crop varieties, planting trees, soil conservation, changing planting dates, and irrigation (Kato et al., 2009). Nevertheless, in spite of the awareness of climate variability, it is not easy to implement these changes. Access to credit, extension services and wealth are obstacles to adaptation for the farmers of both countries. Farmers in Ethiopia also indicated that lack of access to land and information about climate change were barriers to adaptation. Therefore, policymakers should pay attention to small-scale subsistence farmers and enhance adaptation by providing access to information, credit and markets (Kato et al., 2009).

Deressa et al. (2009) undertook a micro-analysis using data from cross-sectional household survey data collected from 1,000 households during the 2004/2005 production season in the Nile Basin of Ethiopia. This survey was the first one to address climate change explicitly. One section of this survey did indeed ask farmers about their perceptions of climate change and their adaptation strategies. More specifically, farm households were asked questions about their observations of the patterns of temperature and rainfall over the past 20 years. The results indicate that 50.6% of the surveyed farmers have observed increasing temperature over the past 20 years, whereas 53% of them have observed decreasing rainfall over the past 20 years. The perception of the farmers in this part of Africa is matched with the climatic observation of temperature. Less unequivocal is the evidence on rainfall. In microeconomic studies undertaken in Ethiopia and South Africa by Kato et al. (2009) and in Kenya by Bryan et al. (2013), farmers perceive changes in precipitation and temperature but just a small portion of them made management adjustments to tackle climate change. For instance, in Ethiopia, 83% of farmers perceive variation in temperature but 56% of them do not use adaptation strategies. Among those who choose adaptation, the most frequent methods are using different crops or different varieties, planting trees, irrigation or changes in planting period. Less frequent is the use of new technologies or migration to urban areas. Similar conclusions describe South Africa's situation.

The fact that only a small percentage of farmers undertake adaptation means that long-term changes in climate alone do not determine farmers' decision making. Extreme weather events; timing, duration and frequency of precipitation; socioeconomic status; household characteristics; and distance from the market determine the outcome.

The study by Kato et al. (2009) reveals that farmers are more educated in South Africa than in Ethiopia, on average seven years against two; in both countries, families are quite large, on average six members; and farmers in South Africa are better off than in Ethiopia, with higher incomes. In South Africa, farmers receive farm support input, such as seeds, tools, machinery and subsidies, instead of food aid, as in Ethiopia. In term of farm characteristics, Ethiopia has more access to formal and informal credit even though farmers in South Africa borrow more money; farmers in South Africa are closer to markets and are less affected by extreme events.

The study on Kenya is an extension of the previous study conducted in 2009 that underlines the differences in adaptation choices across agro-ecological zones. Farmers' perceptions of climate change and climate risk are a decisive variable in adaptation decision-making. Varying by district, farmers confirmed that they perceived an average increase in temperature and decrease in rainfall and also an increase in rainfall variability during the last 20 years. Generally, farmers from the humid zone are more sensitive to a decrease of precipitation than those who live in arid zones. Generally, farmers with more experience and with more access to extension services are more likely to perceive climate changes (Bryan et al., 2013).

Deressa et al. (2009) defined adaptation as crop switching, late planting, soil conservation and tree planting. They showed that the level of education, gender, age and wealth of the head of household; access to extension and credit; information on climate, social capital, agro-ecological settings; and temperature all influence farmers' choices. The main barriers include financial constraints and lack of information on adaptation methods.

Changing crop mix has been found to be a key strategy in a number of studies on adaptation to climate in Africa. Studies were undertaken across different scales. For instance, Maddison (2007), Kurukulasuriya and Mendelsohn (2008a), Seo and Mendelsohn (2008) and Hassan and Nhemachena (2008) provide evidence at the aggregate level that changing crops is the most likely adaptation strategy. Aggregate studies, however, can mask spatial heterogeneity. Changing the crop mix, or crop switching, is a strategy that farmers have implemented for a long time. Farmers, in fact, match crops to soils and environmental conditions – including climate. Moreover, greater use of different crops could be associated with lower expense and ease of access by farmers (Deressa et al., 2009). Implementing more structural adaptation measures (i.e., irrigation) requires more resources and public investment.

Di Falco and Chavas (2009) documented that growing a combination of different barley landraces (e.g., crop genetic diversity) was associated with less production risk exposure in the highlands of Ethiopia. These results showed that maintaining a higher biodiversity regime can be an important asset for sub-Saharan agriculture. In this case, a study of the highlands of Ethiopia showed that conserving landraces in the field delivered important productive services and allowed farmers to mitigate some of the negative effects of harsh weather and agro-ecological conditions. Therefore, *in situ* conservation of crop biological diversity is one of the strategies that can help improve Ethiopia's poor agricultural performance and alleviate food insecurity. The analysis also showed that the beneficial effects of this diversity become of greater value in degraded land. When the land is less fertile, the contribution of crop biodiversity toward reducing crop failure becomes stronger. This underlines the potential role that crop genetic diversification can play as an adaptation strategy. It also can be considered to be one of the cheapest adaptation strategies, when compared to more labour-intensive activities, such as building soil conservation measures or water harvesting methods. These findings also highlight the importance of the genetic traits of African crops. This can be extremely valuable in providing raw material for future genetic improvement of existing crops. Crop mix or crop switching implies taking advantage of the fact that different crop species have different genetic traits. Genetic variability within and between species confers the potential to resist biotic and abiotic stresses, both in the short and the long term (Giller et al., 1997). Growing more crop species enhances the possibility of producing in years when rainfall regimes or environmental conditions are more challenging. Thus, having functionally similar plants that respond differently to weather and temperature randomness contributes to resilience (Holling, 1973) and ensures that "whatever the environmental conditions there will be plants of given functional types that thrive under those conditions" (Heal, 2000). In the African context, for instance, it has been found that more diverse cropping systems provide a wider range of productive responses to weather and climatic shocks.

Besides diversification of crops, diversification of farm activities and household income is relevant. In principle, obtaining income from non-farm (less climate-sensitive) sources is seen as crucial. In the literature, the evidence supporting this as an adaptation strategy is very thin. Moreover, Deressa et al. (2009) found that non-farm income also significantly increases the likelihood of planting trees, changing planting dates and using irrigation as adaptation options. In other words, the extra amount of resources is actually reinvested in the farm.

Moving into a mixed crop-livestock system is also related to diversification and adaptation (e.g., Kurukulasuriya and Mendelsohn, 2008b; Seo and Mendelsohn, 2008; Hassan and Nhemachena, 2008). Livestock choice is also climate-sensitive. Farmers are more likely to raise sheep and goats as temperatures rise and less likely to raise dairy and beef cattle (Seo and Mendelsohn, 2008). Whether they increase or decrease their reliance on chickens depends on their current climate.

Modelling adaptation and its implications: a structural approach

From a policy perspective, understanding adaptation to climate change is of paramount importance. Besides determining the impact of climatic variables on welfare, it is necessary to understand how the set of strategies implemented in the field by farmers (e.g., changing crops, adopting new technologies or soil conservation measures) is chosen in response to long-term changes in environmental conditions and how they affect productivity or revenues (Di Falco et al., 2011).[7] The standard Ricardian approach assumes optimal adaptation to climate by the farmers in the past; the regression coefficients estimate the marginal impact on outputs of future temperature or rainfall changes incorporating farmers' adaptive response. It thus does not provide any insight into how farmers adapt (Seo and Mendelsohn, 2008). To overcome this issue, Seo and Mendelsohn (2008), Kurukulasuriya and Mendelsohn (2008a) and Di Falco et al. (2011) developed the so-called structural Ricardian model, which explicitly models the underlying endogenous decisions by farmers.

The climate change adaptation decision and its implications in terms of an outcome of interest (e.g., productivity, food security, revenue) can be modelled in the setting of a two-stage framework. In the first stage, I use a selection model for climate change adaptation where a representative risk-averse farm household chooses to implement climate change adaptation strategies if they generate net benefits.[8] Let A* be the latent variable that captures the expected benefits from the adaptation choice with respect to not adapting. We specify the latent variable as

$$A_i^* = \mathbf{Z}_i\alpha + \eta_i \ \text{with} \ A_i = \begin{cases} 1 & \text{if } A_i^* > 0 \\ 0 & \text{otherwise} \end{cases} \tag{1}$$

That is, farm household i will choose to adapt ($A_i = 1$), through the implementation of some strategies in response to long-term changes in mean temperature and rainfall, if A* > 0, and 0 otherwise.

The vector Z represents variables that affect the expected benefits of adaptation. These factors can be classified in different groups. First, we consider characteristics of the operating farm (e.g., soil fertility and erosion). For instance, farms characterized by more fertile soil might be less affected by climate change and therefore relatively less likely to implement adaptation strategies. Then, climatic factors (e.g., rainfall and temperature) as well as the experience of previous extreme events, such as droughts and flood, can also play a role in determining the probability of adaptation. To account for selection bias, I adopt an endogenous switching regression model where farmers face two regimes, (1) to adapt and (2) not to adapt, defined as follows:

Regime 1: $y_1i = \mathbf{X}_{1i}\boldsymbol{\beta}_1 + \varepsilon_{1i} \text{if } A_i = 1$ (2a)
Regime 2: $y_{2i} = \mathbf{X}_{2i}\boldsymbol{\beta}_2 + \varepsilon_{2i} \text{if } A_i = 0$ (2b)

where y_i is the quantity produced or revenues per hectare in Regimes 1 and 2, and X_i represents, for instance, a vector of inputs (e.g., seeds, fertilizers, manure and labour), the farm household's characteristics, soil characteristics, assets and the climatic factors included in Z. Finally, the error terms in Equations (1), (2a) and (2b) are assumed to have a trivariate normal distribution, with zero mean and covariance matrix Σ – that is, $(\eta, \varepsilon_1, \varepsilon_2)' \sim N(0, \Sigma)$

$$\text{with} \ \pounds = \begin{bmatrix} \sigma_\eta^2 & \sigma_{\eta 1} & \sigma_{\eta 2} \\ \sigma_{1\eta} & \sigma_1^2 & \cdot \\ \sigma_{2\eta} & \cdot & \sigma_2^2 \end{bmatrix}$$

where σ_n^2 is the variance of the error term in the selection Equation (1), which can be assumed to be equal to 1 because the coefficients are estimable only up to a scale factor (Maddala, 1983, p. 223), σ_1^2 and σ_2^2 are the variances of the error terms in the productivity functions (2a) and (2b), and $\sigma_{1\eta}$ and $\sigma_{2\eta}$ represent the covariance of η_i and ε_{1i} and ε_{2i}.[10] Because y_{1i} and y_{2i} are not observed simultaneously, the covariance between ε_{1i} and ε_{2i} is not defined (reported as dots in the covariance matrix Σ; Maddala, 1983, p. 224). An important implication of the error structure is that, because the error term of the selection equation (1), η_i, is correlated with the error terms of the productivity functions (2a) and (2b) (ε_{1i} and ε_{2i}), the expected values of ε_{1i} and ε_{2i} conditional on the sample selection are non-zero:

$$E\left[\varepsilon_{1i} \mid A_i = 1\right] = \sigma_{1\eta} \frac{\phi(\mathbf{Z}_i\alpha)}{\Phi(\mathbf{Z}_i\alpha)} = \sigma_{1\eta}\lambda_{1i}, \text{ and}$$

$$E\left[\varepsilon_{2i} \mid A_i = 0\right] = -\sigma_{2\eta} \frac{\phi(\mathbf{Z}_i-)}{1-\Phi(\mathbf{Z}_i-)} = \sigma_{2\eta}\lambda_{2i},$$

where $\phi(.)$ is the standard normal probability density function, $\Phi(.)$ the standard normal cumulative density function, and $\lambda_{1i} = \frac{\phi(\mathbf{Z}_i\alpha)}{\Phi(\mathbf{Z}_i\alpha)}$, and $\lambda_{2i} = -\frac{\phi(\mathbf{Z}_i\alpha)}{1-\Phi(\mathbf{Z}_i\alpha)}$.

If the estimated covariances $\hat{\sigma}_{1\eta}$ and $\hat{\sigma}_{2\eta}$ are statistically significant, then the decision to adapt and the quantity produced per hectare are correlated; that is, if one finds evidence of endogenous switching, then we can reject the null hypothesis of the absence of sample selectivity bias. This model is defined as a "switching regression model with endogenous switching" (Maddala and Nelson, 1975). An efficient method to estimate endogenous switching regression models is full information maximum likelihood estimation (Lee and Trost, 1978). The logarithmic likelihood function, given the previous assumptions regarding the distribution of the error terms, is:

$$\ln L_i = \sum_{i=1}^{N} A_i \left[\ln \phi\left(\frac{\varepsilon_{1i}}{\sigma_1}\right) - \ln \sigma_1 + \ln \Phi(\theta_{1i}) \right]$$
$$+ (1 - A_i)\left[\ln \phi\left(\frac{\varepsilon_{2i}}{\sigma_2}\right) - \ln \sigma_2 + \ln\left(1 - \Phi(\theta_{2i})\right) \right],$$

where $\theta_{ji} = \frac{(\mathbf{Z}_i\alpha + \rho_j\varepsilon_{ji}/\sigma_j)}{\sqrt{1-\rho_j^2}}$, $j = 1,2$, with ρ_j denoting the correlation coefficient between the error term η_i of the selection equation (1) and the error term ε_{ji} of Equations (2a) and (2b), respectively.[11] The model can easily be expanded in the context of multiple adaptation strategies and multiple outcomes.

Stage I – selection model of multiple climate change adaptation strategies

In the first stage, let A^* be the latent variable that captures the expected net revenues from implementing strategy j ($j = 1 \ldots M$) with respect to implementing any other strategy k. We specify the latent variable as

$$A_{ij}^* = \bar{V}_{ij} + \eta_{ij} = \mathbf{Z}_i-_j + \eta_{ij} \tag{3}$$

$$\text{with } A_i = \begin{cases} 1 & \textit{iff} \quad A_{i1}^* > \max_{k \neq 1}(A_{ik}^*) \text{ or } \varepsilon_{i1} < 0 \\ \vdots & \quad \vdots \qquad\qquad \vdots \\ M & \textit{iff} \quad A_{iM}^* > \max_{k \neq M}(A_{ik}^*) \text{ or } \varepsilon_{iM} < 0 \end{cases}$$

That is, farm household i will choose strategy j in response to long-term changes in mean temperature and rainfall if strategy j provides expected net revenues greater than any other strategy $k \neq j$ – that is, if $\varepsilon_{ij} = \max_{k \neq j}(A_{ik}^* - A_{ij}^*) < 0$. Equation (3) includes a deterministic component ($\overline{V}_{ij} = Z_i\alpha_j$), and an idiosyncratic unobserved stochastic component η_{ij}. The latter captures all the variables that are relevant to the farm household's decision maker but are unknown to the researcher, such as skills or motivation. It can be interpreted as the unobserved individual propensity to adapt.

The deterministic component $\overline{V}_{ij}$ depends on factors Z_i, as defined earlier, that affect the likelihood of choosing strategy j. It is assumed that the covariate vector Z_i is uncorrelated with the idiosyncratic unobserved stochastic component η_{ij} – that is, $E(\eta_{ij} \mid Z_i) = 0$. Under the assumption that η_{ij} are independent and identically Gumbel distributed – that is, under the independence of irrelevant alternatives (IIA) hypothesis – selection model (1) leads to a multinomial logit model (McFadden, 1973), where the probability of choosing strategy j (P_{ij}) is

$$P_{ij} = P(\varepsilon_{ij} < 0 \mid Z_i) = \frac{\exp(Z_i\alpha_j)}{\sum_{k=1}^{M} \exp(Z_i\alpha_k)} \tag{3}$$

Stage II – multinomial endogenous switching regression model

In the second stage, a multinomial endogenous switching regression model to investigate the impact of each strategy on net revenues can be estimated by applying the selection bias correction model of Bourguignon et al. (2007). The model implies that farm households face a total of M regimes (one regime per strategy, where $j=1$ is the reference category "non-adapting"). A net revenue equation corresponding to each possible j strategy is defined as m regime:

Regime 1: $y_{i1} = X_i\beta_1 + u_{i1}$ if $A_i = 1$ (3a)
Regime M: $y_{iM} = X_i\beta_M + u_{iM}$ if $A_i = M$ (3m)

where y_{ji} is the net revenue per hectare of farm household i in Regime j, ($j = 1 \ldots M$), and X_i represents a vector of inputs (e.g., seeds, fertilizers, manure and labour), household head's and farm household's characteristics, soil characteristics and the past climatic factors included in Z_i; u_{ij} represents the unobserved stochastic component, which verifies $E(u_{ij} \mid X_i, Z_i) = 0$ and $V(u_{ij} \mid X_i, Z_i) = \sigma_j^2$. For each sample observation, only one among the M dependent variables net revenues is observed.[13] To correct for the potential inconsistency, one can take into account the correlation between the error terms η_{ij} from the multinomial logit model estimated in the first stage and the error terms from each net revenue equation u_{ij}. I refer to this model as a "multinomial endogenous switching regression model."

Bourguignon et al. (2007, p. 179) show that consistent estimates of β_j in the outcome equations $(3a)$–$(3m)$ can be obtained by estimating the following selection bias-corrected net revenues equations:

$$\text{Regime 1:} \quad y_{i1} = \mathbf{X_i}\boldsymbol{\beta_1} + \sigma_1 \left[\rho_1 m(P_{i1}) + \sum_j \rho_j m(P_{ij}) \frac{P_{ij}}{(P_{ij}-1)} \right] + \nu_{i1} \text{ if } A_i = 1 \quad (4a)$$

$$\vdots \qquad \vdots$$

$$\text{Regime M:} \quad y_{iM} = \mathbf{X_i}\boldsymbol{\beta_M} + \sigma_M \left[\rho_M m(P_{iM}) + \sum_j \rho_j m(P_{ij}) \frac{P_{ij}}{(P_{ij}-1)} \right] \quad (4m)$$

$$+ \nu_{iM} \text{ if } A_i = M$$

where P_{ij} represents the probability that farm household i chooses strategy j as defined in (2), ρ_j is the correlation between u_{ij} and η_{ij}, and $m(P_{ij}) = \int J(v - \log P_j)g(v)dv$ with $J(.)$ being the inverse transformation for the normal distribution function, $g(.)$ the unconditional density for the Gumbel distribution, and $V_{ij} = \eta_{ij} + \log P_j$. This implies that the number of bias correction terms in each equation is equal to the number of multinomial logit choices M.

If panel data are at hand, one can specify a fixed effect version of the foregoing models. Di Falco and Veronesi (2013) exploit plot-level information to deal with the issue of farmers' unobservable characteristics, such as their skills. Plot-level information can be used to construct panel data and control for farm-specific effects (Udry, 1996). I follow Mundlak (1978) and Wooldridge (2002) to control for unobservable characteristics. We exploit the plot-level information, and insert in the net revenues equations $(4a)$–$(4m)$ the average of plot-variant variables $\bar{S}_i$ – for instance, the inputs used (seeds, manure, fertilizer and labour). This approach relies on the assumption that the unobservable characteristics v_i are a linear function of the averages of the plot-variant explanatory variables $\bar{S}_i$ – that is, $v_i = \bar{S}_i\boldsymbol{\pi} + \psi_i$ with $\psi_i \sim IIN(0, \sigma_\psi^2)$ and $E(\psi_i / \bar{S}_i) = 0$, where π is the corresponding vector of coefficients, and ψ_i is a normal error term uncorrelated with $\bar{S}_i$.

Building up a counterfactual analysis

Switching regression models allows estimating counterfactuals. One can estimate the treatment effects (Heckman et al., 2001), and thus the effect of the treatment "adoption of strategy j" on the net revenues of the farm households that adopted strategy j. In the absence of a self-selection problem, it would be appropriate to assign to farm households that adapted a counterfactual net revenue equal to the average net revenue of non-adapters with the same observable characteristics. Unobserved heterogeneity in the propensity to choose an adaptation strategy also affects net revenues and creates a selection bias in the net revenue equation that cannot be ignored. The multinomial endogenous switching regression model can be applied to produce selection-corrected predictions of counterfactual net revenues.

In particular, one can follow Bourguignon et al. (2007, p. 179 and pp. 201–203), and derive the expected net revenues or land values of farm households that adapted, which in our study means $j = 2 \ldots M$ ($j = 1$ is the reference category "non-adapting"), as

$$E(y_{i2} \mid A_i = 2) = \mathbf{X_i}\boldsymbol{\beta}_2 + \sigma_2 \left[\rho_2 m(P_{i2}) + \sum_{k \neq 2}^{M} \rho_k m(P_{ik}) \frac{P_{ik}}{(P_{ik} - 1)} \right]$$

$$E(y_{i2} \mid A_i = 2) = \mathbf{X_i}\boldsymbol{\beta}_2 + \sigma_2 \left[\rho_2 m(P_{i2}) + \sum_{k \neq 2}^{M} \rho_k m(P_{ik}) \frac{P_{ik}}{(P_{ik} - 1)} \right] \qquad (5a)$$

$$\vdots$$

$$E(y_{iM} \mid A_i = M) = \mathbf{X_i}\boldsymbol{\beta}_M + \sigma_M \left[\rho_M m(P_{iM}) + \sum_{k=1 \ldots M-1} \rho_k m(P_{ik}) \frac{P_{ik}}{(P_{ik} - 1)} \right] \qquad (5m)$$

Then, one can obtain the expected net revenues or land values of farm households that adopted strategy j in the counterfactual hypothetical case that they did not adapt ($j = 1$) as

$$E(y_{i1} \mid A_i = 2) = \mathbf{X_i}\boldsymbol{\beta}_1 + \sigma_1 \left[\rho_1 m(P_{i2}) + \rho_2 m(P_{i1}) \frac{P_{i1}}{P_{i1} - 1} \right.$$

$$\left. + \sum_{k=3 \ldots M} \rho_k m(P_{ik}) \frac{P_{ik}}{(P_{ik} - 1)} \right] \qquad (6a)$$

$$\vdots$$

$$E(y_{i1} \mid A_i = M) = \mathbf{X_i}\boldsymbol{\beta}_1 + \sigma_1 \left[\rho_1 m(P_{iM}) + \sum_{k=2 \ldots M} \rho_k m(P_{ik-1}) \frac{P_{ik-1}}{(P_{ik-1} - 1)} \right] \qquad (6m)$$

This allows us to calculate the treatment effects (TT), for example, as the difference between Equations (5a) and (6a) or (5m) and (6m).

What have we learned from the structural approach?

Di Falco and Veronesi (2013) used the foregoing multinomial framework to answer the following questions. What are the factors affecting the adoption of strategies in isolation or in combination? What are the "best" strategies that can be implemented to deal with climatic change in the field? In particular, what are the economic implications of different strategies? They used plot-level farm data and found that the choice of which adaptation strategy to adopt is crucial to support farm revenues. They found that strategies adopted in combination with other strategies rather than in isolation are more effective. Adaptation is, therefore, more effective when it is composed of a portfolio of actions rather than one single action. More specifically, it is found that the positive impact of changing crops is significant when coupled with water and soil conservation strategies. This highlights the importance of not implementing water or soil conservation programs in isolation.

With regard to the drivers of adaptation, the first-stage analysis highlighted the role of tenure security. The estimated coefficient is positively correlated with all the strategies. The dissemination of information on changing crops and implementing soil conservation strategies also is found to be important. Extension services are, for instance, significant in determining the implementation of adaptation strategies, which could result in more food security for all farmers, irrespective of their unobservable characteristics. Moreover, the availability of information on climate change may raise farmers' awareness of the threats posed by the changing climatic conditions. This is consistent with the finding of Deressa et al. (2009).

The behavioural dimensions of adaptation to climate change: risk aversion

From the results reported earlier, a set of different institutional drivers (e.g., tenure security, extension services) and market drivers (e.g., missing credit markets) have been identified in connection with the issue of lack of adaptation. Lately, attention to climatic effects on different outcomes has been increasing. A large body of literature has used the exogenous variation in climatic factors to identify the causal effects of climate on different outcomes. For instance, some researchers focused on economic outcomes, such as land values, income and growth (e.g., Mendelsohn et al., 1994; Deschenes and Greenstone, 2007, 2011; Dell et al., 2009; Fisher et al., 2012; Graff-Zivin and Neidell, 2010; Graff-Zivin et al., 2013; Hsiang, 2010; Schlenker and Roberts, 2009). Others have paid attention to other crucial impacts of climatic variables, such as conflicts (Hsiang et al., 2011), education, health, migration (Barrios et al., 2006) and social norms (Miguel, 2005).

In this section, I examine the causal effect of climatic conditions on behavioural parameters that can have crucial implications for the choice of adaptation strategies: specifically, farmers' risk aversion. If farmers are averse to risk, they may also be more reluctant to undertake potentially profitable investments if these entail more risk. In this case, higher variability leads to a higher risk premium and lower investment. This finding is well established in the risk literature (e.g., Just and Pope, 1979; Chavas, 2004; Rosenzweig and Binswanger, 1993; Dercon and Christiansen, 2011). I address this issue directly and analyze how the first and the second moments of the long-run distribution of rainfall affect risk preferences.

More specifically, I ask the following research question: are people who are exposed to more variable rainfall more likely to display higher risk aversion? To my knowledge, the estimation of the effect of climatic factors on behavioural parameters is novel.

I use a series of economic experiments where payoffs vary in terms of both riskiness and payoff. The experiments were carried out in the highlands of Ethiopia in 2007. In order to elicit each participant's risk preference, the respondents were presented with a hypothetical farming scenario involving alternative levels of output depending on the weather. The hypothetical agricultural scenario consisted of two plots, the productivity of which differs depending on whether the rains are good or bad, each at a 50% probability. As can be seen in Table 5.1, a series of six choices were presented to the respondents, with each choice consisting of a payment with a higher spread and a higher payoff versus a choice with a lower spread and a lower payoff (see Yesuf and Bluffstone, 2009, for a full description of the experiment). I consider whether farmers choose scenarios that qualify them as risk-averse. We therefore assign a dummy that takes the value of 1 if risk-averse and 0 otherwise. We regress this against the first two moments of the long-run distribution of rainfall. These are calculated over a 30-year period. Table 5.2 provides the summary statistics and Table 5.3 reports the regression results. We find that the probability of being classified as risk-averse is determined by these climatic factors. More specifically, higher rainfall is negatively correlated with the probability of being risk-averse. The second moment of the distribution of rainfall (captured by the coefficient of variation) is, however, positively correlated with the probability of being risk-averse. We extend the analysis with different controls. The results are consistent from both a qualitative and quantitative point of view.

Increasing long-term rainfall variability is therefore associated with higher risk aversion. The result underscores the potential importance of behavioural parameters in climate change adaptation. These parameters are crucially affected by climate. This may uncover a mechanism through which climate may also affect investment decisions that are central in the adaptation process.

Table 5.1 Choice Sets for the Risk Preference Experiment

	Bad harvest	Good harvest	Expected mean	Spread	CPRA* coefficient	Risk classification
Choice set 1	100	100	100	0	∞ – 7.5	Extreme
Choice set 2	90	180	105	90	7.5–2.0	Severe
Choice set 3	80	240	160	160	2.0–0.812	Intermediate
Choice set 4	60	300	180	240	0.812–0.316	Moderate
Choice set 5	20	360	190	360	0.316–0.0	Slight
Choice set 6	0	400	200	400	0.0 – ∞	Neutral

Note: Numbers represent kilogram output.
* Constant partial risk aversion.

Table 5.2 Summary Statistics

Variable	Description	Mean	Std. dev.	Min	Max
Risk averter	HH head classified as risk-averse (see Table 5.1)	33%		0	1
Rainfall	Rainfall average (1970–2004) in mm/year	1173.6	89.4	269.76	1550.2
Rainfall CV	Rainfall coefficient of variation (1970–2004)	0.27	0.052	0.22	1.039
Distance to plots	Average walking distance from the homestead to the plots in minutes	6.25	12.5	0	150
Distance to town	Average walking distance to the nearest market town in minutes	63.39	42.3	0	240
Tenure insecurity	Expect no changes in land holdings (1 = yes; 0 = otherwise)	41%		0	1
Household size	Number of members of household	6.495	2.38	1	19
Livestock	Livestock units	4.367	3.207	0	18.6
Gender	Gender of HH head (1 = female; 0 = male)	17%		0	1
Age	Age of the HH head	52.31	15.06	18	105
Illiterate	Household head unable to write or read (1 = yes; 0 = otherwise)	60%	0.489	0	1
Temperature	Long-run temperature average in degrees Celsius	10.38	4.734	2.78	19.64

Table 5.3 Risk Aversion and Climate Change: Probit Marginal Effects

Dependent variable: risk-averse

	(1)	(2)	(3)	(4)
Rainfall	−0.000815***	−0.000652***	−0.000550***	−0.000564***
	(0.000119)	(0.000110)	(0.000106)	(0.000105)
Rainfall CV	0.940***	1.007***	1.114***	1.115***
	(0.355)	(0.366)	(0.313)	(0.312)
Distance to the plots		−0.00312***	−0.00326***	−0.00310***
		(0.000338)	(0.000251)	(0.000281)
Distance to town		−0.0000175	−0.000171	−0.000171
		(0.000306)	(0.000324)	(0.000319)
Tenure insecurity		0.202***	0.209***	0.208***
		(0.0237)	(0.0188)	(0.0183)
HH size			0.0357***	0.0356***
			(0.00955)	(0.00968)

Dependent variable: risk-averse

	(1)	*(2)*	*(3)*	*(4)*
Livestock			0.00780**	0.00865**
			(0.00341)	(0.00348)
Gender			0.161***	0.165***
			(0.0319)	(0.0323)
Age			0.000706*	0.000633
			(0.000405)	(0.000422)
Illiterate			−0.0368	−0.0341
			(0.0390)	(0.0386)
Temperature				0.00427***
				(0.000900)
N	763	626	626	626

Note: Standard errors in parentheses *$p < 0.10$,**
$p < 0.05$,***
$p < 0.01$; constant not reported.

Concluding remarks

In this study, I reviewed some of the existing evidence on climate change impact and adaptation, with a focus on sub-Saharan Africa. Research published to date highlights that adaptation based upon a portfolio of strategies significantly increases farm net revenues. Changing crop varieties has a positive and significant impact on net revenues when coupled with water conservation strategies or soil conservation strategies but not when implemented in isolation. It is also found that tenure security and access to extension services are key determinants of the decision to adapt. Finally, I combined climatic data and experimentally elicited risk preferences to analyze the impact of climatic factors on behaviour. More rainfall variability is associated with more risk aversion. This may uncover an important behavioural dimension of the impact of climate change in agriculture. More variable rainfall may make farmers more risk-averse. This could also imply a lower propensity to undertake investment. Future research on the behavioural dimensions of climate change is necessary to uncover mechanisms and psychological impacts.

At this stage, other considerations are appropriate. Most of the results published and reported in this chapter rely on cross-sectional and plot-level data. More and better data (e.g., panel data with long time dimension) should be made available to provide more robust evidence on both the role of adaptation and its implications for agriculture. The dynamics of the problem should be also explicated. Some adaptation strategies can be effective in the short run while others may deliver a payoff in the long term.

Notes

1 This includes the ecological and agricultural characteristics (agro-climatic condition, livestock raising conditions, land resource conditions) by which the landscape is classified.
2 This index is based on annual rainfall expressed as a percentage of potential evaporation.
3 More recently, the body of literature on impacts has been expanding quickly. Please refer to the following surveys to appreciate the evolution of this strand of literature: Dell et al. (2014) and Carleton and Hsiang (2016).
4 Mendelsohn et al. (1994, p. 754).
5 The first economic studies on the impact of climate change in developing countries using this approach came from India and Brazil (Mendelsohn and Dinar, 1999; Kumar and Parikh, 2001). These studies confirm earlier predictions that even a modest level of warming would affect agricultural productivity in these countries, although the impact may not be uniform in all areas, with some regions benefiting and the vast majority being affected adversely.
6 Schlenker and Roberts (2009) dealt with the omitted variables issue in a production function set-up. They combine historical crop production and weather data in sub-Saharan Africa into a panel analysis. This approach can take care of the omitted variable problem by using fixed effects capturing all time-invariant effects for which data is not available (e.g., soil texture). A similar approach to deal with unobservables is presented by Fisher et al. (2012).
7 It should be noted that the original Ricardian approach assumes that markets function properly. Access to inputs, credit or technology may, however, be "imperfect."
8 A more comprehensive model of climate change adaptation is provided by Mendelsohn (2000).
9 For notational simplicity, the covariance matrix Σ does not reflect the clustering implemented in the empirical analysis. In addition, constraining the variance term in a single equation to equal 1 is not the same as deriving the proper form of the posterior or even the sampling distribution of the cross-equation correlation matrix.
10 An alternative estimation method is the two-step procedure (see Maddala, 1983, p. 224, for details).
11 When estimating an OLS model, the net revenues equations (3a)–(3m) are estimated separately. However, if the error terms of the selection model (1), η_{ij}, are correlated with the error terms u_{ij} of the net revenues functions (3a)–(3m), the expected values of u_{ij} conditional on the sample selection are nonzero, and the OLS estimates will be inconsistent.

References

Barnwal, P., and K. Kotani. 2010. Impact of Variation in Climatic Factors on Crop Yield: A Case of Rice Crop in Andhra Pradesh, India. Working Paper No. EMS-2010–17. IUJ. www.iuj.ac.jp/research/workingpapers/EMS_2010_17.pdf (accessed January 7, 2014).
Barrios, S., B. Ouattara and E. Strobl. 2006. Climatic Change and Rural-Urban Migration: The Case of Sub-Saharan Africa. *Journal of Urban Economics* 60(3): 357–371.
Barrios, S., B. Ouattara and E. Strobl. 2008. The Impact of Climatic Change on Agricultural Production: Is It Different for Africa? *Food Policy* 33: 287–298.
Below, T.B., K.D. Mutabazi, D. Kirschke, C. Franke, S. Sieber, R. Siebert and K. Tscherning. 2012. Can Farmers' Adaptation to Climate Change Be Explained by Socio-Economic Household-Level Variables? *Global Environmental Change* 22: 223–235.

Benhin, J.K.A. 2008. South African Crop Farming and Climate Change: An Economic Assessment of Impacts. *Global Environmental Change* 18: 666–678.

Bourguignon, F., M. Fournier and M. Gurgand. 2007. Selection Bias Corrections Based on the Multinomial Logit Model: Monte Carlo Comparisons. *Journal of Economic Surveys* 21(1): 174–205.

Bryan, E., T.T. Deressa, G.A. Gbetibouo and C. Ringler. 2009. Adaptation to Climate Change in Ethiopia and South Africa: Options and Constraints. *Environmental Science and Policy* 12(4): 413–426.

Bryan, E., C. Ringler, B. Okoba, C. Roncoli, S. Silvestri and M. Herrero. 2013. Adapting Agriculture to Climate Change in Kenya: Household Strategies and Determinants. *Journal of Environmental Management* 114: 26–35.

Carleton, T.A., and S.M. Hsiang. 2016. Social and Economic Impacts of Climate Change. *Science* 353(6304): 1112.

Chavas, J.-P. 2004. *Risk Analysis in Theory and Practice*. London: Elsevier.

Christiaensen, L., L. Demery and J. Kuhl. 2011. The (Evolving) Role of Agriculture in Poverty Reduction: An Empirical Perspective. *Journal of Development Economics* 96(2): 239–254.

Cline, W.R. 1996. The Impact of Global Warming on Agriculture: Comment. *American Economic Review* 86: 1309–1312.

Cline, W.R. 2007. *Global Warming and Agriculture: Impact Estimates by Country*. Washington, DC: Center for Global Development and Peter G. Peterson Institute for International Economics.

Dell, M., B.F. Jones and B.A. Olken. 2009. Temperature and Income: Reconciling New Cross-Sectional and Panel Estimates. *American Economic Review: Papers and Proceedings* 99(2): 198–204.

Dell, M., B.F. Jones and B.A. Olken. 2014. What Do We Learn from the Weather? The New Climate-Economy Literature. *Journal of Economic Literature* 52(3): 740–798.

Dercon, S., and L. Christiansen. 2011. Consumption Risk, Technology Adoption and Poverty Traps: Evidence from Ethiopia. *Journal of Development Economics* 96(2): 159–173.

Deressa, T.T., and R. Hassan. 2010. Economic Impact of Climate Change on Crop Production in Ethiopia: Evidence from Cross-Section Measures. *Journal of African Economies* 18(4): 529–554.

Deressa, T.T., R. Hassan, T. Alemu, M. Yesuf and C. Ringler. 2009. Analyzing the Determinants of Farmers' Choice of Adaptation Measures and Perceptions of Climate Change in the Nile Basin of Ethiopia. International Food Policy Research Institute (IFPRI) Discussion Paper No. 00798. Washington, DC: IFPRI.

Deressa, T.T., R.M. Hassan and C. Ringler. 2010. Perception of and Adaptation to Climate Change by Farmers in the Nile Basin of Ethiopia. *Journal of Agricultural Science* 149(1): 23–31.

Deschenes, O., and M. Greenstone. 2007. The Economic Impacts of Climate Change: Evidence from Agricultural Output and Random Fluctuations in Weather. *American Economic Review* 97(1): 354–385.

Deschenes, O., and M. Greenstone. 2011. Climate Change, Mortality, and Adaptation: Evidence from Annual Fluctuations in Weather in the US. *American Economic Journal: Applied Economics* 3(4): 152–185.

Diao, X.A., P. Hazell and J. Thurlow. 2010. The Role of Agriculture in African Development. *World Development* 38(10): 1375–1383.

Diao, X.A., and N. Pratt. 2007. Growth Options and Poverty Reduction in Ethiopia: An Economy-Wide Model Analysis. *Food Policy* 32(2): 205–228.

Di Falco, S., and J.-P. Chavas. 2009. On Crop Biodiversity, Risk Exposure and Food Security in the Highlands of Ethiopia. *American Journal of Agricultural Economics* 91(3): 599–611.

Di Falco, S., and M. Veronesi. 2013. How Can African Agriculture Adapt to Climate Change? A Counterfactual Analysis from Ethiopia. *Land Economics* 89(4): 743–766.

Di Falco, S., M. Veronesi and M. Yesuf. 2011. Does Adaptation to Climate Change Provide Food Security? A Micro-Perspective from Ethiopia. *American Journal of Agricultural Economics* 93(3): 825–842.

Di Falco, S., M. Yesuf, G. Köhlin and C. Ringler. 2012. Estimating the Impact of Climate Change on Agriculture in Low-Income Countries: Household-Level Evidence from the Nile Basin, Ethiopia. *Environmental and Resource Economics* 52: 457–478.

Dinar, A., R. Hassan, R. Mendelsohn and J. Benhin. 2008. *Climate Change and Agriculture in Africa: Impact Assessment and Adaptation Strategies*. London: EarthScan.

Fisher, A.C., W.M. Hanemann, M.J. Roberts and W. Schlenker. 2012. The Economic Impacts of Climate Change: Evidence from Agricultural Output and Random Fluctuations in Weather: Comment. *American Economic Review* 102(7): 3749–3760.

Giller, K.E., M.H. Beare, P. Lavelle, A.M.N. Izac and M.J. Swift. 1997. Agricultural Intensification, Soil Biodiversity and Agro-Ecosystem Function. *Applied Soil Ecology* 6: 3–16.

Giorgis, K., A. Tadege and D. Tibebe. 2006. Estimating Crop Water Use and Simulating Yield Reduction for Maize and Sorghum in Adama and Miesso Districts Using the CROPWAT Model. Centre for Environmental Economics and Policy in Africa (CEEPA) Discussion Paper No. 31, p. 14. Pretoria: University of Pretoria.

Graff-Zivin, J., S. Hsiang and M. Neidell. 2013. Climate, Human Capital and Adaptation. Mimeo.

Graff-Zivin, J., and M. Neidell. 2010. Temperature and the Allocation of Time: Implications for Climate Change. U.S. National Bureau of Economic Research Working Paper No. 15717. Cambridge, MA: NBER.

Hassan, R., and C. Nhemachena. 2008. Determinants of Climate Adaptation Strategies of African Farmers: Multinomial Choice Analysis. *African Journal of Agricultural and Resource Economics* 2(1): 83–104.

Heal, G. 2000. *Nature and the Marketplace: Capturing the Value of Ecosystem Services*. New York: Island Press.

Heckman, J.J., J.L. Tobias and E.J. Vytlacil. 2001. Four Parameters of Interest in the Evaluation of Social Programs. *Southern Economic Journal* 68(2): 210–233.

Holling, C.S. 1973. Resilience and Stability of Ecological Systems. *Annual Review of Ecology and Systematics* 4: 1–23.

Hsiang, S. 2010. Temperatures and Cyclones Strongly Associated with Economic Production in the Caribbean and Central America. *Proceedings of the National Academy of Sciences* 107(35): 15367–15372.

Hsiang, S., K. Meng and M. Cane. 2011. Civil Conflicts Are Associated with the Global Climate. *Nature* 476: 438–441.

Hurni, H. 1998. Agroecological Belts of Ethiopia: Explanatory Notes on Three Maps at a Scale of 1:1,000,000. Soil Conservation Research Programme, Research Report No. 43. Addis Ababa and Bern: Centre for Development and Environment (CDE).

Intergovernmental Panel on Climate Change (IPCC). 2007. Summary for Policymakers: Climate Change 2007: The Physical Science Basis. Working Group I Contribution to IPCC Fourth Assessment Report: Climate Change 2007. Geneva: IPCC.

International Assessment of Agricultural Knowledge, Science and Technology for Development (IAASTD). 2009. *Agriculture at a Crossroad – Global Report.* 1718 Connecticut Avenue, NW, Suite 300, Washington, DC, 20009: Island Press. www.unep.org/dewa/ agassessment/reports/IAASTD/EN/Agriculture%20at%20a%20Crossroads_Global%20 Report%20(English).pdf (accessed December 2013).

Just, R.E., and R.D. Pope. 1979. Production Function Estimation and Related Risk Considerations. *American Journal of Agricultural Economics* 61: 276–284.

Kabubo-Mariara, J., and F.K. Karanja. 2007. The Economic Impact of Climate Change on Kenyan Crop Agriculture: A Ricardian Approach. *Global and Planetary Change* 57: 319–330.

Kumar, K.S.K., and J. Parikh. 2001. Indian Agriculture and Climate Sensitivity. *Global Environmental Change* 11: 147–154.

Kurukulasuriya, P., R. Mendelsohn, R. Hassan, J. Benhin, M. Diop, H.M. Eid, K.Y. Fosu, G. Gbetibouo, S. Jain, A. Mahamadou, S. El-Marsafawy, S. Ouda, M. Ouedraogo, I. Sène, N. Seo, D. Maddison and A. Dinar. 2006. Will African Agriculture Survive Climate Change? *World Bank Economic Review* 20(3): 367–388.

Kurukulasuriya, P., and R. Mendelsohn. 2008a. Crop Switching as an Adaptation Strategy to Climate Change. *African Journal Agriculture and Resource Economics* 2: 105–125.

Kurukulasuriya, P., and R. Mendelsohn. 2008b. A Ricardian Analysis of the Impact of Climate Change on African Cropland. *African Journal of Agricultural and Resource Economics* 2: 1–23.

Lautze, S., Y. Aklilu, A. Raven-Roberts, H. Young, G. Kebede and J. Learning. 2003. Risk and Vulnerability in Ethiopia: Learning from the Past, Responding to the Present, Preparing for the Future. Report for the U.S. Agency for International Development. Addis Ababa, Ethiopia. Inter-University Initiative on Humanitarian Studies and Field Practice, Harvard University, Cambridge, MA, U.S., and Feinstein International Famine Center, Gerald J. and Dorothy R. Friedman School of Nutrition Science and Policy, Tufts University, Medford, MA, U.S.

Lee, L.F., and R.P. Trost. 1978. Estimation of Some Limited Dependent Variable Models with Application to Housing Demand. *Journal of Econometrics* 8: 357–382.

Lobell, D.B., M.B. Burke, C. Tebaldi, M.M. Mastrandrea, W.P. Falcon and R.L. Naylor. 2008. Prioritizing Climate Change Adaptation Needs for Food Security in 2030. *Science* 319: 607–610.

Maddala, G.S. 1983. *Limited Dependent and Qualitative Variables in Econometrics.* Cambridge, UK: Cambridge University Press.

Maddala, G.S., and F.D. Nelson. 1975. Switching Regression Models with Exogenous and Endogenous Switching. In *Proceedings of the American Statistical Association* (Business and Economics Section), pp. 423–426, ASA.

Maddison, D. 2007. The Perception of and Adaptation to Climate Change in Africa. Policy Research Working Paper No. 4308. The World Bank, Development Research Group, Sustainable Rural and Urban Development Team. Washington, DC: The World Bank.

Mary, L., and A.E. Majule. 2009. Impacts of Climate Change, Variability and Adaptation Strategies on Agriculture in Semi-Arid Areas of Tanzania: The Case of Manyoni District in Singida Region, Tanzania. *African Journal of Environmental Science and Technology* 3(8): 206–218.

Massetti, E., and R. Mendelsohn. 2011. Estimating Ricardian Functions with Panel Data. *Climate Change Economics* 2: 301–319.

McFadden, D.L. 1973. Conditional Logit Analysis of Qualitative Choice Behavior. In *Frontiers in Econometrics*, edited by Paul Zarembka. New York: Academic Press, pp. 105–142.

McIntyre, B.D., H.R. Herren, J. Wakhungu and R.T. Watson. 2009. Agriculture at a Cross Road. International Assessment of Agricultural Knowledge, Science and Technology for Development (IAASTD), Sub-Saharan Africa (SSA) Report. Washington, DC: The World Bank.

Mendelsohn, R., and A. Dinar. 1999. Climate Change, Agriculture, and Developing Countries: Does Adaptation Matter? *The World Bank Research Observer* 14(2): 277–293.

Mendelsohn, R., and A. Dinar. 2003. Climate, Water, and Agriculture. *Land Economics* 79: 328–341.

Mendelsohn, R., and A. Dinar. 2009. *Climate Change and Agriculture: An Economic Analysis of Global Impacts, Adaptation, and Distributional Effects*. Cheltenham, UK: Edward Elgar.

Mendelsohn, R., W. Nordhaus and D. Shaw. 1994. The Impact of Global Warming on Agriculture: A Ricardian Analysis. *American Economic Review* 84: 753–771.

Miguel, E. 2005. Poverty and Witch Killing. *Review of Economic Studies* 72: 1153–1172.

Mundlak, Y. 1978. On the Pooling of Time Series and Cross Section Data. *Econometrica* 46(1): 69–85.

Parry, M., C. Rosenzweig and M. Livermore. 2005. Climate Change, Global Food Supply and Risk of Hunger. *Philosophical Transactions of the Royal Society B* 360: 2125–2138.

Rosenzweig, C., and M. Parry. 1994. Potential Impact of Climate Change on World Food Supply. *Nature* 367: 133–138.

Rosenzweig, M., and H. Binswanger. 1993. Wealth, Weather Risk and the Composition and Profitability of Agricultural Investments. *Economic Journal* 103(416): 56–78.

Schlenker, W., and D.B. Lobell. 2010. Robust Negative Impacts of Climate Change on African Agriculture. *Environmental Research Letters* 5: 1–8.

Schlenker, W., and M.J. Roberts. 2009. Non-Linear Temperature Effects Indicate Severe Damages to U.S. Crop Yields under Climate Change. *PNAS* 106(37): 15594–15598.

Seo, S.N., and R. Mendelsohn. 2008. Measuring Impacts and Adaptations to Climate Change: A Structural Ricardian Model of African Livestock Management. *Agricultural Economics* 38(2): 151–165.

Stage, J. 2010. Economic Valuation of Climate Change Adaptation in Developing Countries. *Annals of the New York Academy of Sciences* 1185: 150–163.

Udry, C. 1996. Gender, Agricultural Production, and the Theory of the Household. *Journal of Political Economy* 104(5): 1010–1046.

UNEP. 2002. *Africa Environment Outlook: Past, Present and Future Perspectives*. Nairobi: UNEP, p. 422.

Wooldridge, J.M. 2002. *Econometric Analysis of Cross Section and Panel Data*. Cambridge, MA: MIT Press.

World Bank. 2010. *The Economics of Adaptation to Climate Change*. Washington, DC: The World Bank Group. http://climatechange.worldbank.org/sites/default/files/documents/EACC_Ethiopia.pdf (accessed November 2013).

Yesuf, M., and R. Bluffstone. 2009. Poverty, Risk Aversion, and Path Dependence in Low-Income Countries: Experimental Evidence from Ethiopia. *American Journal of Agricultural Economics* 91(4): 1022–1037.

6 Climate, shocks, weather and maize intensification decisions in rural Kenya*

*Martina Bozzola, Melinda Smale and
Salvatore Di Falco*

Introduction

Farmers in Africa are among the most vulnerable to climate change. On the African continent, multiple stresses occur at multiple scales; African smallholder farmers, who are among the world's poorest, have limited capacity for adaptation (IPCC, 2007). Kenya is heavily exposed to changing climatic patterns, with serious repercussions for the well-being of farming households (Oremo, 2013). Many areas of the country have registered rising seasonal mean temperatures over the last 50 years. Regional climate model studies suggest drying over most parts of Kenya in August and September, although climate impacts are likely to be unevenly distributed across the country (Niang et al., 2014).

Smallholder farm families pursue various adaptive strategies to cope with climate change, but intensification of production through increased use of hybrid seed and mineral fertilizer is not generally one of them. Since the Green Revolution in Asia, researchers have debated whether the yields of improved varieties and hybrids are higher but also more variable, exposing poor farmers to greater risk (e.g., Anderson and Hazell, 1989). In general, empirical evidence on this point depends on the counterfactual (which varieties/hybrids are compared) and

* The farm household data used in this work was collected and made available by the Tegemeo Institute of Agricultural Policy and Development of Egerton University, Kenya. However, the specific findings and recommendations remain solely the authors' and do not necessarily reflect those of Tegemeo Institute.

    We are grateful to Jean-Paul Chavas, Robert Finger, Nick Hanley, Mary K. Mathenge, Michele Redi, Timothy Swanson and Cédric Tille for their thoughtful comments. We are also indebted to Simone Fatichi for assistance with the climate and soil quality data, to Xavier Vollenweider for his support in preparing the SPEI index and to Mark Schaffer for sharing the user-written stata command xtivreg3.

    The research leading to these results has received funding from the European Union's Seventh Framework Program FP7/2007–2011 under Grant Agreement Number 290693 FOODSECURE, as well as the US Agency for International Development and Michigan State University. The authors alone are responsible for any omissions or deficiencies. Neither the FOODSECURE project nor any of its partner organizations, nor any organization of the European Union or European Commission, the US Agency for International Development nor Michigan State University is accountable for the content of papers.

the geographical scale of analysis. In the major agricultural regions of Kenya, farm families depend on maize as a food staple and ready source of cash. Maize growers have high adoption rates and a history of growing maize hybrids with and without fertilizer (Mathenge et al., 2014). They have limited access to credit and no access to insurance, so they have a strong incentive to plant seed that reduces the variance of yields and limits their exposure to downside risk. Preliminary research by Jones et al. (2012), who considered several of the major maize-growing agroecologies, suggested that the use of hybrids in maize production not only enhanced mean yields but also reduced downside risk, with no significant effect on yield variance.

Smallholder agricultural production in rain-fed agriculture, like that found in Kenya, relies on environmental production conditions that are "exogenously" determined – that is, largely outside the control of farm families (Sherlund et al., 2002). Ochieng et al. (2016) estimated the effects of climate variability and change in crop revenue on maize and tea revenues earned by smallholder farmers in Kenya, finding differences between the two crops; temperature affected crop revenues negatively in maize but positively in tea production, while rainfall had a negative effect on income from tea. An analysis by Wineman et al. (2016) explored the channels through which exposure to extreme weather in Kenya affects the well-being of smallholder farm households, based on longitudinal and spatial analysis of income- and calorie-based measures of welfare. The authors found that extreme weather generally affects household welfare via crop production, recommending the development of new varieties with enhanced tolerance of dry and moist extremes.

Here, we focus on variety choice. Climate and soil characteristics are rarely incorporated into microeconomic analysis of variety choice. Mutiso (1996) showed that farmers in southeastern Kenya follow local knowledge systems when choosing the time to prepare land and plant. Other agronomic factors also guide planting decisions, especially in areas with sparse rainfall (Sacks et al., 2010). Thus, it is important to account not only for environmental conditions, including climate and soil quality, but also for factors influencing farm management choices and adaptation options, such as human capital (labor supply and quality) and financial and physical capital (assets, access to credit, farm size and tenure).

We address two research questions in this chapter. First, we ask how climatic shocks, weather and climate change affect smallholder decisions to intensify maize production. We measure intensification as the share of maize area per farm allocated to hybrid seeds. While controlling for relevant covariates as noted earlier, we differentiate and test the separate effects of climatic shocks, climate and weather on hybrid area shares. *Climate shocks* refer to the number of times during the previous decade that each village experienced a serious drought. The term *climatology* refers to climate normals. Climate normals are measured as average weather conditions over a 30-year period (1971–2010). *Weather* indicates the rainfall and temperature registered during the main rainfall season of the corresponding data collection year.

Second, we ask whether and how allocating a higher share of maize area in hybrid seeds per farm affects the vulnerability of smallholder income, expressed in terms of expected crop income, crop income variability and downside risk. Maize is the primary staple food grown by all smallholder farmers in the sample, across a wide range of livelihood types and farming systems. Mathenge et al. (2014) report that maize accounts for about 28% of gross farm output in the small-scale farm sector and that outside the semiarid areas 98% of households grow maize. Tegemeo Institute data, used by Mathenge et al. (2014) and here, shows that the share of crop income in household income averages 45%, varying only between 44% and 48% over the years of the survey.

With reference to the portfolio theory of decision making, we address the second question by using Antle's (1983) method of moments. Our econometric strategy reflects the structure of our data-generating process and conceptual frame. We apply our model to four waves of panel data collected in the major agricultural regions of Kenya, controlling for time-invariant heterogeneity by applying the Mundlak-Chamberlain procedure. In our two-stage econometric modeling, we are interested in examining both the maize intensification decision and the relationship of maize intensification to income vulnerability. We also consider the potential endogeneity of input choices in crop income outcomes.

We contribute to the existing literature in several directions. First, from a methodological perspective, we differentiate the roles of climatology, climate shock and weather on input choices in a microeconomic context. Second, we explore the intensification of staple food production through the dimension of hybrid seed. The Boserupian hypothesis (Boserup, 1975) suggests that population pressures on a declining environmental base generate incentives for intensifying food production. In a volatile environmental context, input intensification could aggravate smallholder vulnerability. We test this hypothesis.

Third, in the second stage of the estimation procedure, we include, along with intensification variables, climatology, climate shocks and weather as explanatory factors, gauging the impact on crop income and risk across agro-regional zones. The inclusion of both climate and weather variables allows us to capture the full extent of underlying adaptation decisions (Bezabih et al., 2014). Thus, our work contributes to illuminating an ongoing debate concerning the appropriate measurement methods in adaptation studies.

Finally, we include detailed information on environmental production conditions, such as climate and soil characteristics at the village level, and separate the main rainfall season and short-season rainfall. The incidence of seasons and their length vary across Kenya's agro-regional zones, and across years. Sherlund et al. (2002) have demonstrated the potential bias in production models of failing to control for soil quality. In terms of measurement techniques, we utilize the Standardized Precipitation Evapotranspiration Index (SPEI), which is the most advanced drought index. The SPEI is a multiscalar drought index that accounts for the fact that the impact of rainfall on the growing cycle of a plant depends on the extent to which water can be retained by the soil.

Theoretical basis

A leading paradigm in models of seed-fertilizer adoption since the Green Revolution has been the portfolio theory of investment attributed to Markowitz (1952), articulated by Just and Zilberman (1983) in terms of trade-offs between the mean and variance of yield distributions, where the choice variable was the crop area share allocated to new techniques (here, hybrid seed) with higher mean yields as compared to more traditional farmers' techniques (local maize, no mineral fertilizer).

However, the approach has been far less commonly applied in the analysis of natural resource management. We apply it in the context of intensification choices made under climate change and extend it to include skewness effects, following Antle's method of moments (1983). Recent research has demonstrated the importance of the third moment in analyzing climate-related risk in agricultural production (Koundouri et al., 2006; Di Falco and Chavas, 2009). Notably, we assume that a farm family will maximize the following function:

$$\max_{\alpha_t} \alpha_t (E_t R_{t+1} - R_{f,t+1}) - \frac{k_1}{2} \alpha_t^2 \sigma_t^2 + \frac{k_2}{3} \alpha_t^3 \gamma_t \qquad (1)$$

where the family chooses to allocate the maize area share of α_t to the riskier hybrid seeds at time t and the other terms are defined as follows: R_{t+1} is the return to hybrid maize, from time t to time t+1; $R_{f,t+1}$ is the return to local maize, from time t to time t+1; $E_t R_{t+1}$ is the conditional mean (conditional on the farmer's information at time t; thus they are written with the t subscript) of the maize area planted with hybrid seeds and σ_t^2 is the conditional variance of the maize area planted with hybrid seeds.

The terms k_1 and k_2 are coefficients representing farmers' risk aversion to yield variance and skewness respectively. Higher terms k_1 and k_2 indicate more conservative farmers who hold fewer hybrid seeds.

We extend the standard mean-variance model by adding skewness, defined as:

$$\gamma_t = E\left(\frac{X-\mu}{\sigma}\right)^3 = \frac{\mu^3}{\sigma^3} \qquad (2)$$

Given the definition of α_t, we can redefine the overall variance of (maize) yields as: $\sigma_{pt}^2 = \alpha_t^2 \sigma_t^2$. Similarly, we define the skewness as: $\gamma_{pt}^2 = \alpha_t^3 \gamma_t$. The return on the input mix (the seeds portfolio) is a linear combination of the simple returns of the riskier and less risky inputs. By definition, the input that generates higher means (in this case, the intensifying input, hybrid seeds) also generates greater variance. One of the factors that can contribute to increasing the vulnerability of modern agricultural production systems is the utilization of a narrow range of genetic material in plant breeding, so that the different varieties grown by farmers are in fact closely related; another is the slow turnover of modern varieties in farmers' fields, although we have no evidence of such a situation in Kenya (Smale and Olwande, 2014).

We assume that the benchmark input set has sufficiently low risk, so that the solution of the problems is almost identical to the standard mean-variance model with a riskless asset. The farm family prefers a high mean and low variance of returns on the mix. As in the standard mean-variance model, we assume that the farm family maximizes a linear combination of mean and variance of returns from inputs, with a positive weight on the mean and a negative weight on the variance. The farm family is averse to results skewed in a specific direction ($\gamma_t < 0$).

By solving the first-order condition of Equation 1, we can find the optimal share of maize land to be farmed with the riskier input set.

$$E_t(R_{t+1} - R_{f,t+1}) - k_1 \alpha_t \sigma_t^2 + k_2 \alpha_t^2 \gamma_t = 0 \tag{3}$$

Defining:

$$\Delta = E_t R_{t+1} - R_{f,t+1} \; ; \quad x = \frac{\Delta}{k_1 \sigma_t^2} \; ; \quad y = \frac{k_2 \gamma_t}{\Delta}$$

The solution of the maximization problem also can be written as:

$$\alpha_t = \min\left(\alpha_t = \frac{1 - \sqrt{1 - 4xy}}{2xy}, \; 1 \right) \tag{4}$$

In cases where the yield skewness plays a small part, an approximate solution, up to the first order in γ, is:

$$\alpha_t = \frac{\Delta}{k_1 \sigma_t^2} + \frac{\gamma_t \Delta^2 k_2}{k_1^3 \sigma_t^6} + \ldots 0\left(\gamma_t^2\right) \tag{5}$$

where $\frac{\Delta}{k_1 \sigma_t^2} > \frac{|\gamma_t| \Delta^2 k_2}{k_1^3 \sigma_t^6}$ and higher orders are expected to be negligible, because of the way we defined the approximation.

Equation 5 indicates the share of (maize) land on which the farmer is willing to plant riskier inputs (hybrid seeds). This share is equal to the risk premium divided by the conditional variance times a coefficient representing the risk aversion of the farm family plus a term capturing aversion to negative outcomes of the distribution of skewness. By including skewness in the model, we can approximate downside risk exposure. Increased skewness of yield (income) implies lower exposure to downside risk. Downside risk refers to the probability of zero or negative crop income for a smallholder farming family, which is potentially disastrous (Di Falco and Chavas, 2009). Reducing downside risk decreases the asymmetry of the yield (income) distribution by shifting it toward higher outcomes, holding both means and variances constant (Menezes et al., 1980; Di Falco and Chavas, 2009). We can view the short-term decision of a farming family as intended to avert negative outcomes or yield fluctuations in a specific direction.

Figure 6.1 illustrates the result of the farmer's maximization problem presented in Equation 4. The figure shows the optimal share of (maize) land planted to intensified inputs, for given ranges of variance, skewness and expected returns. Figure 6.1 has some interesting features. First, we notice that, for high values of

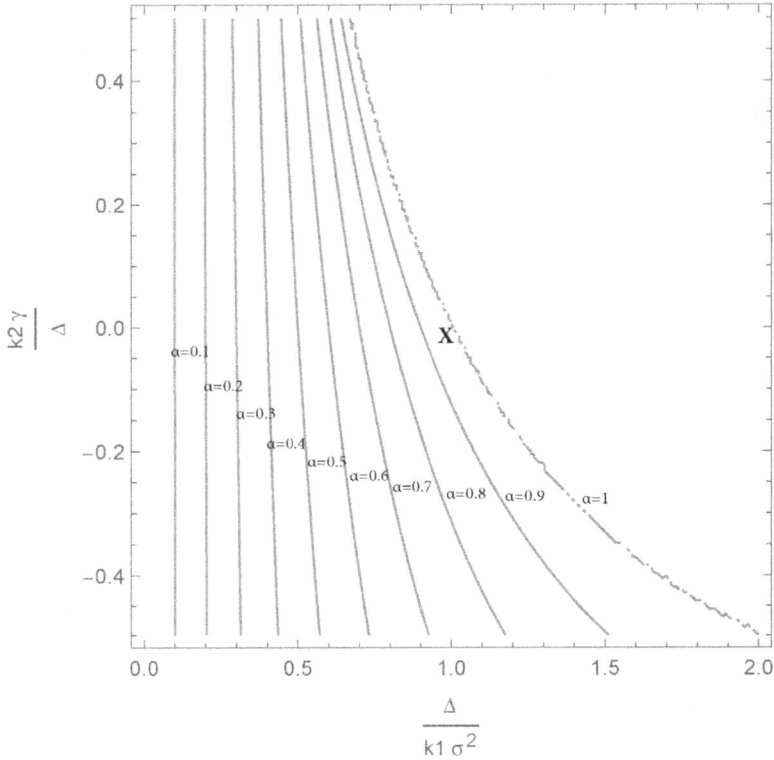

Figure 6.1 The Farmer's Maximization Problem

Note: Figure 6.1 is a visual representation of the framer's maximization problem presented in Equation (4).

the variance term σ_t^2, the distribution of the skewness of the yield does not matter in defining the share of land allocated to the risky inputs α_t, as indicated by the almost vertical boundary lines for values in the range from α_t=0.1 to α_t=0.4. However, as variance decreases or expected crop income increases (i.e., as we move to the right along the horizontal axis), the distribution of yield skewness (captured by γ_t) becomes increasingly relevant in determining the family's allocation to the risky input, up to a point where the variance is very low and only extremely adverse distributions of outcomes matter.

For example, we can consider an initial land allocation between risky and less risky input combinations, defined by the cross in the graph, where 90% of the area is allocated to intensified inputs and 10% to the other input set. We assume that the distribution of skewness equals zero and that the ratio $\frac{\Delta}{k_1\sigma_t^2}$ equals almost 1.1 for this input choice.

If skewness increases to high enough positive outcomes, holding the variance constant will cause the share α_t to increase to 1. However, as the distribution of skewness assumes negative outcomes (and the farm family fears crop failure), the share of cropland allocated to the intensified input decreases. Under downside risk aversion, farm families are adversely affected by downside risk (e.g., risk of negative crop income). We expect that a downside-averse decision maker will invest in adaptation strategies to reduce such risk (Menezes et al., 1980; Antle, 1983, 1987; Di Falco and Chavas, 2009). Our research interest in capturing how inputs contribute to the skewness of the crop income distribution is greatest when variance is neither too low (and thus there is very little risk associated with the second moment of the distribution of crop income) nor extremely high.

The precise shape of the cutting lines in Figure 6.1 depends on the range of values attributed to the expected yields, as well as their variance and skewness. These are determined by the types of crops grown and local environmental conditions. Figure 6.1 provides intuition concerning why, under some conditions, the third moment of the distribution of agricultural yields does not seem to be a key determinant of farmer input choices, although some literature has found it to be crucial (Koundouri et al., 2006; Di Falco and Chavas, 2009; Groom et al., 2008; Di Falco and Veronesi, 2014). For example, skewness and variance effects might be very different across a country like Kenya, which is characterized by the presence of various agro-regional zones. Notably, hybrid seed would be preferred in areas where the marginal productivity is higher.

Empirical approach

Data sources

We draw on three comprehensive data sources in our analysis. The first source is household survey data collected by Egerton University's Tegemeo Institute of Agricultural Policy and Development, with technical support from Michigan State University, in four rounds (2000, 2004, 2007 and 2010). Argwings-Kodhek et al. (1999) provides a detailed description of the sample design, which was implemented in consultation with the Kenya National Bureau of Statistics (KNBS). Survey instruments are available online (www.tegemeo.org). All non-urban divisions in the selected districts were assigned to one or more agro-regional zones based on agronomic information from secondary data. The panel dataset comprises eight agro-regional zones. Within each division, villages and households (in that order) were randomly selected. The sample excluded large farms with over 50 acres and two pastoral areas. The final dataset used in this study contains detailed farm-level data from 1,243 agricultural households in 22 districts. Certain village-level covariates, such as population density and agro-regional zones, are included in these data and our analysis.

Second, we associate climate variables developed from the monthly average maximum, minimum and average temperature and monthly cumulative precipitation for 107 villages across Kenya from 1971. These climate data are from the Climatic Research Unit (CRU) TS3.21 dataset (Harris et al., 2014). We compile climate data to match the main rainy season and the short-rains season, taking into account local differences in the length and timing of these two seasons.

These data were used to calculate the SPEI index; as discussed earlier, this multi-scalar drought index accounts for the impact of rainfall on plant growth in the context of the soil's capacity to retain water. This in turn depends on the characteristics of the soil and on the extent to which sunshine induces evaporation (Harari and La Ferrara, 2014). The indicator, developed by Vicente-Serrano et al. (2010), considers the joint effects of precipitation, potential evapotranspiration (PET) and temperature in determining droughts.

The SPEI index is an extension of the widely used Standardized Precipitation Index (SPI) (McKee et al., 1993), and can be used for determining the onset, duration and magnitude of drought conditions with respect to normal local conditions. Increasingly, the SPEI index is considered an improved measure over similar indexes previously used[1] because it provides a better measure of the effective amount of moisture received by the soil (Vicente-Serrano et al., 2010; Harari and La Ferrara, 2014). We employ a three-month SPEI index (SPEI3), determined for the last month of the main rainfall season and comprising also the two preceding months, taking into account differences between agro-climatic zones in establishing the reference month.

Third, we draw on soils data at the village scale from the Harmonized World Soil Database, a partnership between the Food and Agriculture Organization (FAO) and the European Soil Bureau Network (FAO/IIASA/ISRIC/ISS-CAS/JRC, 2012).

Estimation strategy

Our analysis is conducted in two stages. In the first stage, we analyze the determinants of maize intensification, paying particular attention to the role of past climatic shocks (captured by the SPEI3 index), access to markets (captured by population density) and the price per kg of fertilizer used on dry maize from hybrid seeds during the main rainfall season.[2] Second, we probe how maize intensification, along with climate and weather, affects farmers' welfare under uncertainty, taking into account the heterogeneity in agro-regional conditions within Kenya. In this second step, we model the production technology as a stochastic production function, assessing its probability distribution using the sequential estimation procedure (Antle, 1983; Kim and Chavas, 2003). The dependent variables in the second step of the estimation procedure are expected crop income and variance and skewness of crop income.

We test and control for the potential endogeneity of maize intensification decisions by estimating two-stage least squares. This is a robust estimation method that provides a standard starting point for applying instrumental variables (Angrist and Krueger, 2001). In the first stage of the estimation procedure, we use the frequency of climatic shocks, the logarithm of population density at the village level, and the price of fertilizer as instrumental variables for the decision about maize intensification. For identification purposes, some of the variables in the equation determining maize intensification (Equation 5) will later be excluded from the crop income and risk equations (Equations 7a, b and c).

In the first stage of our estimation, we use Equation 6 to represent the optimal strategy undertaken by the representative farm household:

$$\alpha_{it} = \alpha\left(x_{it}^h, x_r^s, x_{rt}^v, x_{ikt}^p, x_{it}^f, x_{rt}^{CR}, x_{rk}^c, x_{rkt}^w, Z_a; \gamma\right) + \varepsilon_{it} \tag{6}$$

The subscript *it* denotes the *ith* farm household in year *t*, while the subscript *r* is used for village-level observations and *k* indexes the rainfall seasons. Terms γ and λ are vectors of parameters, and ε_{it} is the household-specific random error term. The dependent variable α_{it} is a continuous variable indicating the share of land planted with hybrid seeds of dry maize.

Explanatory variables

Definitions for each variable are presented in Table 6.1. Descriptive statistics of the variables used in this study are presented in Table 6.2.

Table 6.1 Variables Definitions

Variable	Description
Farm-specific variables (source: Tegemeo)	
Crop income	Value of crop production minus input and land preparation costs (labor and seeds costs excluded).
Price fertilizer (KES)	Farm price fertilizer for dry maize from purchased hybrids. If the household did not buy fertilizer for this crop category, the village's average is used.
Educated men	No. of adult men with secondary education
Educated women	No. of adult women with secondary education
Livestock assets (KES)	Total nominal value (KES) of livestock assets
Credit village	Proportion of village households that received credit, by year
Salaries & remittances	Share of salaries and remittance earnings in total household income
Land	Total household land area (ha)
Land title deed	=1 if land owned with no title deed, 0 otherwise
Village-specific climate characteristics (source: CRU TS3.21)	
SPEI3 Index	Three months Standardized Precipitation Evapotranspiration Index (SPEI3) for the last month of the main rainfall season (January, July or August, depending on the division and agro-regional zone to which each village belongs) and the two preceding months. We calculated the SPEI index manually, using the R routines developed by Vicente-Serrano et al. (2010). SPEI index for each location is based on monthly precipitation and rainfall at village level, downloaded from the CRU TS3.21 dataset (Harris et al., 2014) for the period 1971–2012.
Droughts_165	Number of times in the last decade[*] the value of the SPEI3 was < -1.65 in the last month of the main rainfall season.

(*Continued*)

Table 6.1 (Continued)

Variable	Description
Temperature max (°C)* Temperature min (°C)* Temperature average (°C)* Rainfall (mm/mo)*	Monthly average maximum air temperature (°C) during the major rainfall season Monthly average minimum air temperature (°C) during the major rainfall season Monthly average air temperature (°C) during the major rainfall season Cumulated rainfall (mm/mo) during the major rainfall season
Temperature average climatologies* Rainfall climatologies (mm/mo)*	Average air temperature (°C) 1971–2010 during the major rainfall season Cumulated rainfall (mm/mo) 1971–2010 during the major rainfall season
Village-specific soil characteristics (source: World Soil database)	
AWC_mm Ph topsoil (-log(H+)) Gravel topsoil (%vol)	Available water storage capacity class of the soil unit, measured in mm/m pH measured in a soil-water solution. It is a measure of the acidity/alkalinity of the soil. Volume % gravel (materials in soil larger than 2mm) in the topsoil (i.e., 0–30 cm) (%vol)
Village-level socioeconomic variables & agro-regional zones	
Population density Agro-regional zone	Village population density (cap/km²) HPMZ high-potential maize zone (26.6%); CHI central highlands (19.4%), WLO western lowlands (12%); WTR western transitional (11.7%); ELO eastern lowlands (11.3%); WHI western highlands (10.3%); CLO coastal lowlands (5.9%); MRS marginal rain shadow (2.7%). Percentages indicate the frequency of farms in our sample in each agro-regional zone.

Note: The exchange rate in 2017 for 1 Kenyan Shilling (KES) is USD .0097.

* We take into account the relevant cropping season: e.g., for villages in the Rift Valley, the reference period is March (year-1) to (August year-1).
# Reference Decades: 1989–1999 for 2000; 1993–2003 for 2004; 1996–2006 for 2007; 1999–2009 for 2010.

Vectors x_{it}^{h}, x_{it}^{f}, include household and other farm characteristics, respectively. The human capital resources of the household are measured as the number of adult men and adult women in the household with a secondary education. Financial capital is measured in terms of livestock wealth, access to credit at the village level, and salaries and remittances, which provide liquidity that is uncorrelated with crop income. Farm physical capital is represented by scale of land cultivated, with a dummy variable indicating ownership by deed.

Vector x_{ikt}^{p} includes the farm price for fertilizer applied to hybrid maize grown during the main rainfall season,[3] while the vector x_{rt}^{s} includes soil quality

Table 6.2 Descriptive Statistics

Variables	Mean	Std. dev.	Min	Max
Farm-specific variables				
Crop income	87,911	142,264	0	3,883,123
Hybrid seeds (maize land share)	0.338	0.323	0	1
Price fertilizer (KES)	37.08	18.03	0.32	700
Educated men	0.89	0.98	0	10
Educated women	0.71	0.81	0	6
Livestock assets (KES)	81,366	217,682	0	8,679,900
Credit village	0.47	0.30	0	1
Salaries & remittances income share	0.18	0.24	0	1
Land	5.80	8.72	0	157
No land title deed dummy	0.36	0.48	0	1
Village-specific climatic variables				
Temperature max (°C)	26.56	3.63	19.12	33.47
Temperature min (°C)	14.04	3.76	7.5	23.95
Temperature min (°C)	14.04	3.76	7.5	23.95
Temperature average (°C)	20.27	3.62	13.3	28.67
Rainfall (mm/mo)	708.75	209.29	145	1154.1
Temperature average climatologies (°C)	19.57	3.69	13.61	27.89
Rainfall climatologies (mm/mo)	708.95	186.32	184.58	946.44
SPEI3 Index	−0.18	1.01	−2.28	2.24
Droughts_165	0.74	0.72	0	2
Village-specific soil characteristics				
AWC_mm	149.42	3.77	125	150
Gravel top soil (%vol)	1.25	4.09	0	28
Ph top soil (-log(H+))	5.75	1.04	4.5	8.9
Village-specific socioeconomic variables				
Population density	363.47	214.88	16.43	1,245.11

information at the village level. Vector x_{rt}^{v} includes the logarithm of population density at village level.

Of special interest is vector x_{rt}^{CR}, capturing how climatic risk affects intensification decisions. This vector includes a climate risk proxy stemming from the SPEI3 index, determined for the last month of the main rainfall season, taking into account differences between agro-climatic zones in establishing the reference month. Notably, we include the number of times during the previous decade that the value of the SPEI3 was lower than −1.65. This value indicates the exposure of the village to serious drought stress. The SPEI3 drought index expresses the incidence of past droughts (climatic shocks) as determinants of input choices. Vectors x_{rt}^{c}, x_{it}^{w} include climate and weather information. Vector Z_{a} contains agro-regional zones fixed effects. We include them in the analysis, as we believe that being in a specific agro-regional zone significantly affects farm management decisions. For example, the way farmers adapt to climate change might differ significantly depending on whether the farm is located in a zone with a bimodal or a unimodal rainfall regime.

The role of variable α_{it}, representing farmers' decisions on the intensification of production, enters the second stage of our estimation strategy via the predictions from Equation 6. Through this second step, we investigate how intensification affects farmers' expected crop income under risk.

Next, in order to capture the full extent of risk exposure, we assess the impact of intensification strategies on the distribution of expected crop income (Equation 7a) and its variance (Equation 7b) and skewness (Equation 7c). To do this, we follow Antle's moment-based approach to specify the stochastic structure of the model.

Accordingly, the estimated relationship between crop income risk equations, climatic variables, maize intensification decisions and other covariates is given by:

$$\ln y_{it} = -+ \beta_w w_{rkt} + \beta_C c_{rk} + \mu\, x_{i,t} + \varphi s_r + \vartheta \alpha_{it} + Z_a + \varepsilon_{it} \tag{7a}$$

$$\hat{\varepsilon}_{it}^2 = -+ \beta_w w_{rkt} + \beta_C c_{rk} + \mu\, x_{i,t} + \varphi s_r + \vartheta \hat{\alpha}_{it} + Z_a + \check{\varepsilon}_{it} \tag{7b}$$

$$\hat{\varepsilon}_{it}^3 = -+ \beta_w w_{rkt} + \beta_C c_{rk} + \mu\, x_{i,t} + \varphi s_r + \vartheta \hat{\alpha}_{it} + Z_a + \tilde{\varepsilon}_{it} \tag{7c}$$

The subscripts i, t, r and k are defined as in Equation 8. The dependent variable $\ln y_{it}$ denotes the logarithm of crop income for the *ith* farm household at year t.[4] We incorporate both weather and climate measures as determinants of farm-level crop income and risk, as presented in Equations 7 a, b and c. Therefore, w_{rkt} is a vector of weather variables: temperature (minimum and maximum) and precipitation (monthly cumulative) in year t, while c_{rk} is a vector of climate normals for the mean temperature and cumulative rainfall. Both vectors refer to village r, for the main rainfall season ($k=1$). Vector x_{it} includes socioeconomic and physical farm characteristic variables at time t. Vector s_r contains soil quality variables, available at the village level. Z_a is a set of agro-regional zone fixed effects. These dummy variables can capture exogenous variables that vary by agro-regional zone but have not been measured.

The coefficients β_w, β_c, μ, ϕ, , ϑ and ξ represent the vectors of parameter estimates for each associated vector of variables, while ε_{it} is the error term. The composite error term is composed of a normally distributed random error term, $u_{it} \sim N(0,\sigma_u^2)$, and an unobserved household-specific time-invariant component (q_i), as follows:

$$\varepsilon_{it} = q_i + u_{it} \tag{8}$$

Similarly, $\check{\varepsilon}_{it}$ and $\tilde{\varepsilon}_{it}$ are the composite error terms for the variance, Equation 7b, and the downside risk, Equation 7c, respectively, and have the same distribution properties as ε_{it}.

The panel structure of our dataset requires the use of a fixed effect estimator that permits the time-variant regressors to be correlated with the time-invariant component of the error term, while assuming that these regressors are uncorrelated with the idiosyncratic error. This estimation provides consistent parameters even if there is correlation between the independent variables and time-invariant unobserved heterogeneity, such as soil quality. The estimation of an instrumental variables model with fixed effect methodology would allow us to test and control for potential endogeneity caused by a correlation between decisions regarding intensification and vulnerability outcomes. However, standard fixed effect models rely on a data transformation that removes the individual effect.

We have previously discussed the importance of including in our framework variables that are by their nature time-invariant regressors, such as climatology and soil quality variables. One way to include time-invariant variables while addressing endogeneity is to estimate a random effects model while controlling for unobserved heterogeneity using the Mundlak-Chamberlain approach (referred to as the pseudo-fixed effects model). Following Mundlak (1978) and Chamberlain (1982, 1984), the right-hand side of our regression equation includes the mean value of the time-varying explanatory variables. This approach relies on the assumption that unobserved effects are linearly correlated with explanatory variables. Thus, the unobserved household-specific time-invariant component in Equation 8 can be specified as:

$$q_i = \zeta \bar{x} + v_i \tag{9}$$

where $\bar{x}$ is the mean of the time-varying explanatory variables within each farm household (cluster mean), ζ is the corresponding vector coefficient, and v_i is a random error unrelated to the $\bar{x}$'s. The vector ζ will be equal to zero if the observed explanatory variables are uncorrelated with the random effects. The use of the Mundlak-Chamberlain device also addresses the problem of selection and endogeneity bias where these are due to time-invariant unobserved factors, such as household heterogeneity (Wooldridge, 2002). If we failed to control for these factors, we would not obtain consistent parameter estimates. Moreover, estimation of parameters ζ allows us to test for the relevance and strength of the fixed effects via an F test, performed for the endogenous variable.

Results

First-stage regression results for the potentially endogenous variables are reported in Table 6.3. Frequent past climatic shocks, as manifested by drought incidence, reduce the maize area share per farm allocated to hybrid seeds. Looking at the weather variables, extreme temperature influences maize intensification. Maximum temperature has a negative (insignificant) impact on intensification of production at an increasing rate (significant), while minimum temperatures have the opposite signs with the same significance. Higher rainfall has a significant, positive correlation with hybrid seed use, at a decreasing rate. Farmers in areas

Table 6.3 Estimation Results – First-Stage Regressions

	Share acreage under purchased hybrids	
Droughts_165	−0.0565***	[0.0085]
ln population density	0.3534***	[0.0536]
ln fertilizers price	−0.0163	[0.0104]
Temperature max	−0.2010	[0.1514]
Temperature max squared	0.0059**	[0.0029]
Temperature min	0.0581	[0.0657]
Temperature min squared	−0.0042*	[0.0022]
Rainfall	0.0004*	[0.0002]
Rainfall squared	−2.81e-07**	[1.41e-07]
Temperature average climatologies	−8.2869***	[1.1446]
Temperature average climatologies squared	0.1481***	[0.0257]
Rainfall climatologies	−0.0133***	[0.0039]
Rainfall climatologies squared	6.02e-06**	[2.36e-06]
AWC_mm	0.0008	[0.0030]
Ph top soil	0.0892***	[0.0117]
Gravel top soil	0.0087***	[0.0019]
ln livestock assets	0.0060*	[0.0031]
ln credit village	0.0468	[0.0323]
No land title deed dummy	−0.0045	[0.0092]
ln educated men	0.0267**	[0.0120]
ln educated women	0.0012	[0.0121]
ln salaries & remittances income share	0.0558	[0.0347]
ln land	−0.0466***	[0.0125]
Agro-regional FE	Yes	
F test of excluded instruments	$F(5, 4041)=30.58$	
Observations	4,085	

Note: Pseudo-fixed effect estimation. Robust standard errors in brackets. *** $p<0.01$,

** $p<0.05$,
* $p<0.1$.

where the weather is more favorable tend to allocate more maize area to hybrid seeds; temperatures are lower in the areas with the highest historical adoption rates.

Population density has a positive correlation with maize intensification, consistent with the Boserupian hypothesis. The presence of educated men in the household has a positive impact on the adoption of maize hybrid seeds, since education provides access to information, services and communication, as well as the potential to utilize these more effectively.

Fertilizer price has the expected negative sign but is not a statistically significant determinant of maize area shares planted to hybrid seed. Other research in Kenya has suggested that Kenyan farmers tend to apply nonoptimal quantities of mineral fertilizers. Ogada et al. (2010) found that most Kenyan farm households apply insufficient quantities of mineral fertilizers; Sheahan (2011) and Marenya and Barrett (2009) found the opposite. Our empirical model does not enable us to draw conclusions regarding the quantitative response of fertilizer application rates on maize to price variation, but only to observe the jointness of use of mineral fertilizers and hybrid seed. In about 90% of the plots in our dataset where maize hybrid seeds are planted during the main rainfall season, mineral fertilizers were also applied.[5] Also using the Tegemeo dataset, Smale et al. (2015) found a strongly significant and expected negative sign for the relationship between nitrogen nutrient kgs per ha of maize and the fertilizer price.

Wealthier farm families and families with higher human capital resources are more likely to plant a larger share of their land with hybrids. There is no statistically significant evidence that those farmers living in villages with less binding expenditure constraints are planting larger land shares with hybrid seeds. Credit is not provided directly for maize production in Kenya, but farmers who obtain credit for other purposes (e.g., tea growers in the highlands) may also be more likely to plant hybrids. A larger land endowment is negatively associated with the land share allocated to maize hybrid seed planted, suggesting that larger landowners might allocate a larger land share to other crops, such as cash crops, instead of staple crops. Soil quality strongly affects intensification decisions.

Table 6.4 reports the results for the second-stage regressions. We address the issue of the instruments' relevance using an F test of the joint significance of the excluded instruments, reported at the bottom of Table 6.3. The F statistic is greater than 10. This result indicates the strength of the chosen instruments (Staiger and Stock, 1997). The choice of instruments seems appropriate and we turn to discussing our main regression results.

In Table 6.4, Column 1 reports the impact of intensification of production on expected crop income. Consistent with previous research, a larger share of maize area per farm allocated to hybrid seed tends to positively affect expected crop income (Jones et al., 2012; Mathenge et al. 2014). To capture the full extent of how these management decisions determine risk exposure, we also report both the farm-specific variance function (Column 2) and the skewness function (Column 3) for crop income. We find no evidence that a larger share of land allocated to dry maize hybrid seeds (intensification) increases risk, in terms of either variance (this finding is consistent with Jones et al., 2012) or skewness of the distribution of crop income. The share of maize area allocated to hybrids has no significant impact on either the variance or the skewness of crop income. Thus, planting a greater area with maize hybrid seed contributes positively to mean crop income, with no statistically significant impact on the other risk equations.

In general, long-term impacts are larger than short-term effects, a result also found in Bezabih et al. (2014). Looking at weather, the squared temperature and precipitation coefficients are generally significant. This finding implies that the model is nonlinear. The fact that the squared terms are positive or negative reveals that seasonal effects are convex or concave, respectively. The maximum diurnal

Table 6.4 Estimation Results – Second-Stage Regressions (Pseudo Fixed Effects Estimation)

	(1) ln crop income		(2) Variance		(3) Skewness	
Share acreage under purchased hybrids	0.8873***	[0.2615]	-0.4921	[0.4843]	-1.8270	[60.4570]
Temperature max	-3.3191***	[0.4660]	0.9754	[1.0438]	-12.4135**	[6.2528]
Temperature max squared	0.0637***	[0.0091]	-0.0175	[0.0200]	0.2433*	[0.1661]
Temperature min	1.9952***	[0.2060]	-0.4800	[0.4717]	5.3069*	[3.0216]
Temperature min squared	-0.0659***	[0.0071]	0.0146	[0.0150]	-0.1723*	[0.0934]
Rainfall	0.0068***	[0.0006]	-0.0005	[0.0007]	0.0065	[0.0208]
Rainfall squared	-0.000004***	[0.0000]	0.0000	[0.0000]	0.000004*	[0.0000]
Temperature average climatologies	-2.6787	[8.1524]	-9.1297	[30.8128]	6.2084	[4944.3]
Temperature average climatologies sq.	0.0826	[0.1686]	0.1458	[0.6034]	0.0967	[100.828]
Rainfall climatologies	-0.0037	[0.0088]	0.0270	[0.0271]	-0.0289	[3.9933]
Rainfall climatologies sq.	-0.00001	[0.00001]	-0.00002*	[0.00002]	0.00003	[0.0029]
AWC_mm	-0.0008	[0.0133]	0.0593**	[0.0273]	0.0228	[0.3569]
Ph top soil	-0.4427***	[0.0447]	0.0831	[0.0689]	0.1851	[1.4567]
Gravel top soil	-0.0449***	[0.0087]	-0.0083	[0.0170]	-0.0250	[1.0743]
ln livestock assets	0.0456***	[0.0096]	-0.0268*	[0.0159]	0.2112	[0.3059]
ln credit village	0.3046***	[0.1029]	0.3764*	[0.2040]	-2.1129	[2.8854]
No land title deed dummy	-0.0365	[0.0273]	-0.0742**	[0.0308]	0.2418	[0.1665]
ln educated men	0.0093	[0.0386]	0.0813	[0.0504]	0.0534	[1.3603]
ln educated women	0.0992**	[0.0462]	-0.1979	[0.1217]	1.2878*	[0.6849]
ln salaries & remittances income share	-1.0519***	[0.1429]	0.5615	[0.4673]	-4.9736	[6.7722]
ln land	0.4596***	[0.0464]	0.1290*	[0.0731]	-0.7405	[2.9065]
Agro-regional FE	Yes		Yes		Yes	
Observations	4,085		4,085		4,085	
Number of hhid	1,166		1,166		1,166	

Note: Pseudo-fixed effect estimation. Robust standard errors in brackets. *** p<0.01, ** p<0.05, * p<0.1.

temperature correlates negatively with expected crop income, whereas higher minimum temperature is beneficial. Furthermore, higher diurnal temperature is associated with crop failure. Several agronomic studies confirm that maize reacts differently to maximum and minimum temperature (Harrison et al., 2011). Rainfall during the current main rain season has a bell-shaped relationship with crop income. Looking at the crop income equation (Column 1), we also notice that the coefficients associated with temperatures are much larger than the coefficients associated with rainfall. This result confirms those of Kabubo-Mariara and Karanja (2007), who concluded that in Kenya, the temperature component of global warming is much more important than precipitation. Interestingly, weather, but not climate, has an impact on the third moment of the distribution of crop income.

The impacts of climate normals on expected crop income are very similar, generally larger than the impacts of weather, but not statistically significant. An increase in rainfall climatologies enhances the risk associated with the variance of the distribution of agricultural yields, as well as the risk of crop failure. This result is probably related to the fact that most of the agriculture in the country (and in our sample) is rain-fed and depends strongly on the quantity and distribution of rainfall across space and time.

Soil quality is an important determinant of farm crop income. Higher values associated with the gravel variable indicate a higher percentage of materials that are larger than 2 mm in the soil. In areas where this type of soil is predominant, farming is more difficult and plant life is sparser. Notably, the higher the value associated with gravel soil, the lower the ability of the soil to retain moisture, and the lower the presence of mineral nutrients. Henceforth, the negative coefficient associated with this variable complies with our expectations. pH is a measure for the acidity and alkalinity of the soil, measured in concentration levels $(-\log(H+))$. pH between 5.5 and 7.2 (acid to neutral soil) offers the best growing conditions, and the mean value of the sample is in this range. Higher pH (associated with alkaline soils) is negatively correlated with crop income. Furthermore, farmers in Kenya tend to apply diammonium phosphate (DAP) as the main fertilizer type on dry maize from hybrid seeds when they need to increase the soil pH. This application might, however, also increase acidity of the land over the medium to long term.

Farm size, as expected, plays an important, positive role in determining crop income, as does the value of livestock assets. Higher shares of remittances and other salaries in total household income negatively affect crop income, probably because farmers with outside options in terms of income diversification have lower incentives to take management and investment decisions to improve maize farming conditions.

Whether the family has a deed title over the land it operates is not statistically significant for expected crop income. However, the associated coefficient is negative, indicating that land tenure insecurity could be detrimental to crop income. Since the ratification of the new constitution in Kenya, land tenure and entitlement have been a prominent concern. This finding suggests that land certification could be an effective policy instrument to buffer against climate anomalies. The presence in the household of women with secondary education is positively correlated with crop income and reduces the risk of crop failure. This result highlights the importance of human capital in efficiently managing agricultural technology.

Conclusions and implications

In this chapter, we have analyzed two major research questions. First, we have explored how climatology, weather and climate shocks affect maize intensification, other factors held constant. Second, we have tested whether maize intensification affects the vulnerability of smallholder farmers to crop income variability and downside risk, in the presence of these factors. We have defined maize intensification as the maize area share per farm allocated to hybrid seed (much of which is fertilized). Drawing from and extending the portfolio theory of investment choice, we estimated a two-stage model to identify the determinants of input use and assess the effects of input use on the mean, variance and skewness of crop income among smallholder farmers in Kenya. We focus on maize production, considering the importance of this crop as not only a food staple but also an income source in Kenya. In order to include time-invariant variables, such as soils and agro-regional fixed effects, while addressing endogeneity, we estimate a random effects model while controlling for unobserved heterogeneity using the Mundlak-Chamberlain approach (referred to as the pseudo-fixed effects model). We extend the portfolio investment approach previously applied to the analysis of input use decisions by incorporating and differentiating the effects of weather, climate change and climatic shocks.

Our approach enables us first to demonstrate that maize intensification is strongly affected by weather, climate shocks and climatology, in addition to commonly cited, household-farm characteristics, such as education, wealth, access to credit, and off-farm earnings. Next, we find that maize intensification has a positive effect on expected crop income but has no significant effect on crop income variability or downside risk. Moreover, relying on a higher proportion of hybrid seed use, which is negatively associated with persistent climatic shocks, is not enough to statistically significantly reduce the likelihood that crop income falls below a given threshold (downside risk). Importantly, cropping system decisions are related to longer-term investment choices, while decisions about specific hybrid types are, rather, annual decisions.

Thus, maize intensification is not in and of itself an effective strategy in the face of climate change and climate shocks. Further, our results suggest that farmers are not adapting optimally to climate change. Suboptimal choices might reflect multiple market failures, such as credit constraints, poor access to input and output markets, and information asymmetries.

In addition to these major findings, our empirical analysis confirms the need to account for agro-regional zones and soil quality variables in microeconomic models of input use and adaptation strategies. Omission of these factors could cause biased estimates of included coefficients. Regression results support the Boserupian theory that rising population density provides incentives for the shift toward more intensive farming systems. Finally, we find trade-offs between nonfarm employment and crop income, indicating changing dynamics of income in rural communities as Kenya urbanizes.

Our findings lead us to recommend that the government of Kenya play an active role in encouraging smallholder adaptation to changing climate patterns and climate shocks. In Kenya, multiple market failures include poor or nonexistent insurance, so that farmers use other risk-coping mechanisms, which can be weak (Fafchamps, 1992; Kurosaki and Fafchamps, 2002). Safety nets typically provide only limited support (Dercon and Krishnan, 2000; Dercon, 2004), while off-farm income that is not covariant with agricultural shocks is limited in more remote rural areas. In this context, smallholder maize growers need adaptation mechanisms other than the use of hybrid seed as a strategy to buffer against downside risk. Smallholders need continued improvement of access not only to well-adapted hybrid seed and other inputs through decentralized, competitive markets but also to effective, widely diffused market information services and other insurance mechanisms. Helping farmers learn about weather, climate, production and postharvest handling, as well as other adaptation strategies, would be beneficial.

Notes

1 Examples include the SPI, which is based on rainfall only, and on the assumption that temperature and potential evapotranspiration have negligible variability compared to precipitation, as well as the Palmer Drought Severity Index (PDSI) (Palmer, 1965), which is based on the soil-water balance equation on a fixed temporal scale between 9 and 12 months.
2 By dry maize, we refer to maize grown and harvested dry, rather than green. Price is averaged over fertilizer types, of which the dominant type applied to hybrid maize was diammonium phosphate (DAP).
3 If the household did not buy fertilizer for this crop during the main rainfall season, the village's average is used.
4 In order to treat the zero values in the sample, which would result in reduction of the sample size, we add the constant 1 to each variable before taking the natural logarithm – that is, ln(variable)=ln[1 + (variable)]. By doing this, we ensure that all of the logarithms will exist.
5 The remaining 10% of observations have missing values regarding the application of mineral fertilizers on the plot; thus, we cannot exclude the possibility that mineral fertilizer was applied on an even higher percentage of the plot farmed with hybrid seeds.

References

Anderson, J.R., and P. Hazell. 1989. *Variability in Grain Yields*. London: The John Hopkins University Press.

Angrist, J.A., and A.B. Krueger. 2001. Instrumental Variables and the Search for Identification: From Supply and Demand to Natural Experiments. *Journal of Economic Perspectives* (15): 69–85.

Antle, J.M. 1983. Testing the Stochastic Structure of Production: A Flexible Moment-Based Approach. *Journal of Business and Economic Statistics* 1: 192–201.

Antle, J.M. 1987. Econometric Estimation of Producers' Risk Attitudes. *American Journal of Agricultural Economics* 69: 509–522.

Argwings-Kodhek, G., T.S. Jayne, G. Nyambane, T. Awuor and T. Yamano. 1999. *How Can Micro-Level Household Information Make a Difference for Agricultural Policy Making? Selected Examples from the KAMPAP Survey of Smallholder Agriculture and Non-Farm Activities for Selected Districts in Kenya.* Nairobi, Kenya: Tegemeo Institute, Egerton University.

Bezabih, M., S. Di Falco and A. Mekonnen. 2014. Is It the Climate or the Weather? Differential Economic Impacts of Climatic Factors in Ethiopia. Grantham Research Institute on Climate Change and the Environment Working Paper No. 148. London, UK: London School of Economics.

Boserup, E. 1975. The Impact of Population Growth on Agricultural Output. *The Quarterly Journal of Economics* 89(2): 257–270.

Chamberlain, G. 1982. Multivariate Regression Models for Panel Data. *Journal of Econometrics* 1: 5–46.

Chamberlain, G. 1984. Panel Data. In *Handbook of Econometrics*, Volume 2, edited by Z. Griliches and M.D. Intriligator. Amsterdam: North Holland, pp. 1248–1318.

Dercon, S. 2004. *Insurance against Poverty.* Oxford: Oxford University Press.

Dercon, S., and P. Krishnan. 2000. Vulnerability, Poverty and Seasonality in Ethiopia. *Journal of Development Studies* 36: 25–53.

Di Falco, S., and J.P. Chavas. 2009. On Crop Biodiversity, Risk Exposure and Food Security in the Highlands of Ethiopia. *American Journal of Agricultural Economics* 91: 599–611.

Di Falco, S., and M. Veronesi. 2014. Managing Environmental Risk in the Presence of Climate Change: The Role of Adaptation in the Nile Basin of Ethiopia. *Environmental and Resource Economics* 57: 553–577.

Fafchamps, M. 1992. Cash Crop Production, Food Price Volatility and Rural Market Integration in the Third World. *American Journal of Agricultural Economics* 74: 90–99.

FAO/IIASA/ISRIC/ISS-CAS/JRC. 2012. *Harmonized World Soil Database (Version 1.2).* Rome, Italy and Laxenburg, Austria: FAO and IIASA.

Groom, B., P. Koundouri, C. Nauges and A. Thomas. 2008. The Story of the Moment: Risk-Averse Cypriot Farmers Respond to Drought Management. *Applied Economics* 40: 315–326.

Harari, M., and E. La Ferrara. 2014. Conflict, Climate and Cells: A Disaggregated Analysis. IGIER Working Paper No. 461, 2012 (Updated Version: February 2014). Milan, Italy: Bocconi University.

Harris, I., P.D. Jones, T.J. Osborn and D.H. Lister. 2014. Updated High-Resolution Grids of Monthly Climatic Observations. *International Journal of Climatology* 34: 623–642.

Harrison, L., J. Michaelsen, C. Funk and G. Husak. 2011. Effects of Temperature Changes on Maize Production in Mozambique. *Climate Research* 46: 211–222.

IPCC. 2007. Climate Change 2007: Impacts, Adaptation and Vulnerability. Contribution of Working Group II to the Fourth Assessment Report of the Intergovernmental Panel on Climate Change. Cambridge, UK: Cambridge University Press, pp. 433–467.

Jones, A., T. Dalton and M. Smale. 2012. A Stochastic Production Function Analysis of Maize Hybrids and Yield Variability in Drought-Prone Areas of Kenya. Tegemeo Working Paper No. 49. Nairobi: Tegemeo Institute of Agricultural Policy and Development.

Just, R.E., and D. Zilberman. 1983. Stochastic Structure, Farm Size and Technology Adoption in Developing Agriculture. *Oxford Economic Papers* 35(2): 307–328.

Kabubo-Mariara, J., and F.K. Karanja. 2007. The Economic Impact of Climate Change on Kenyan Crop Agriculture: A Ricardian Approach. *Global and Planetary Change* 57(3–4): 319–330.

Kim, K., and J.-P. Chavas. 2003. Technological Change and Risk Management: An Application to the Economics of Corn Production. *Agricultural Economics* 29: 125–142.

Koundouri, P., C. Nauges and V. Tzouvelekas. 2006. Technology Adoption under Production Uncertainty: Theory and Application to Irrigation Technology. *American Journal of Agricultural Economics* 88: 657–670.

Kurosaki, T., and M. Fafchamps. 2002. Insurance Market Efficiency and Crop Choices in Pakistan. *Journal of Development Economics* 67: 419–453.

Marenya, P., and C. Barrett. 2009. Soil Quality and Fertilizer Use Rates among Smallholder Farmers in Western Kenya. *Agricultural Economics* 40: 561–572.

Markowitz, H. 1952. Portfolio Selection. *Journal of Finance* 7: 77–91.

Mathenge, M.K., M. Smale and J. Olwande. 2014. The Impacts of Hybrid Maize Seed on the Welfare of Farming Households in Kenya. *Food Policy* 44: 262–271.

McKee, T.B.N., J. Doesken and J. Kleist. 1993. The Relationship of Drought Frequency and Duration to Time Scales. Eighth Conference on Applied Climatology, American Meteorological Society, Anaheim, CA, pp. 179–184.

Menezes, C., C. Geiss and J. Tressler. 1980. Increasing Downside Risk. *American Economic Review* 70: 921–932.

Mundlak, Y. 1978. On the Pooling of Time Series and Cross Section Data. *Econometrica* 46: 69–85.

Mutiso, S.K. 1996. Indigenous Knowledge in Drought and Famine Forecasting. In *Machakos District, Kenya: Indigenous Knowledge and Change in African Agriculture*, edited by W.M. Adams and L.J. Slikkerveer. Ames, IA: Technology and Social Change Program, Iowa State University, pp. 67–86.

Niang, I., O.C. Ruppel, M.A. Abdrabo, A. Essel, C. Lennard, J. Padgham and P. Urquhart. 2014. Africa. In *Climate Change 2014: Impacts, Adaptation, and Vulnerability, Part B: Regional Aspects*. Contribution of Working Group II to the Fifth Assessment Report of the Intergovernmental Panel on Climate Change, edited by V.R. Barros, C.B. Field, D.J. Dokken, M.D. Mastrandrea, K.J. Mach, T.E. Bilir, M. Chatterjee, K.L. Ebi, Y.O. Estrada, R.C. Genova, B. Girma, E.S. Kissel, A.N. Levy, S. MacCracken, P.R. Mastrandrea and L.L. White. Cambridge and New York: Cambridge University Press.

Ochieng, J., L. Kirimi and M. Mathenge. 2016. Effects of Climate Variability and Change on Agricultural Production: The Case of Small-Scale Farmers in Kenya. WPS 55/2016. Nairobi, Kenya: Tegemeo Institute of Agricultural Policy and Development.

Ogada, J.M., W. Nyangena and M. Yesuf. 2010. Production Risk and Farm Technology Adoption in the Rain-Fed Semi-Arid Lands of Kenya. *African Journal of Agricultural and Resource Economics* 4: 159–174.

Oremo, F.O. 2013. Small-Scale Farmers' Perceptions and Adaptation Measures to Climate Change in Kitui County, Kenya. MA Thesis. University of Nairobi, Nairobi, Kenya.

Palmer, W.C. 1965. Meteorological Drought. U.S. Department of Commerce Weather Bureau Research Paper No. 45. Washington, DC: U.S. Weather Bureau.

Sacks, W.J., D. Deryng, J.A. Foley and N. Ramankutty. 2010. Crop Planting Dates: An Analysis of Global Patterns. *Global Ecology and Biogeography* 19(5): 607–620.

Sheahan, M.B. 2011. Analysis of Fertilizer Profitability and Use in Kenya. MSc. Thesis. Department of Agriculture, Food and Resource Economics. Michigan State University, East Lansing.

Sherlund, S.M., C.B. Barrett and A.A. Adesina. 2002. Smallholder Technical Efficiency Controlling for Environmental Production Conditions. *Journal of Development Economics* 69: 85–101.

Smale, M., M.K. Mathenge and J. Opiyo. 2015. Off-Farm Work and Fertilizer Intensification among Smallholder Farmers in Kenya: A Cross-Crop Comparison. Tegemeo Working Paper Series No. 53. Nairobi: Tegemeo Institute.

Smale, M., and J. Olwande. 2014. Demand for Maize Hybrids and Hybrid Change on Smallholder Farms in Kenya. *Agricultural Economics* 45: 409–420.

Staiger, D., and J.H. Stock. 1997. Instrumental Variables Regression with Weak Instruments. *Econometrica* 65: 557–586.

Vicente-Serrano, S.M., S. Beguerìa, J.I. López-Moreno, M. Angulo and A. Kenawy. 2010. A Global 0.5 Gridded Dataset (1901–2006) of a Multiscalar Drought Index Considering the Joint Effects of Precipitation and Temperature. *Journal of Hydrometeorology* 11(4): 1033–1043.

Wineman, A., N.M. Mason, J. Ochieng and L. Kirimi. 2016. Let It Rain: Weather Extremes and Household Welfare in Rural Kenya. WPS 57/2016. Nairobi, Kenya: Tegemeo Institute of Agricultural Policy and Development.

Wooldridge, J.M. 2002. *Econometric Analysis of Cross Section and Panel Data*. Cambridge, MA: MIT Press.

7 Adaptation to climate change by smallholder farmers in Tanzania

Coretha Komba and Edwin Muchapondwa

Introduction

Agriculture is the most important sector in sub-Saharan Africa (SSA) and is set to be hit the hardest by climate change. Indeed, this is confirmed by several studies (see, e.g., Deressa, 2006; Moussa and Amadou, 2006; Jain, 2006; Hassan and Nhemachena, 2008; Molua and Lambi, 2006; Mano and Nhemachena, 2006). Although climate change may affect the agricultural sectors of different countries in different ways, what is clear is that these changes will bring about substantial welfare losses, especially for smallholders whose main source of livelihood derives from agriculture. There is a need for nations to neutralize the potential adverse effects if welfare losses to this vulnerable segment of society are to be avoided. Adaptation seems to be the most efficient way for farmers to reduce these negative impacts (Füssel and Klein, 2006). This can be achieved through the smallholder farmers themselves taking adaptive actions or by governments implementing policies aimed at promoting appropriate and effective adaptation measures.

In order to implement appropriate interventions, governments need to understand the opportunities (or lack thereof) for adaptation and the key drivers behind voluntary adaptation by vulnerable smallholder farmers. Some studies report that agricultural measures such as the use of improved crop varieties, the planting of trees, soil conservation, changing planting dates, and irrigation are the most used adaptation strategies in African countries. Other studies have pointed out several socioeconomic, environmental and institutional factors, as well as economic structures, as key drivers influencing farmers to choose specific methods in Africa as a whole and in some specific SSA countries (Deressa et al., 2009; Kabubo-Mariara, 2008; Mideksa, 2009; Bryan et al., 2009). Thus, there is a need for each nation to understand the scope of climate change and the drivers of adaptation, particularly among its smallholder farmers, in order to craft appropriate policy responses, as the vulnerability and sensitivity of each country differ, as does the accessibility of the different adaptation methods.

Tanzania is one of the SSA countries in which agriculture is the backbone of the economy. Thus, agriculture remains the largest sector in the economy and hence its performance has a significant effect on output and corresponding income and poverty levels (United Republic of Tanzania, 2003). Tanzanian agriculture is the major source of food, and accounts for about 45% of GDP, 60% of merchandise exports, 75% of rural household income and 80% of employment (Andersson et al., 2005). Furthermore, agriculture stimulates economic growth indirectly, through larger consumption linkages than other sectors have with the rest of the economy. Higher and sustained agricultural growth is needed to meet Tanzania's National Strategy for Growth and Reduction of Poverty (NSGRP, also called MKUKUTA in Kiswahili) and Millennium Development Goals of halving poverty and food insecurity by 2015 (United Republic of Tanzania, 2005).

Key constraints to achieving Tanzania's agricultural growth targets include: (1) high transaction costs due to poor state infrastructure or its absence; (2) underinvestment in productivity-enhancing technologies; (3) limited access to technology demand and delivery channels, with 60%–75% of households estimated to have no contact with agricultural research and extension services; (4) limited access to financing for the uptake of technologies; (5) unmanaged risks, with significant exposure to variability in weather patterns with periodic droughts, the impact of which is amplified by the dependency on rain-fed agriculture and the limited capacity to manage land and water resources; and (6) weak capacity in policy formulation and implementation of public interventions, along with weak coordination among the various actors in the public sector (United Republic of Tanzania, 2003). Recently, the Tanzanian government adopted the Agricultural Sector Development Strategy (ASDS) and its operational program (ASDP). The intention of this strategy is to achieve sustained agricultural growth of about 5% annually, primarily through transformation from subsistence to commercial agriculture. However, the agricultural development strategy also needs to address the serious challenges posed by climate change, which can become a crucial limiting factor for agricultural growth in the medium to long term. To date, insufficient attention has been paid to the issue of climate change in relation to agriculture. Accordingly, this study will attempt to gather evidence which can form the basis for mainstreaming climate change in discussions surrounding the agricultural sector.

It is important to know whether farmers respond to their perceptions of events. If they do, and if they recognize that climate change is occurring, then the state would simply need to help them overcome the constraints they face in implementing appropriate adaptation methods. On the other hand, if they do respond to their perceptions about events but do not recognize the role of climate change, then the state would need to ensure increased awareness. However, if farmers do not respond at all to their perceptions, then the state would need to be proactively involved in ensuring that farmers undertake appropriate adaptation if the impending welfare losses to this vulnerable group are to be abated.

The main purpose of this study is threefold: (1) to investigate whether smallholder farmers in Tanzania perceive climate change, (2) to investigate whether, as a consequence, they adapt at all in their agricultural activities and (3) to investigate the factors influencing their choice of particular adaptation methods. This study collected data from 556 randomly selected smallholder farming households from four representative administrative regions representing six of the seven agro-ecological regions of the country.

The rest of the chapter is arranged as follows. The next section reviews relevant previous studies on adaptation to climate change by individual farmers. The following section discusses the methods, variables and data used in this study. The fourth section presents and discusses the results. The concluding section draws policy implications.

Individual farmers' adaptation to climate change

Research has been undertaken by scholars around understanding farmers' awareness of climate change, options for adaptation to these changes, and the factors influencing choice of adaptation methods. Mixed evidence has been presented as to whether farmers are aware that the climate is changing in their areas. For example, Ishaya and Abaje (2008) report a lack of awareness and knowledge by farmers in Jema'a, Nigeria. On the other hand, working in the Nile Basin of Ethiopia, Deressa et al. (2009) report that 50.6% of the surveyed farmers had observed increasing temperatures over the past 20 years, whereas 53% of them had observed decreasing rainfall over the past 20 years. Thus, in line with the current definition of climate change, the majority of the surveyed Ethiopian farmers demonstrated awareness. According to Deressa et al. (2009), it appears that the easiest way of assessing this awareness is to inquire from a sample whether they have observed a change in the climate across two adjacent decades (e.g., between the 1990s and the 2000s, in terms of both the means and the variances of precipitation and temperature). With that goal, our study will use that approach in its investigation.

It might be expected that farmers who recognize climate change will take some actions to cushion themselves against its adverse effects. In the Ethiopian study, 58% of farmers who claimed to have observed changes in climate over the past 20 years had responded to it by undertaking some adaptation measures. In fact, several studies report agricultural adaptation measures such as the use of crop varieties, planting trees, soil conservation, changing planting dates, diverging from crop production to livestock keeping, and irrigation as the most used adaptation methods in African countries (Deressa et al., 2009; Kabubo-Mariara, 2008; Mideksa, 2009; Ajao and Ogunniyi, 2011; Bryan et al., 2009). However, it is clear that for various reasons, not all farmers will adapt. In this study, the reasons for failing to adapt mentioned by farmers included lack of funds, shortage of water, poor planning and shortage of seeds.

Several factors have been put forward to explain the presence or absence of adaptation to climate change. Downing et al. (1997) explore fairly standard variables to explain adaptation in Africa. Nhemachena and Hassan (2007) identify the important determinants of adaptation in South Africa, Zambia and Zimbabwe as access to credit and extension, and also awareness. Their study suggests enhancing access to credit and information about climate and agronomy so as to boost adaptation. Ishaya and Abaje (2008) find that lack of awareness and knowledge about climate change and adaptation strategies, lack of capital and improved seeds, and lack of water for irrigation played an important role in hindering adaptation in Jema'a, Nigeria.

Gbetibouo (2009) proposes that the major driver influencing farmers' adaptation in Limpopo Basin, South Africa, is the way that they formulate their expectations of future climate in dealing with the changing weather patterns. According to that study, the major factor restraining farmers' adaptation is inadequate access to credit. The study also argues that, among other things, the main factors that promote adaptive capacity are farmers' income, the size of the household, farmers' experience, and engaging in nonfarm activities. Below et al. (2012) acknowledge the role of public investment in rural infrastructure, a good education system that allows females equal education opportunities, availability of micro-credit services, availability and technically efficient use of agricultural inputs, and availability of agricultural extension in improving adaptation in Mlali and Gairo villages in Tanzania.

While analyzing farmers' perceptions of climate change, governance and adaptation constraints in the Niger Delta region of Nigeria, Nzeadibe et al. (2011) also point out that the factors responsible for hindering adaptation are inadequate information, limited awareness and knowledge about adaptation methods, and poor government attention to the phenomenon of climate change. Deressa et al. (2011) also find that education level and gender of the head of the household, size of the household, livestock ownership, availability of credit, and temperature significantly influence the presence of farmers' adaptation in Ethiopia. Ogalleh et al. (2012), in analyzing perceptions and responses in Kenya, find that smallholders' perceptions are that climatic variability is increasing. In dealing with the negative impacts of this variability, the smallholders in this community use diversification of crop varieties, migration and sale of livestock. In addition, West et al. (2008) analyzed the local perceptions and regional climate trends on the central plateau of Burkina Faso and found that rural households in the study area vary their agricultural practices – for example, integrating different crop varieties in their agricultural activities and implementing a host of soil and water conservation practices in order to respond to drought.

For those farmers who undertake any adaptation at all, the choice of specific method depends on a number of elements, including socioeconomic, environmental and institutional factors, as well as the economic structure of the country. Thus, the choice of adaptation methods depends on a range of variables which are considered important for the availability, accessibility and affordability of particular adaptation procedures. Several studies have identified specific variables which may positively or negatively affect the choice of particular adaptation methods. Deressa et al. (2009) conclude that farmers' education level, access to extension and credit, climate information, social capital and agroecological settings greatly influence their choices, while financial constraints and lack of information hinder farmers' uptake of other adaptation methods. Adesoji and Ayinde (2013), investigating the methods used by arable crop farmers to mitigate the negative impact of climate change in Osun State, Nigeria, suggest that age, household size, income, source of information and farm size are the main determinants of the choice of adaptation strategies implemented by farmers. In that study, the authors mention that the adaptation strategies which are regularly employed are use of different planting dates, multiple cropping and cover cropping.

In analyzing options and constraints in adaptation in Ethiopia and South Africa, Bryan et al. (2009) insist on a better understanding of climate change by farmers as a way of reducing its negative impacts. That study finds that government farm support, farmers' income, and access to fertile land and credit influence the choice of adaptation methods in South Africa, while access to extension and credit, farmers' income and information about climate change influence the choice in Ethiopia. The study further finds that the main barrier to uptake of other adaptation methods in both countries was lack of access to credit.

Each of the studies discussed earlier has something to offer the big picture, as summarized in Table 7.1. However, as mentioned earlier, what is important for the uptake of adaptation methods is the availability, accessibility and affordability of such techniques. Indeed, many socioeconomic variables have been investigated for their impacts on the choice of adaptation methods in different agroecological zones. For example, Downing et al. (1997) explore the standard variables to explain adaptation strategies in Africa but investigate specific factors affecting choice of adaptation strategies in the case of specific countries.

In that respect, the current study examines how socioeconomic, environmental and institutional factors as well as economic structure influence Tanzanian farmers' choice of adaptation methods. Thus, this study includes variables which capture the availability, accessibility and affordability of such techniques for Tanzanian smallholder farmers. The starting points are the following variables identified from the literature: access to credit and extension, farmers' awareness of climate change, knowledge about climate variation and adaptation strategies, availability of capital and improved seeds, availability of water for irrigation, farmers' income, the size of the household, farmers' experience, engaging in nonfarm activities, knowledge about adaptation methods, education and gender of the head of the household, livestock ownership, social capital, agroecological settings, government farm support, and access to fertile land.

Most research cited in this study modeled determinants of the choice of adaptation method using either a probit model or a multinomial logit model (Deressa et al., 2009; Bryan et al., 2009; Gbetibouo, 2009). Using MNL, as in Deressa et al. (2009) and Gbetibouo (2009), could be appropriate, as the farmers can make the choice from among more than two methods. However, this model imposes a very restrictive assumption that the choices of adaptation methods are independent across alternatives – that is, the assumption of independence of irrelevant alternatives (IIA) (Wooldridge, 2001). This assumption is not an easy one because farmers' choice of adaptation methods depends on different factors. In this case, the probability of choosing one method over another may change depending on the influence of the dependent factors. Alternatively, this study employs a multinomial probit model (MNP) which does not impose the independence assumption

Table 7.1 Literature Review Summary Table

Source	Purpose	Sample	Methods	Results
Ishaya and Abaje (2008)	To examine the way indigenous farmers in Jema'a, Nigeria, perceive climate change and their adaptation strategies to climate change	200 households	Analysis of variance (ANOVA) and chi-square	– Indigenous people perceive that the climate has been changing over the years. – The threat of climate change affects health, food supply, biodiversity loss and fuelwood availability. – Lack of improved seeds, access to water for irrigation, current knowledge of modern adaptation strategies, capital, awareness and knowledge of climate change scenarios are factors hindering the adoption of modern techniques of combatting climate change.
Deressa et al. (2009)	To identify the major methods used by farmers to adapt to climate change in the Nile Basin of Ethiopia, the factors that affect their choice of method, and the barriers to adaptation	1000 households	Multinomial logit (MNL) model	The level of education, gender, age, and wealth of the head of household; access to extension and credit; information on climate; social capital; agroecological settings and temperature all influence farmers' choices.
Kabubo-Mariara (2008)	To examine the economic impact of climate change on livestock production in Kenya	722 households	Ricardian model Hadley Centre coupled model (HADCM) and parallel climate model (PCM)	– Modest gains from rising temperatures and losses from increased precipitation. – Livestock farmers in Kenya are likely to incur heavy losses from global warming.
Mideksa (2009)	To quantify the general equilibrium impact of climate change on the GDP of Ethiopia	Macro data using a World Bank 2005 SAM	A multi-sector, multi-product, comparative static small open economy general equilibrium model	Climate change will make the prospect of economic development harder, either by reducing agricultural production in the sectors linked to the agricultural sector through a 10% decrease in GDP, or by increasing the degree of income inequality, in which the Gini coefficient increases by 20%.

Study	Objective	Sample	Method	Findings
Bryan et al. (2009)	To examine farmers' perceptions of climate change, the extent of adaptation, barriers to adaptation, and the factors influencing adaptation and adaptation choices in Ethiopia and South Africa.	1800 farm households	A probit model	– Farm-level adaptation involves more than adopting new agricultural technologies. – The results by country and income terciles suggest that strategies should also be tailored to meet the particular needs and constraints of different countries and groups of farmers.
Nhemachena and Hassan (2007)	To examine farmers' adaptation strategies to climate change in Southern Africa (South Africa, Zambia and Zimbabwe)	1719 households	A multivariate discrete choice model	Access to credit and extension and awareness of climate change are some of the important determinants of farm-level adaptation.
Gbetibouo (2009)	To determine whether the climate has changed, whether farmers perceive climate change and variability, and what characteristics differentiate farmers who perceive changes from those who do not, in South Africa	794 households	– A Heckman probit model – A multinomial logit (MNL) model	– Household size, farming experience, wealth, access to credit, access to water, tenure rights, off-farm activities, and access to extension are the main factors that enhance adaptive capacity. – Lack of access to credit is the main factor inhibiting adaptation.
Nzeadibe et al. (2011)	– To appraise the perception and understanding of Niger Delta farmers of the role of national governments in climate change governance. – To examine grassroots communities' perceptions of constraints to adaptation to changing climate	400 households	Simple descriptive statistics (results presented in tables, figures and charts)	– The major constraints to climate change adaptation by farmers in the Niger Delta are lack of information, low awareness levels, irregularities of extension services, poor government attention to climate problems, inability to access available information, lack of access to improved crop varieties, ineffectiveness of indigenous methods, no subsidies for planting materials, limited knowledge of adaptation measures, low institutional capacity, and absence of government policy on climate change.

(Continued)

Table 7.1 (Continued)

Source	Purpose	Sample	Methods	Results
				– Farmers have a low level of awareness of government policies/programs on climate change. – Farmers have a poor perception of the effectiveness of the policies/programs and low awareness of the existence and impact of Committees on Climate Change in the National Assembly.
Deressa et al. (2011)	To analyze the two-step process of adaptation to climate change, which initially requires Ethiopian farmers' perception that climate is changing prior to responding to changes through adaptation	1000 mixed crop and livestock farmers	Heckman sample selection model	– Farmers' perception of climate change is significantly related to the age of the head of the household, wealth, knowledge of climate change, social capital and agroecological settings. – Factors significantly affecting adaptation to climate change are education of the head of the household, household size, whether the head of the household was male, whether livestock were owned, the use of extension services in crop and livestock production, the availability of credit and the temperature.
West et al. (2008)	To analyze local perceptions and regional climate trends on the central plateau of Burkina Faso	120 people	Ethnographic interviews, focus groups, and participant observation	– Farmers perceive that both overall seasonal rainfall and the number of "big rains" during the rainy season have decreased over the last 30 years. – Rural households respond to drought by changing their agricultural practices.
Adesoji and Ayinde (2013)	– To identify the mitigation strategies being used by the arable crop farmers in Osun State of Nigeria – To determine the factors influencing farmers' mitigation strategies	120 arable crop farmers	Multiple regression analysis	– Arable crop farmers mitigate change in climate mostly with indigenous or ethno-methods, which do not involve importation of technology in order to sustain production. – When planning extension programs for arable crop farmers, their age, household size, income, sources of information, and farm size should be considered.

Ajao and Ogunniyi (2011)	To examine farmers' strategies for adapting to climate change in the Ogbomoso agricultural zone of Oyo State of Nigeria	150 farmers	Probit model	– The types of climate change identified in the study area were delayed onset of rainfall, higher temperature and less rain. – The outcome of climate change was food shortage, decline in livestock yield, decline in crop yield and death of livestock. – The identified actions taken to address climate change are growing a new crop, adoption of drought-tolerant/resistant crop varieties, diversification from crops to livestock production and new land management practices. – The long-term improvement investments commonly adapted in the study area were tree planting/agroforestry, mulching/surface cover, fallowing and improved fallowing.
Ogalleh et al. (2012)	To present empirical evidence that demonstrates local knowledge, perceptions and adaptations to climate change and variability among the smallholders of Laikipia district of Kenya	– 46 transcripts from focus group discussions – 206 farmers	– The Palmer Drought Severity Index (PDSI) – Tabulations and frequency tables	– Climatic variability is increasingly changing. – Local perceptions include decreasing rainfall, increasing temperatures, increasing frosts and increasing hunger. – Coping and adaptation strategies used include diversification of crop varieties, migration and sale of livestock.

and is shown to produce more accurate results than MNL (see, e.g., Alvarez and Nagler, 1998; Schofield et al., 1998; Alvarez et al., 2000; Dow and Endersby, 2004). Because there are some choices involved (e.g., crop choice, income), possible sample selection bias needs to be addressed for the proper analysis of the determinants of the choice of adaptation methods. To address the selection bias, this study employs Heckman's two-stage estimation (Heckman, 1979) in analyzing the likelihood that Tanzanian farmers will adapt to climate change and their choice of adaptation methods.

Methods

This section provides an overview of the methodology used in addressing each of the objectives of this study. To reiterate, this study investigates (1) whether smallholder farmers in Tanzania perceive climate change; (2) whether, as a consequence, they adapt at all in their agricultural activities; and (3) the factors influencing their choice of adaptation methods. In order to determine whether smallholder farmers in Tanzania perceive climate change, a sample of smallholder farmers were asked whether they have observed variation in the climate across two adjacent decades (i.e., between the 1990s and the 2000s, in terms of both the means and variances of precipitation and temperature).

Heckman sample selection model

Farmers make the choice of adaptation methods as they decide to adapt. Because there are some choice variables involved (e.g., crop choice, income) in the farmers' choice, the possible sample selection bias needs to be addressed for the proper analysis of the determinants of this choice. To address possible selection bias, the study employs Heckman's two-stage estimation (Heckman, 1979). This study follows the sample selection methodology of Grilli and Rampichini (2007) in which the outcome equation consists of multiple choices. The difference between this study and that of Grilli and Rampichini is that the outcome equation in this study is a multinomial probit model. The choice of a multinomial probit over a multinomial logit model was explained earlier in this study.

Therefore, the selection equation analyzing the probability that the farmer adapts to climate change is specified by following a probit model. This follows the assumption that the cumulative distribution of ε_i is normal (Wooldridge, 2001):

$$p(Y = 1 \mid X) = \Phi(X'\alpha) = \int_{-\infty}^{X'\alpha} \frac{e^{-\frac{(X'\alpha)^2}{2}}}{\sqrt{2\pi}} d(X'\alpha) \tag{1}$$

where Φ is the normal cumulative distribution function. It is assumed that the probability of a farmer undertaking any adaptation at all (Y=1) depends on a vector of independent variables (X), unknown parameters (α) and the stochastic error term (ε) (Gujarati, 2003). The probability of a farmer undertaking any adaptation at all $P(Y=1 \mid X)$) has been modeled empirically as a function of independent variables, such as experience, gender, education and household income; whether a farmer has observed decadal changes in rainfall and temperature; general availability of information about climate change; agroecological zone; and distance from input markets. This model implies a diminishing magnitude of marginal effects for the independent variables; the coefficients give the signs of the marginal effects of each of the independent variables on the probability that the farmer undertakes any adaptation at all. The corresponding log likelihood function for the probability is:

$$\ln L = \sum_{i=1}^{n} I_i \ln[\Phi(X'\alpha)] + (1 - I_i)\ln[1 - \Phi(X'\alpha)] \tag{2}$$

where I_i is the dummy indicator equal to 1 if farmer i undertakes any adaptation at all to climate change and 0 otherwise. The consistent maximum likelihood parameter estimates are obtained by maximizing the foregoing log likelihood function. The marginal impact for each variable on the probability is given by:

$$\frac{\partial p(Y=1 \mid X)}{\partial X_k} = \frac{\partial \Phi(Y=1 \mid X)}{\partial X_k} = \varphi(X'\alpha)\alpha_k \tag{3}$$

while the marginal effect for a dummy variable, say X_k, is the difference between two derivatives evaluated at the possible values of the dummy – that is, 1 and 0. Thus,

$$\frac{\partial p(Y=1 \mid X)}{\partial X_k} = [\varphi(X'\alpha)]_{j=1} - [\varphi(X'\alpha)]_{j=0} \tag{4}$$

In order to determine the factors influencing the farmer's choice of particular adaptation methods, another probability model is used, where the dependent variable is multinomial, with as many categories as the number of adaptation methods to climate change available in the sampled population. Thus, when it comes to the choice of a particular adaptation method, the model assumes that farmer i maximizes his perceived utility from using a certain adaptation method subject to given factors. In this case, utility is observed through the actions of the farmer in choosing adaptation methods. The farmer's choices are unordered multinomial outcomes. The farmer's choice of one adaptation method from among others is modeled in a random utility framework. The utility function is only partially observed. Following Cameron and Trivedi (2005), the partially observable utility attached to each adaptation method j=0,1, . . . ,J by farmer i can be expressed as:

$$u_0 = \varepsilon_0$$
$$u_1 = X\beta_1 + \varepsilon_1$$
$$u_2 = X\beta_2 + \varepsilon_2$$
$$. . . .$$
$$u_j = X\beta_j + \varepsilon_j$$

where j=0 indicates that the farmer chooses not to adapt and j=1,2,..,J indicates the available suite of adaptation methods from which farmers can choose; X is a vector of farmers' characteristics and other factors that may affect the farmers' choice of particular methods; β are unknown parameters to be estimated;[1] and ε are idiosyncratic factors which are independent from each other. Given the several choices that farmers face, the rule is to choose the adaptation method which gives the highest utility – that is, if option j gives a farmer the highest utility of all the alternatives, then we expect to observe the outcome y = j, provided that:

$$
\begin{aligned}
P(y = j) &= \Pr(U_j \geq U_k), \textit{for all } k \\
&= \Pr(U_k - U_j \leq 0), \textit{for all } k \\
&= \Pr(\varepsilon_k - \varepsilon_j \leq X'_j\beta - X'_k\beta), \textit{for all } k
\end{aligned}
\tag{5}
$$

Farmers choose whether to adapt, but their choice of adaptation method is influenced by many factors. It has been pointed out that, in order to avoid sample selection bias on unobserved variables, the units should be sampled randomly so that the unobserved variables should not correlate with the error terms of the statistical model of interest (Copas and Li, 1997). As noted before, the use of a Heckman sample selection model is ideal. After estimating the selection equation using a probit model, the study now estimates the second part of the Heckman model, which is an outcome equation that involves the farmers' choice of adaptation method. The probability model for examining the factors influencing farmers' choice of different adaptation methods is the multinomial probit (MNP) model. The MNP model is needed because farmers have to choose from many adaptation methods which are unordered and nominal in character (Bartels et al., 1999; Greene, 2000; Wooldridge, 2001; Gujarati, 2003).[2] In MNP, it is assumed that the error term follows a multivariate normal distribution in which each error has a zero mean and the errors are allowed to be correlated. As it is, MNP models' direct evaluation of the likelihood entails a large number of integrals (one for each observation) of moderate dimension. The omitted outcome in the multinomial model is not adapting (not adapting is considered as one of the choices that farmers are expected to make). The assignment of not adapting as an omitted outcome is because (1) it is easy for any farmer to choose not to adapt even though the farmer has the ability and has access to other adaptation methods, and (2) it is the most frequently occurring outcome. From Equation (5), the probability that alternative j is chosen equals

$$
\Pr(y = j) = \Pr\{\varepsilon_k - \varepsilon_j \leq (X_j - X_k)'\beta, \textit{for all } k
\tag{6}
$$

where Xs are alternative-specific regressors and εs are multivariate normally distributed (Cameron and Trivedi, 2005).

The inverse Mill's ratio (IMR) calculated after the first-stage selection equation (the probability of adapting to climate change) is included in the second-stage multinomial probit model as one of the predictors (a correcting term). The significance of IMR indicates the existence of selection bias; however, if it is not significant, this does not necessarily imply that there is no selection bias.

Description of variables

From a review of the relevant literature, a set of variables was identified which might be important in explaining the uptake of adaptation to climate change in general, as well as the choice of specific adaptation methods. These include socioeconomic factors, environmental factors, institutional factors and the economic structure in which the choices occur.[3]

Socioeconomic variables

Key socioeconomic variables are household consumption and household income, which includes both farm income derived from selling farm products and nonfarm income derived from other nonfarm activities, including income from wages and small businesses (e.g., kiosks). Household income is expected to be positively related to undertaking adaptation to climate change – that is, the more income the farmer has, the more likely it is she will undertake adaptation. Nonfarm income is also relevant here because farmers generally finance adaptation from their overall incomes regardless of source.

Another key variable is awareness about climate change and adaptation methods – that is, whether farmers are informed about climate change and various adaptation techniques. Such information may be obtained from media sources – that is, radio, television or newspapers. Being aware of climate change and the different adaptation methods gives farmers a wide range of options for response and allows them to choose those methods which are personally more convenient.

The farming experience of the household head is expected to be positively related to undertaking adaptation. A farmer with more experience would know when climate change is occurring in the area and which methods work well in that specific agroecological zone. The selection of particular crops to be grown as the household's major crop is also an important factor in choosing certain adaptation methods. Large households are expected to offer more of the technical and manual skills required to respond to climate change. Higher educational credentials of the household head increase the knowledge base. In addition, when looking at the member of the household who has the highest level of education (who may or may not be the head), a higher educational credential of that individual increases the household's knowledge.

Environmental variables

The environmental variables used in this study are incidences of droughts and floods, agroecological zones, the farmer's observation of changes in rainfall and temperature, and the average annual rainfall and temperature for the respective regions under study. These variables are important because they help give concrete signs of climate change at the farm level. Farmers who perceive changes in rainfall and temperature, including increasing droughts and floods, are more likely to adapt to climate change. The location of plots in certain agroecological zones influences the adaptation modes used.

Institutional variables

Institutional factors include all social mechanisms of interaction which are used to manage adaptation to climate change. These mechanisms include regulations, enforcement and agricultural extension, all of which determine access to adaptation. Government intervention is of great importance here, especially now that Tanzania is implementing the *Kilimo Kwanza* policy, which seeks to promote sustainable growth in the agricultural sector. However, the presence of social capital within the farming communities is probably more important (Mathijs, 2003; Munasib and Jordan, 2011). Farmers can receive technical support about adaptation to climate change from both the government and community groups.

The economic structure

The national economic structure is an important determinant of the uptake of adaptations to climate change. Here, the economic structure includes the market conditions governing agricultural activity and other economic alternatives. For example, farm size, access to formal and informal credit,[4] and distance from input and output markets will affect agricultural productivity and the uptake of adaptation techniques.

Data sources

This study uses a survey dataset collected from 556 randomly selected farmers' households from December 2010 to January 2011 in four administrative regions of Tanzania – namely, Iringa, Morogoro, Dodoma and Tanga. These four were expressly chosen out of 26 regions in order to include most of the agroecological zones and therefore represent varying climate change impacts in Tanzania. The four selected regions represent six of the seven agroecological zones in Tanzania, as reported in United Republic of Tanzania (2007): coastal, arid, plateau, southern highlands, alluvial plains and semiarid.[5] This is a sample survey with a cross-sectional design. The units of analysis were drawn from the lists of households provided by *Nyumba Kumi* leaders.[6] The sample was randomly selected from the lists of eligible farmers' households as provided by the leaders.

Data was collected from farmers using a structured questionnaire and face-to-face interviews during a two-month field trip to the aforementioned regions. During the process, participation was voluntary and ethical considerations were taken into account, with the farmers being assured of the confidentiality of the information they revealed. The respondents in the study were selected if they fulfilled three main conditions: (1) the household head is a smallholder farmer – that is, they own farming plots of not more than three hectares (Montiflor, 2008; Eicher et al., 2006); (2) the household head is aged 18 years or above;[7] and (3) the household head's major economic activity is agriculture. The interview was carried out in Kiswahili, which is the Tanzanian national language and is spoken by the majority of Tanzanians. Because this study is about perception of climate change in the past 20 years, the 22 households with household heads younger than 30 years were dropped from the sample. Those household heads are assumed to be too young to remember what happened when they were less than 10 years old. In this case, this study uses the information provided by 534 households. The descriptive statistics of the explanatory variables that will be used in the analysis are presented in Table 7.2.

Table 7.2 Descriptive Statistics of Explanatory Variables to Be Used in the Analysis

Variable	Mean	Std. dev.	Min	Max
Annual household income (in '000 TZS*)	5260.92	3016.63	9100	24500
Age of the head of household (years)	46.80	12.25	30	96
Head of household is male (male = 1, female = 0)	0.75	0.40	0	1
Household has access to media (yes = 1, no = 0)	0.79	0.40	0	1
Highest education in the household (years)	10.21	3.08	0	19
Number of years worked as farmer (years)	22.71	12.97	1	70
Size of the household (numbers)	6.47	3.48	1^	17
Farm size (hectares)	1.92	0.75	0.5	3
Frequency experienced floods in the past 20 years (years)	1.42	1.19	0	6
Frequency experienced drought in the past 20 years (years)	2.61	2.07	0	10
Average rainfall in household's neighborhood in 2010 (millimeters)	874.55	250.51	583	1370.7
Average temperature in household's neighborhood in 2010 (degrees centigrade)	24.10	2.34	21	27.07
Has received agricultural technical support from community group or government (yes = 1, no = 0)	0.57	0.49	0	1
Grows rice as the major crop (yes = 1, no = 0)	0.06	0.24	0	1
Grows sorghum as the major crop (yes = 1, no = 0)	0.13	0.33	0	1
Has observed changes in rainfall and temperature (yes = 1, no = 0)	0.99	0.11	0	1
Access to credit (yes = 1, no = 0)	0.49	0.50	0	1
Distance from input markets (kilometers)	5.84	4.34	0.5	11

(*Continued*)

Table 7.2 (Continued)

Variable	Mean	Std. dev.	Min	Max
Located in the Coastal agroecological zone (yes = 1, no = 0)	0.27	0.44	0	1
Located in the Arid agroecological zone (yes = 1, no = 0)	0.06	0.26	0	1
Located in the Alluvial agroecological zone (yes = 1, no = 0)	0.27	0.44	0	1
Located in the Southern Highlands agroecological zone (yes = 1, no = 0)	0.07	0.26	0	1
Located in the Semi-arid agroecological zone (yes = 1, no = 0)	0.09	0.29	0	1
Located in the Plateau agroecological zone (yes = 1, no = 0)	0.23	0.46	0	1

Source: Own survey data, December 2010–January 2011.

* The exchange rate used is 1USD = 1592 Tanzanian Shilling (TZS), January 2012.

^ 15 households have one household member. Most of them are female and unmarried (widowed or not married at all), aged between 45 and 56 years, whose children have started their own families.

Results and discussion

Results

Farmers were asked to compare the climate in the two decades between the 1990s and the 2000s with respect to mean and variance precipitation and temperature. 528 farmers (98.9%) perceived mean and variance changes in both precipitation and temperature. The perception was that mean precipitation had decreased while the variance of precipitation had increased. Both the mean and variance of temperature were perceived to have increased. In fact, 531 farmers (99.46%) perceived climate changes with respect to precipitation or temperature or both. Only 3 farmers (0.54%) did not perceive any climate change. The research therefore indicates overwhelming evidence that Tanzanian smallholder farmers perceive climate change to have occurred over the past two decades.

It is necessary to know whether farmers' perceptions are consistent with reality. If their perceptions deviate from fact, then there is a risk that they might not respond at times when they should be responding. Even though climate change is a rather long-term phenomenon, there seems to be evidence that this has been occurring in the study areas across the two decades in question.[8] Statistical evidence from data provided by the Tanzanian Meteorological Agency shows a decrease in mean decadal rainfall from 847.3 mm in the 1990s to 763.5 mm in the 2000s and an increase in mean decadal temperature from 23.20°C in the 1990s to 23.8°C in the 2000s. This source also shows an increase in the decadal variances of both rainfall and temperature; the rainfall decadal variance rose from 8476.08 in the 1990s to 41934.1 in the 2000s and the temperature decadal variance rose from 7 in the 1990s to 8 in the 2000s.

The rainfall data from TMA is then segmented into two seasons: long rains (*Masika*) and short rains (*Vuli*).[9] Statistical evidence still shows a decrease in decadal mean rainfall in both the Vuli and Masika rain seasons. While the mean rainfall in the Vuli seasons decreased from 274.3 mm in the 1990s to 244.2 mm in the 2000s, the mean rainfall in the Masika seasons decreased from 558.2 mm in the 1990s to 442.5 mm in the 2000s. The surprising result is the decadal rainfall variance in the Vuli season. Generally, the science of climate assumes that precipitation variability increases with an increase in temperature. Statistical evidence shows that the decadal rainfall variance in the Masika seasons increased from 10056.5 in the 1990s to 17149.7 in the 2000s; in the Vuli seasons, the variance decreased from 54190.6 in the 1990s to 20360.1 in the 2000s. The decrease in rainfall variance was also found by Sun et al. (2012) when analyzing global monthly mean precipitation. In their study, they argue that this variability of rainfall patterns leads to a redistribution of rainfall in which dry seasons get wetter and wet seasons get drier. Thus, farmers' perceptions about climate change are consistent with reality and, therefore, a pro-adaptation response to their perceptions would be appropriate and helpful to government efforts to avoid potential agricultural losses.

Now that we have found evidence that Tanzanian smallholder farmers perceive climate change to be occurring in their areas, we proceed to investigate the other two objectives of the study. This includes investigating whether, as a consequence of their perceptions about climate change, they attempt to adapt at all, and investigating the factors influencing their choice of adaptation methods. Multicollinearity tests were performed in order to check whether independent variables in the models to be estimated provide redundant information about the response variables. We tested for the presence of multicollinearity using the variance inflation factor, $VIF_j = 1/(1-R_j^2)$, where R_j^2 is the coefficient of determination of the model which includes all independent variables except the jth variable. Table 7.3 demonstrates the VIF for all variables that are less than 10. This indicates that there is no problem with multicollinearity.

Here we report the probit estimation results for (1) the probability of adapting to climate change in general, and (2) the multinomial probit estimation results for the probability of using short-season crops, using crops resistant to drought, irrigating, changing planting dates and planting trees, relative to not adapting. In both models, the marginal effects of the independent variables are reported. Table 7.4 reports the marginal effects results for the selection and outcome

Table 7.3 VIF Test for Multicollinearity

Variable	VIF	SQRT VIF	Tolerance	Eigenval	Cond index	R-squared
Annual household income	2.02	1.42	0.496	3.871	1.	0.504
Household has access to media	1.04	1.02	0.964	2.403	1.269	0.035
Number of years worked as farmer	1.24	1.11	0.809	1.805	1.464	0.191

(*Continued*)

Table 7.3 (Continued)

Variable	VIF	SQRT VIF	Tolerance	Eigenval	Cond index	R-squared
Head of household is male	1.12	1.06	0.894	1.436	1.642	0.105
Size of the household	1.39	1.18	0.718	1.337	1.701	0.281
Highest education in the household	1.41	1.19	0.707	1.201	1.795	0.293
Farm size	1.08	1.04	0.925	1.127	1.853	0.075
Frequency experienced floods in the past 20 years	1.14	1.07	0.879	0.974	1.993	0.120
Frequency experienced drought in the past 20 years	1.34	1.16	0.747	0.960	2.008	0.253
Average rainfall in household's neighborhood in 2010	6.27	2.50	0.159	0.893	2.082	0.841
Average temperature in household's neighborhood in 2010	4.61	2.15	0.217	0.862	2.118	0.783
Has received technical support	1.57	1.25	0.635	0.772	2.238	0.364
Grows rice as the major crop	1.78	1.33	0.568	0.693	2.362	0.437
Grows sorghum as the major crop	2.03	1.43	0.497	0.688	2.371	0.508
Has observed changes in rainfall and temperature	1.06	1.03	0.949	0.480	2.839	0.058
Access to credit	1.39	1.18	0.725	0.433	2.988	0.279
Distance from input markets	1.97	1.40	0.508	0.392	3.140	0.492
Located in Coastal agroecological zone	7.83	2.80	0.127	0.312	3.517	0.872
Located in Plateau agroecological zone	5.14	2.27	0.197	0.175	4.692	0.805
Located in Alluvial agroecological zone	4.52	2.13	0.223	0.105	6.049	0.779
Located in Southern highlands agroecological zone	2.29	1.51	0.436	0.071	7.367	0.564
Located in Semi-arid agroecological zone	5.83	2.41	0.172	0.032	9.065	0.828

Note: Mean VIF 2.49; condition number 7.3669; determinant of correlation matrix 0.0004.

Table 7.4 Marginal Effects Heckman Sample Selection Model

Explanatory variable	Outcome equation: Choice of adaptation method; a multinomial probit model					Selection equation: probability of adapting
	Method 1 Short-season crops	Method 2 Crops resistant to drought	Method 3 Irrigation	Method 4 Changing planting dates	Method 5 Planting trees	
Annual household income	-0.035 (0.054)	0.059 (0.047)	0.012 (0.019)	0.011 (0.03)	0.002 (0.011)	0.074 (0.054)
Number of years worked as farmer	0.001 (0.002)	-0.003 (0.002)	-0.001 (0.001)	0.001 (0.001)	0.0002 (0.0004)	-0.001 (0.002)
Farm size	0.043 (0.03)	-0.007 (0.024)	-0.004 (0.01)	-0.012 (0.016)	-0.015 (0.012)	-0.04 (0.029)
Highest education in the household	-0.009 (0.01)	0.023** (0.008)	0.005 (0.004)	-0.019*** (0.005)	0.004 (0.003)	0.022*** (0.008)
Size of the household	0.001 (0.007)	-0.001 (0.006)	-0.002 (0.002)	0.002 (0.004)	0.001 (0.001)	-0.0002 (0.007)
Average temperature in the neighborhood in 2010	0.082*** (0.017)	0.049** (0.018)	-0.005 (0.006)	-0.01 (0.009)	-0.011 (0.007)	-0.055*** (0.02)
Average rainfall in the neighborhood in 2010	-0.00001 (0.0002)	0.0002 (0.0002)	0.0002*** (0.0001)	-0.0003** (0.0001)	-0.0001** (0.0001)	-0.001*** (0.0003)
Head of household is male#	-0.004 (0.053)	0.028 (0.044)	-0.009 (0.019)	-0.029 (0.033)	0.01 (0.01)	0.042 (0.051)
Household has access to media#	0.032 (0.053)	0.019 (0.046)	0.017 (0.015)	0.077* (0.042)	0.021 (0.018)	0.061 (0.053)
Access to credit#	0.037 (0.051)	0.041 (0.044)	-0.003 (0.015)	-0.099** (0.028)	0.021 (0.018)	0.039 (0.049)

(Continued)

Table 7.4 (Continued)

	Outcome equation: Choice of adaptation method; a multinomial probit model					Selection equation: probability of adapting
Frequency of experienced drought in the past 20 years#	0.001 (0.013)	0.019* (0.012)	0.003 (0.004)	-0.027*** (0.009)	0.006 (0.004)	0.030*** (0.012)
Frequency experienced flood in the past 20 years#	-0.02 (0.019)	-0.017 (0.019)	0.007 (0.006)	0.024** (0.009)	-0.004 (0.004)	-0.015 (0.019)
Has received technical support#	0.099** (0.051)	-0.019 (0.044)	-0.022 (0.019)	-0.027 (0.031)	-0.002 (0.009)	0.015 (0.052)
Grows rice as the major crop#	-0.271*** (0.032)	-0.059 (0.108)	-0.034** (0.013)	0.079 (0.149)	-0.495** (0.245)	0.31*** (0.058)
Grows sorghum as the major crop#	-0.083 (0.075)	-0.072 (0.067)	0.346 (0.235)	-0.041 (0.036)	-0.001 (0.013)	0.038 (0.087)
Located in the Coastal agroecological zone#	0.022*** (0.109)	-0.014** (0.006)	0.001 (0.002)	-0.003 (0.004)	-0.003 (0.003)	0.541*** (0.082)
Located in the Plateau agroecological zone#	-0.069 (0.094)	-0.089 (0.057)	0.142 (0.126)	-0.04 (0.079)	0.127 (0.13)	0.28*** (0.067)
Located in the Alluvial plains agroecological zone#	-0.028 (0.109)	-0.235*** (0.042)	0.507** (0.173)	-0.037 (0.034)	-0.0003 (0.013)	-0.027 (0.094)
Located in the Southern Highlands agroecological zone#	-0.063 (0.142)	-0.212*** (0.022)	0.604** (0.237)	-0.076*** (0.018)	0.027 (0.045)	0.003 (0.115)
Located in the Semi-arid agroecological zone#	0.129 (0.197)	-0.169*** (0.044)	0.408* (0.246)	-0.08*** (0.023)	-0.009 (0.014)	0.008 (0.135)
Distance from input markets						-0.01* (0.007)

Has observed changes in rainfall and temperature#						0.439** (0.182)
Inverse Mill's ratio	-0.569*** (0.172)	0.326** (0.138)	0.039 (0.071)	-0.546*** (0.109)	-0.101 (0.063)	
Number of observations (543)	131	93	31	60	37	534
Base rate	0.2559	0.186	0.0351	0.0784	0.0141	0.67341149

Note:
- Base category for adaptation methods is "No adaptation."
- Base category for agroecological zone is Arid.
- Standard errors are in brackets; *, **, *** imply significance level at 10%, 5% and 1% respectively.
- (#) dy/dx is for discrete change of dummy variable from 0 to 1.

equations.[10] The results for the binary probit model (selection equation) are reported in Column 7, while the results of outcome equations that represent each dominant adaptation method chosen by farmers are reported in Columns 2 to 6. The log-likelihood ratios test in all the equations strongly rejects the null hypothesis; we therefore conclude that the variables included in the model explain the variation in the regressand. Finally, the results on the inverse Mill's ratios suggest a strong selection mechanism in choosing short-season crops, choosing crops resistant to drought, and changing planting dates. It was important, therefore, to address the sample selection issue. The coefficients −0.569 and −0.546 suggest that, on average, unobservable factors that increase the probability of farmers adapting to climate change decrease smallholder farmers' likelihood of choosing to plant short-season crops and change planting dates. Moreover, the coefficient 0.326 implies that unobservable factors that increase the probability of farmers adapting to climate change increase their likelihood of planting crops resistant to drought.

The Heckman sample selection model has the limitation that different variables might determine participation and outcomes. The independent variables in selection and outcome equations are not mutually exclusive; there are some variables that are included in both equations but there are some variables that are not included in the outcome equation. This is because the outcome equation is performed after the selection equation and the variables that are necessary in the participation equation might not be necessary determinants in the outcome equation because the household is already participating. In this study, the dummies for the fact that the farmer has observed changes in rainfall and temperature and for distance from input markets are excluded from the outcome equation. It is important to include the variable that captures the impact of observing changes in rainfall and temperature to determine the probability of a farmer adapting to climate change, but observing changes does not necessarily determine the adaptation method implemented.

The results of the selection probit model (Column 7) suggest that the probability of a typical Tanzanian farmer adapting to climate change increases with education levels of household members; farmers observing climate change with respect to precipitation and temperature across the two decades; the frequency of drought[11] experienced during the past 20 years; and growing rice as the major crop. The results also suggest that the probability of adapting to climate change decreases with temperature and rainfall levels in the farming area and distance from input market. Farmers located in the coastal and plateau agroecological zones tend to use adaptation strategies more than those located in the arid agroecological zone.

The probit model parameters are estimable up to a scaling factor. The coefficients of the probit model give the change in the mean of the probability distribution of the dependent variable associated with the change in one of the explanatory variables, but these effects are usually not of primary interest. The marginal effects on the probability of possessing the characteristic can be of more use. The marginal effects vary across individuals and, in this case, indicate by how much the probability of a farmer using adaptation measures alters with changes in the explanatory variables.

The marginal effect for having observed changes in rainfall and temperature across the two decades is 43.9%. This implies that farmers who have observed climate change with respect to precipitation and temperature across the past two decades have a 43.9% higher probability of adapting to climate change above the base case. This result is largely expected because respondents were asked about the adaptation which was undertaken in response to observing climate change. It is nevertheless necessary to test this variable, because the model in Table 7.4 is run using data from all respondents, a few of whom did not perceive change. The results seen so far with respect to this variable are very important because they provide two confirmations: first, farmers perceive that climate change is occurring; and, second, farmers respond to their perceptions of this phenomenon by undertaking adaptation measures. Therefore, the major role with which the Tanzanian government needs to occupy itself, relating to the effects of climate change on smallholder agriculture, is simply to assist farmers to overcome the constraints they face – namely, shortage of water, funds and seeds, and poor planning by farmers.[12]

With respect to education, farmers with more education or in the households with more educated members are more likely to pursue adaptation strategies related to climate change than are farmers with lower education levels or in households with members with lower levels of education. On average, one more year of schooling of the household member with the most years of education increases the probability of adapting to climate change by 2.2%. Similar results have been reported by the empirical studies of Deressa et al. (2009), Goulden et al. (2009) and Iglesias et al. (2011).

On average, a 1 degree increase in the average annual temperature in the farmer's neighborhood decreases the probability of farmers adapting by 5.5%. This is a plausible result for crops requiring a higher temperature. At the same time, a 1 mm increase in average annual rainfall in the farmer's neighborhood decreases the probability of adaptation by 0.1%. This seems plausible because most of the adaptation methods that Tanzanian farmers adopt are aimed at dealing with insufficient rainfall. This means that, when there is shortage of rainfall, there is a need for smallholder farmers to adapt to the decreasing rainfall availability by either implementing water conservation technologies or planting crops that do not need much rainfall – for example, sorghum, potatoes and cassava.

The probability of adaptation by farmers who grow rice as their major crop is 31% higher than for those who grow other major crops, including maize. This might be because rice is among the most popular cereal crops in Tanzania and is the preferred foodstuff for many people with medium and high income. In this case, farmers who grow rice as their major crop might take active steps to adapt to climate change so as to ensure good yields. Distance from input markets reduces the probability of farmers adapting. The results show that a 1 kilometer increase in distance from input markets reduces this probability by 1%. This is because, when input markets are located far from farming plots, it is difficult for farmers to access the inputs necessary for adaptation. Farmers who reported experiencing one additional year of drought have a 3% higher probability of adapting. Farmers located in the coastal and plateau agroecological zones are 54.1 and 28% more likely to adapt, respectively, than those who reside in the arid zone.

The results from the multinomial probit model show the direction and the magnitude of the effect of different factors influencing farmers' choice of a particular adaptation method from up to five alternative adaptation methods used by Tanzanian farmers.

Short-season crops

The results for Method 1 suggest that the probability of using short-season crops relative to no adaptation increases with temperature intensity, agricultural technical support from community groups and/or government, and location in the coastal agroecological zone, while the probability decreases if farmers grow rice as their major crop.

Farmers generally use short-season crops when temperatures increase. An increase in the average annual temperature has an impact on farmers' adaptation to climate change using short-season crops; this is shown by our finding that a 1 degree centigrade increase in average annual temperature leads to an 8.2% increase in the use of short-season crops. Receiving agricultural technical support from the government and/or community groups increases the farmers' probability of using short-season crops by 9.9%. Empirical studies recognize the importance of agricultural extension services to farmers; see, for example, Ziervogel et al. (2006), Cooper et al. (2008), Keil et al. (2008), Deressa et al. (2009) and Below et al. (2012). These studies confirm the importance of agricultural extension services provided by government and community groups.

Farmers who grow rice as their major crop have a 27.1% lower likelihood of using short-season crops compared to their peers growing other major crops. Farmers located in the coastal zone are 2.2% more likely to use short-season crops compared to their peers in the arid agroecological zone.

Crops resistant to drought

The results for Method 2 imply that the probability of using crops which are drought-resistant, relative to no adaptation, increases with an increase in the level of education of the household, temperature intensity and incidence of drought, and decreases with location of the plot in agroecological zones other than arid.

An increase in average annual temperature appears to impact on the decision of farmers to plant drought-resistant crops; the study indicates that a 1 degree centigrade increase in average annual temperature above the 2010 level leads to a 4.9% increase in the use of such crops. This result is plausible; it is expected farmers will choose to plant more drought-resistant crops when temperatures are high because those crops can tolerate the high temperature. Farmers who are more educated and those households with more highly educated members tend to use drought-resistant crops more often. On average, an increase in one more year of education increases the probability of farmers using drought-resistant crops as opposed to not undertaking any adaptation measures. It is expected that farmers who have reported experiencing a greater incidence of drought in the past 20 years would want to plant drought-resistant crops. The results tell us that an increase in the number of years that a farmer reported experiencing drought increases the probability of using such crops by 1.9%.

Being located in the coastal, alluvial plains, southern highlands and semiarid zones decreases the likelihood of farmers' using crops which are resistant to drought by 1.4%, 23.5%, 21.2% and 16.9%, respectively, compared to farmers located in the arid agroecological zone.

Irrigation

The results from Method 3 show that the likelihood of using irrigation relative to no adaptation increases with rainfall intensity and being located in alluvial plains, southern highlands and semiarid agroecological zones, and decreases for farmers growing rice as the major crop.

Our results confirm that smallholder farmers in the lowland areas of Tanzania grow rice because they do not need to irrigate their plots in these areas. Farmers who grow rice as the major crop are 3.4% less likely to irrigate their plots.

An increase in average annual rainfall does not considerably impact farmers' adaptation to climate change using irrigation because a 1 millimeter increase in average annual rainfall above the 2010 level leads to only a 0.02% increase in the use of irrigation. Being located in alluvial plains, southern highlands and semiarid agroecological zones increases the probability of the use of irrigation by 50.7%, 60.4% and 40.8%, respectively, compared to farmers located in the arid agroecological zone. This may be simply explained by the fact that water for irrigation is more easily available in any other agroecological zone than in the arid zone. In this case, farmers who are capable of irrigating their plots can easily use this adaptation method, providing they are not residing in an arid agroecological zone.

Changing planting dates

The results from Method 4 suggest that the likelihood of changing planting dates relative to no adaptation increases with incidences of flood but decreases with highest education in the household, rainfall intensity, access to information, access to credit, incidence of drought and being located in the semiarid and southern highlands agroecological zones.

The probability of farmers changing planting dates decreases in relation to education in the household. An additional year of education for the household member with the highest education level decreases the probability of the household's changing planting dates as their adaptation method by almost 2% compared to the base category. The reason for the negative relationship might be that farmers who rely on rainfall in their agricultural activities plant their seeds when rain starts. They do not need to be educated to see that the rainfall season has started. Rainfall intensity does not have much impact on the probability of farmers changing planting dates. The results show that a 1 millimeter increase in rainfall decreases the likelihood of farmers changing planting dates by 0.03%. The results reveal that farmers who have access to the media are 7.7% more likely to change planting dates compared to those who do not have access to the media. Farmers who have access to the media receive information from weather forecasts to aid their decisions on when to plant their crops.

The marginal effect of −0.099 for credit suggests that changing planting dates is an adaptation method predominantly suitable for those lacking access to credit. Access to credit increases the probability of farmers switching away from changing planting dates by almost 10%. Presumably, with access to capital, farmers would use other capital-intensive adaptation methods. This implies that lack of access to credit is a significant constraint preventing some farmers from using methods other than shifting planting dates. Financial institutions such as banks, savings and credit cooperative society (SACOS) and village community banks (VICOBA) are therefore potentially effective institutions in empowering farmers to reduce the impact of climate change by using adaptation methods they deem suitable. In the same way, this also suggests the importance of informal networks, including relatives, friends and neighbors, in credit provision for agricultural investments.

Farmers who reported experiencing more incidence of drought have a 2.7% lower probability of changing planting dates. However, farmers who reported experiencing less incidence of flood are 2.4% more likely to shift dates. When there is drought, changing planting dates might not be a favorable choice for farmers. Whether the plants are planted early or later might not change the fact that the area is dry and therefore not conducive to agriculture. Being located in southern highland and semiarid zones decreases the likelihood of farmers changing planting dates by 7.6% and 8.7%, respectively, compared to those located in the arid agroecological zone.

Planting trees

The results from Method 5 show that the probability of planting trees as an adaptation method relative to no adaptation decreases with growing rice as a major crop and with rainfall intensity.

The results reveal that farmers who grow rice as their major crop have a 49.5% lower probability of planting trees as their adaptation method. This can be explained by the fact that trees attract birds, which may then eat the rice in the fields, thus endangering the crop yield. The results further reveal that a 1 millimeter increase in average annual rainfall decreases the probability of farmers planting trees by 0.01%. Planting trees is associated with attracting rainfall in the area; thus, it is logical that when rainfall increases in a certain area, the farmers might not choose to use that adaptation technique.

Discussion

Undertaking some adaptation to climate change is a step in the right direction by farmers in Tanzania. However, some adaptation techniques are more effective than others. Particular adaptation methods might be more appropriate for particular crops or agroecological zones. The government can play a significant role by promoting adaptation methods appropriate for particular circumstances. In order for this to occur, the government would require information about the key drivers of the current choice of adaptation methods. This information gives two useful hints: the social characteristics of farmers who are likely to voluntarily adopt particular adaptation methods, and the environmental, institutional and economic conditions influencing their adoption of particular methods. The first type of information gives guidance in targeting farmers' recruitment into initiatives aimed at enhancing adaptation by using particular methods. The second set of information provides guidance about the environmental, institutional and economic conditions which need to be changed to promote particular adaptation methods. On the basis of the foregoing information about the drivers of specific adaptation methods, the government can play a significant role by promoting adaptation methods appropriate for particular circumstances. The foregoing results assist in targeting farmers' recruitment into initiatives aimed at enhancing adaptation using particular methods as well as guidance about the environmental, institutional and economic conditions which need to be targeted to promote these specific methods.

As shown in Table 7.5, about 34% of surveyed farmers did not undertake any adaptation at all, even though these adaptations are not necessary for only about 10% of the surveyed farmers. Thus, a sizeable number of farmers who are currently not making changes ought to be doing so. In many cases, farmers are constrained from undertaking these adaptation measures. The reasons given by farmers for not using adaptation methods perceived to be the best in dealing with climate change include lack of funds (144 farmers, 25.9%), shortage of water (152 farmers, 27.3%), poor planning (42 farmers, 7.6%) and shortage of seeds (18 farmers, 3.2%), as shown in Figure 7.1.

Table 7.5 Perceived Best and Implemented Adaptation Methods to Climate Change

Adaptation method	Perceived best by	Implemented by
Irrigation	156 farmers, 28.1%	31 farmers, 5.6%
Short-season crops	147 farmers, 27.0%	131 farmers, 24.1%
Crops resistant to drought	83 farmers, 15.5%	93 farmers, 17.3
Planting trees	61 farmers, 11.7%	37 farmers, 7.4%
Changing planting dates	38 farmers, 7.4%	60 farmers, 11.3%
No adaptation	49 farmers, 10.4%	182 farmers, 34.4%

Source: Own survey data, December 2010–January 2011.

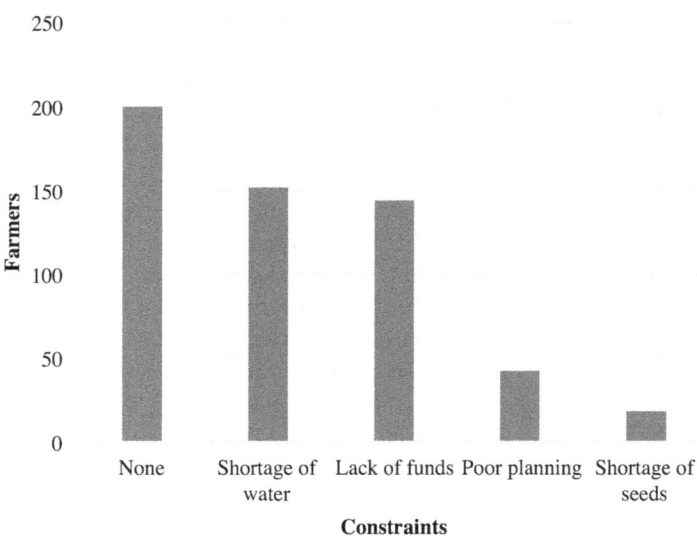

Figure 7.1 Constraints to Implementing the Best Perceived Adaptation Methods
Source: Own survey data, December 2010–January 2011.

In the absence of constraints, more farmers would opt for irrigation (28.1% instead of the current 5.6%), planting short-season crops (27% instead of the current 24.1%) and planting trees (11.7% instead of the current 7.4%). Thus, irrigation is the dominant adaptation method that farmers would ideally want to use to respond to observed climate change; currently, however, they are constrained by circumstances.

Policy implications and conclusions

The main purpose of this study was threefold: (1) to investigate whether smallholder farmers in Tanzania perceive climate change; (2) to investigate whether, as a consequence, they adapt at all in their agricultural activities; and (3) to investigate the factors influencing their choice of particular adaptation methods. The study collected data from 556 randomly selected smallholder farming households from four representative administrative regions representing six of the seven agroecological regions of the country. Farmers were asked to compare their perceptions of the climate in the decade between the 1990s and the 2000s with respect to mean and variance of precipitation and temperature. There is overwhelming evidence that Tanzanian smallholder farmers perceive climate change to have occurred over the past two decades (i.e., 1990s–2000s). Even though climate change is a long-term phenomenon, statistical evidence from data provided by the Tanzania Meteorological Agency provides evidence that climate change has indeed been occurring in the study areas across the two decades in question. Thus, farmers' perceptions about climate change are consistent with reality and, therefore, a pro-adaptation response to their perceptions would be appropriate and helpful to government efforts to avoid potential losses from the effects of climate change on this vulnerable group.

Those farmers who perceive climate change adapt to it in their agricultural activities. The results show that farmers who perceived climate variation with respect to precipitation and temperature across the past two decades have a 43.9% higher probability of adapting. The results of the binary probit model used as a selection equation in the Heckman sample selection model of a famer's decision to use adaptation measures suggest that the probability of undertaking any adaptation increases with household education levels; having observed climate change with respect to precipitation and temperature across the two decades; the frequency of drought experienced during the past 20 years; growing rice as the major crop; and the agroecological zone of the farm. The results also suggest that the probability of undertaking adaptation decreases with temperature and rainfall levels in the farming area, and with the distance from input markets. Farmers located in the coastal and plateau agroecological zones tend to undertake more adaptation compared to those located in the arid agroecological zone.

Farmers mentioned planting short-season crops and drought-resistant crops, using irrigation, changing planting dates, and planting trees as the methods they use to deal with the change. The study used a multinomial probit model as the outcome equation in the Heckman sample selection model to investigate the factors influencing farmers' choice of specific adaptation methods. The probability of using short-season crops increases with temperature intensity, having received agricultural technical support from community groups and/or government, and being located in the coastal agroecological zone; the probability decreases with growing rice as the major crop. The probability of using drought-resistant crops increases with household education levels, temperature intensity and incidence of drought, and decreases with location in the coastal, alluvial plains, southern highlands and semiarid agroecological zones. The probability of using irrigation increases with rainfall intensity and residing in alluvial plains, southern highlands and semiarid agroecological zones, and decreases with growing rice as the major crop. The likelihood of changing planting dates increases with the incidence of floods but decreases with household education levels, rainfall intensity, access to the media, incidence of drought, access to credit, and location in semiarid or southern highland agroecological zones. The probability of planting trees as an adaptation method decreases with growing rice as the major crop and with rainfall intensity. The inverse Mill's ratio shows that there is sample selection bias in three adaptation choices. In this case, we addressed the possibility of endogeneity bias by employing a Heckman sample selection model in our analysis.

The first and foremost role with which the Tanzanian government needs to occupy itself surrounding the effects of climate change on smallholder agriculture is to assist smallholder farmers to overcome the constraints they face. The results offer guidance with respect to the environmental, institutional and economic conditions which need to be reformed to encourage farmers to adapt to climate change and to promote particular adaptation methods. With regard to education, it is important for the Tanzanian government to make sure that young household members are provided with suitable education so that they are able to provide relevant advice to their elders about modern and appropriate adaptation approaches.

Because 36% and 55% of farmers located in the arid and semiarid agroecological zones, respectively, reported shortage of water for irrigation as a major constraint to adaptation, the government should encourage the farmers to concentrate on farming drought-resistant crops instead of planting crops that require more water, while at the same time developing irrigation infrastructure in areas where water is available.

The smallholder farmers identified lack of funds, shortage of water for irrigation, poor planning and shortage of the seeds recommended by agricultural experts as the main constraints in undertaking adaptation. In the case of lack of funds, the Tanzanian government should assist the farmers who are not yet in the SACOS and/or VICOBA credit organizations in forming groups so that they can be considered for low-interest agricultural loans. To diminish the problem of seed shortage, the government should ensure that agricultural officers and agents provide the appropriate amount of required subsidized seeds at the appropriate time. As for poor planning, farmers should be empowered to consider suitable and appropriate activities given the climate conditions; that is, they should be supported to develop long-term adaptation plans, even if this means switching crops completely or engaging in activities other than agriculture.

Furthermore, on the basis of the results revealed in this study on key drivers of specific adaptation methods, the government can play a significant role by promoting adaptation methods appropriate for particular circumstances – for example, particular crops or agroecological zones. The results also contribute guidance for targeting farmers' recruitment into initiatives aimed at enhancing adaptation to climate change using particular methods. For example, the probability of farmers in the arid agroecological zone using short-season crops and irrigation as their adaptation strategies is very low. Thus, in these cases, the government can promote the use of drought-resistant crops because they do not require plentiful water. In the coastal agroecological zone (Tanga administrative region), farmers are most likely to grow short-season crops. This is one of the bimodal areas – that is, the regions that receive two rainy seasons – namely, a long rainfall season (Masika: March to May) and a short rainfall season (Vuli: October to December). During the Vuli rainfall season, farmers in Tanga are reported to grow composite maize, which does not require a long period and plentiful rain to mature (USDA, 2003). In this case, therefore, the government is advised to invest in research and development (R&D) for short-season crop varieties.

Appendix

Results for farmers' choice of adaptation methods

Table 7A1 Heckman Sample Selection Model

Explanatory variable	Outcome equation: choice of adaptation method					Selection equation: probability of adapting
	Method 1 Short-season crops	Method 2 Crops resistant to drought	Method 3 Irrigation	Method 4 Changing planting dates	Method 5 Planting trees	
Annual household income	-0.047 (0.234)	0.269 (0.245)	0.229 (0.336)	-0.024 (0.28)	0.133 (0.409)	0.204 (0.149)
Number of years worked as farmer	0.001 (0.007)	-0.011 (0.008)	-0.01 (0.014)	0.007 (0.009)	0.006 (0.01)	-0.002 (0.005)
Farm size	0.131 (0.133)	-0.017 (0.127)	-0.059 (0.161)	-0.083 (0.144)	-0.464** (0.202)	-0.109 (0.079)
Highest education in the household	-0.02 (0.041)	0.091** (0.043)	0.081 (0.061)	-0.141*** (0.048)	0.131** (0.063)	0.061*** (0.023)
Size of the household	0.001 (0.03)	-0.005 (0.029)	-0.025 (0.039)	0.012 (0.035)	0.022 (0.041)	0.001 (0.018)
Average temperature in the neighborhood in 2010	0.248*** (0.071)	0.167* (0.089)	-0.06 (0.101)	-0.063 (0.079)	-0.337** (0.154)	-0.153*** (0.056)
Average rainfall in the neighborhood in 2010	-0.0001 (0.001)	0.001 (0.0008)	0.004*** (0.001)	-0.002 (0.001)	-0.003 (0.003)	-0.002*** (0.001)

(Continued)

Table 7A1 (Continued)

Explanatory variable	Outcome equation: choice of adaptation method					Selection equation: probability of adapting
	Method 1 Short-season crops	Method 2 Crops resistant to drought	Method 3 Irrigation	Method 4 Changing planting dates	Method 5 Planting trees	
Head of household is male#	-0.089 (0.233)	0.097 (0.239)	-0.131 (0.289)	-0.213 (0.267)	0.386** (0.306)	0.114 (0.139)
Household has access to media#	0.097 (0.238)	0.077 (0.234)	0.289 (0.313)	0.466* (0.276)	0.368 (0.334)	0.165 (0.143)
Access to credit#	0.099 (0.226)	0.143 (0.233)	-0.059 (0.271)	-0.74*** (0.26)	0.641** (0.281)	0.106 (0.138)
Frequency experienced drought in the past 20 years#	0.012 (0.058)	0.081 (0.063)	0.051 (0.074)	-0.19** (0.083)	0.207* (0.107)	0.083** (0.032)
Frequency experienced flood in the past 20 years#	-0.076 (0.086)	-0.082 (0.098)	0.078 (0.11)	0.158* (0.091)	-0.129 (0.117)	-0.041 (0.052)
Has received technical support#	0.343 (0.233)	-0.015 (0.231)	-0.259 (0.322)	-0.146 (0.275)	-0.006 (0.312)	0.042 (0.143)
Grows rice as the major crop#	-1.332* (0.69)	0.258 (0.701)	-0.634 (0.95)	0.983 (0.872)	-2.899** (1.232)	1.313*** (0.50)
Grows sorghum as the major crop#	0.055 (0.472)	0.015 (0.357)	2.061** (0.935)	-0.055 (0.435)	0.288 (0.485)	0.106 (0.251)
Located in the Coastal agroecological zone#	0.07** (0.026)	-0.049 (0.031)	-0.006 (0.037)	-0.019 (0.032)	-0.106** (0.048)	2.233*** (0.583)
Located in the Plateau agroecological zone#	0.113 (0.447)	-0.07 (0.414)	1.345** (0.667)	0.599 (0.54)	1.664 (1.053)	1.074** (0.429)
Located in the Alluvial plains agroecological zone#	0.409 (0.473)	-0.894** (0.425)	3.091*** (0.818)	0.125 (0.441)	0.487 (0.572)	-0.075 (0.258)

	(1)	(2)	(3)	(4)	(5)	(6)
Located in the Southern highlands agroecological zone#	0.566 (0.561)	-1.466** (0.661)	2.947*** (0.921)	-0.502 (0.565)	1.337** (0.601)	-0.007 (0.319)
Located in the Semi-arid agroecological zone#	1.056** (0.623)	-0.375 (0.624)	2.556*** (0.929)	-0.749 (0.787)	0.211 (0.688)	-0.023 (0.373)
Distance from input markets						-0.036* (0.019)
Has observed changes in rainfall and temperature#						1.174** (0.586)
Constant	-4.898 (3.697)	-1.414 (4.209)	-8.473 (5.318)	8.226* (4.028)	3.714 (5.961)	1.099 (2.43)
Inverse Mill's ratio	-2.889*** (0.75)	-0.064 (0.693)	-0.709 (1.12)	-5.329*** (0.935)	-1.937 (1.644)	
Number of observations (543)	131	93	31	60	37	534
Log likelihood	-715.63154					-323.59193
Wald chi2 (p-value)	440.46 (0.0000)					45.96 (0.002)

Note:
- Base category for adaptation methods is "No adaptation."
- Base category for agroecological zone is Arid.
- Standard errors are in brackets; *, ** and *** imply significance level at 10%, 5% and 1% respectively.

Table 7A2 Marginal Effects Heckman Sample Selection Model using MNL as Outcome Equation

Explanatory variable	Outcome equation: choice of adaptation method					Selection equation: probability of adapting
	Method 1 Short-season crops	Method 2 Crops resistant to drought	Method 3 Irrigation	Method 4 Changing planting dates	Method 5 Planting trees	
Annual household income	0.035 (0.057)	0.027 (0.05)	0.003 (0.004)	-0.035 (0.037)	-0.001 (0.001)	0.061 (0.052)
Number of years worked as farmer	0.001 (0.002)	-0.002 (0.002)	-0.0002 (0.0002)	0.001 (0.001)	0.00002 (0.0002)	-0.001 (0.002)
Farm size	0.009 (0.033)	0.016 (0.025)	-0.001 (0.002)	-0.009 (0.018)	-0.001 (0.0004)	-0.034 (0.028)
Highest education in the household	0.009 (0.012)	0.012 (0.01)	0.001 (0.001)	-0.02** (0.008)	0.0001 (0.002)	0.023*** (0.008)
Size of the household	-0.001 (0.008)	-0.003 (0.006)	-0.0003 (0.001)	0.003 (0.004)	0.0001 (0.001)	0.0001 (0.006)
Average temperature in the neighborhood in 2010	0.018 (0.026)	-0.006 (0.028)	-0.001 (0.002)	-0.03 (0.017)	-0.0001 (0.0004)	-0.056*** (0.02)
Average rainfall in the neighborhood in 2010	-0.001** (0.0003)	0.001** (0.0003)	0.0001** (0.00003)	-0.0002 (0.0003)	2.33e-06 (0.0001)	-0.001*** (0.0003)
Head of household is male#	0.035 (0.054)	-0.004 (0.048)	-0.002 (0.005)	-0.023 (0.035)	0.0003 (0.001)	0.042 (0.051)
Household has access to media#	0.081 (0.05)	-0.013 (0.048)	0.003 (0.004)	-0.077 (0.049)	0.0001 (0.001)	0.054 (0.053)
Access to credit#	0.064 (0.056)	0.024 (0.044)	-0.001 (0.004)	-0.099*** (0.032)	0.001* (0.001)	0.049 (0.049)
Frequency experienced drought in the past 20 years#	0.023 (0.015)	0.006 (0.013)	0.001 (0.001)	-0.026** (0.011)	0.0001 (0.0002)	0.030*** (0.012)
Frequency experienced flood in the past 20 years#	-0.036* (0.019)	-0.002 (0.019)	0.002 (0.001)	0.022** (0.01)	-0.0001 (0.0002)	-0.016 (0.019)

	(1)	(2)	(3)	(4)	(5)	(6)
Has received technical support#	0.111** (0.052)	-0.039 (0.046)	-0.006 (0.005)	-0.027 (0.032)	-0.0004 (0.001)	0.015 (0.052)
Grows rice as the major crop#	-0.041 (0.239)	-0.142** (0.072)	-0.006 (0.007)	0.113 (0.275)	0.001 (0.006)	0.307*** (0.062)
Grows sorghum as the major crop#	-0.243*** (0.033)	-0.179*** (0.025)	0.999*** (0.001)	-0.092*** (0.018)	-0.001*** (0.0003)	0.075 (0.081)
Located in the Coastal agroecological zone#	0.0002 (0.157)	-0.34*** (0.113)	0.914*** (0.159)	-0.098 (0.064)	-0.091 (0.09)	0.555*** (0.079)
Located in the Plateau agroecological zone#	-0.219*** (0.028)	-0.203*** (0.027)	0.996*** (0.187)	-0.087*** (0.017)	-0.001*** (0.0003)	0.298*** (0.06)
Located in the Alluvial plains agroecological zone#	-0.188*** (0.04)	-0.249*** (0.036)	0.999*** (0.038)	-0.088*** (0.02)	-0.001*** (0.0004)	0.011 (0.09)
Located in the Southern Highlands agroecological zone#	-0.229*** (0.028)	-0.205*** (0.024)	0.997*** (0.121)	-0.091*** (0.017)	-0.001*** (0.0004)	0.026 (0.11)
Located in the Semi-arid agroecological zone#	-0.216*** (0.03)	-0.193*** (0.027)	0.998*** (0.118)	-0.099*** (0.019)	-0.001*** (0.0004)	0.05 (0.126)
Distance from input markets						-0.012* (0.007)
Has observed changes in rainfall and temperature#						0.43** (0.184)
Inverse Mill's ratio	-0.012 (0.229)	-0.073 (0.202)	0.002 (0.019)	-0.558** (0.198)	-0.002 (0.005)	
Number of observations (556)	131	93	31	60	37	534
Base rate	0.25298	0.1763	0.0089	0.0863	0.0013	0.6685295

Note:

- Base category for adaptation methods is "No adaptation."
- Base category for agroecological zone is Arid.
- Standard errors are in brackets; *, ** and *** imply significance level at 10%, 5% and and 1% respectively.
- (#) dy/dx is for discrete change of dummy variable from 0 to 1.

Table 7A3 Hausman Test for Independence of Irrelevant Alternatives (IIA)

Omitted	Chi-square	Prob (chi-square)	Evidence
Plant short-season crops	0.07	1.0000	For H0
Plant crops which are resistant to drought	0.68	0.9985	For H0
Irrigation	0.62	1.0000	For H0
Change planting dates	1.20	0.9771	For H0
Plant trees	0.87	0.8217	For H0

Note: The Hausman test was conducted to determine whether one of the key assumptions underlying the multinomial logit specification is fulfilled (i.e., the assumption of independence of irrelevant alternatives (IIA)). The assumption holds when, under the null hypothesis, there is no misspecification of the estimation. The results in this table show that the IIA assumption holds in all categories (i.e., the H0 that there is IIA is not rejected).

Notes

1 $(\beta J - \beta 5)$, for example, shows the net influence of farmers' characteristics and other factors in the choice of adaptation methods.
2 The realities that define the farmers' needs and aspirations (i.e., contextual background) shape their decisions on how to adapt to climate change. Thus, the choice of a particular adaptation method is subject to contextual background. For this study, the contextual background includes socioeconomic factors, environmental factors, institutional factors and the economic structure.
3 Our starting point was the following variables: access to credit and extension, farmers' awareness about climate change, knowledge about climate change and adaptation strategies, availability of capital and improved seeds, availability of water for irrigation, farmers' income, the size of the household, farmers' experience, engaging in non-farm activities, knowledge about adaptation methods, education and gender of the head of the household, livestock ownership, social capital, agroecological settings, government farm support and access to fertile land.
4 Informal credit here refers to borrowing from relatives or neighbors.
5 There is a need for diversity in order to get a good proxy for climate change so that the results obtained from the study can be generalized to the rest of the country.
6 In Tanzania there is a "Nyumba Kumi" concept whereby households within ten neighboring houses group together under one leadership which is recognized by the government. The leaders in the groups know almost all members of their groups – their ages, activities and so on.
7 A household with a household head less than 18 years of age is not included because those household heads are minors. Moreover, since this study is about climate change and it requires household heads to remember the changing climatic variables for the past 20 years, we thought it would be very difficult for household heads under age 30 to remember what happened during their childhood.
8 Increases in temperature affects crop yield. Watson et al. (1998) point out that when the temperature is already close to the crop's maximum tolerance, a small increase in temperature will have a substantial negative effect on yield. In line with temperature, an increase/decrease in rainfall above/below the required amount leads to reduction in yields.
9 According to our sampled agroecological areas, only Tanga and Morogoro regions have bimodal rainy seasons. They have short rainy seasons in October to December and long rainy seasons in March to June.

10 The coefficient results for the Heckman sample selection model are reported in Table 7A1 (see Appendix). We also performed Heckman sample selection using a multinomial logit model as an outcome equation. We wanted to compare the MNL model with that of MNP and reach a conclusion about which model should be used. As explained before, our MNL also passed the IIA assumption but, for the reasons explained earlier in this chapter, it was decided to use MNP for our analysis. The results for both the Heckman model with MNL and the IIA test are provided in the appendix (Tables 7A2 and 7A3).

11 In this study, drought means experiencing less rain than usual.

12 The government might also want to promote specific adaptation methods and not others. This issue will be picked up later on during a discussion about specific adaptation methods.

References

Adesoji, S.A., and J.O. Ayinde. 2013. Ethno-Practices and Adaptation Strategies for the Mitigation of Climate Change by Arable Crop Farmers in Osun State: Implications for Extension Policy Formulation in Nigeria. *Journal of Agricultural and Food Information* 14(1): 66–76.

Ajao, A.O., and L.T. Ogunniyi. 2011. Farmers' Strategies for Adapting to Climate Change in Ogbomoso Agricultural Zone of Oyo State. *Agris On-line Papers in Economics and Informatics* 3(3): 3–13. http://ageconsearch.umn.edu/bitstream/116378/2/agris_online_2011_3_ajao_ogunniyi.pdf (accessed November 3, 2014).

Alvarez, R.M., and J. Nagler. 1998. When Politics and Models Collide: Estimating Models of Multiparty Elections. *American Journal of Political Science* 42: 55–96.

Alvarez, R.M., J. Nagler and S. Bowler. 2000. Issues, Economics, and the Dynamics of Multiparty Elections: The British 1987 General Election. *American Political Science Review* 94(1): 131–149.

Andersson, J., D. Slunge and M. Berlekom. 2005. Tanzania-Environmental Policy Brief. MIMEO. http://sidaenvironmenthelpdesk.se/wordpress3/wp-content/uploads/2013/04/Env-Policy-Brief-Tanzania-2005.pdf (accessed October 28, 2014).

Bartels, K., Y. Boztug and M. Müller. 1999. Testing the Multinomial Logit Model. Sonderforschungsbereich 373 Humboldt Universitaet Berlin. http://edoc.hu-berlin.de/series/sfb-373-papers/1999-19/PDF/19.pdf (accessed November 3, 2014).

Below, T.B., K.D. Mutabazi, D. Kirschke, C. Franke, S. Sieber, R. Siebert and K. Tscherning. 2012. Can Farmers' Adaptation to Climate Change Be Explained by Socio-Economic Household-Level Variables? *Global Environmental Change* 22(1): 223–235.

Bryan, E., T.T. Deressa, G.A. Gbetibouo and C. Ringler. 2009. Adaptation to Climate Change in Ethiopia and South Africa: Options and Constraints. *Environmental Science and Policy* 12: 413–426.

Cameron, A.C., and P. Trivedi. 2005. *Microeconometrics: Methods and Applications*. Cambridge: Cambridge University Press.

Cooper, P.J.M., J. Dimes, K.P.C. Rao, B. Shapiro, B. Shiferaw and S. Twomlow. 2008. Coping Better with Current Climatic Variability in the Rain-Fed Farming Systems of Sub-Saharan Africa: An Essential First Step in Adapting to Future Climate Change? *Agriculture, Ecosystems and Environment* 126: 24–35.

Copas, B.J., and H.G. Li. 1997. Inference for Non-Random Samples (with Discussion). *Royal Statistical Society: Series B* 59: 55–95.

Deressa, T.T. 2006. Measuring the Economic Impact of Climate Change on Ethiopian Agriculture: Ricardian Approach. CEEPA DP25. Pretoria, South Africa: University of Pretoria.

Deressa, T.T., R.M. Hassan and C. Ringler. 2011. Perception of and Adaptation to Climate Change by Farmers in the Nile Basin of Ethiopia. *The Journal of Agricultural Science* 149: 23–31.

Deressa, T.T., R.M. Hassan, C. Ringler, T. Alemu and M. Yesuf. 2009. Determinants of Farmers' Choice of Adaptation Methods to Climate Change in the Nile Basin of Ethiopia. *Global Environmental Change* 19: 248–255.

Dow, J.K., and J.W. Endersby. 2004. Multinomial Probit and Multinomial Logit: A Comparison of Choice Models for Voting Research. *Electoral Studies* 23: 107–122.

Downing, T.E., L. Ringius, M. Hulme and D. Waughray. 1997. Adapting to Climate Change in Africa. In *Mitigation and Adaptation Strategies for Global Change*. Liege, Belgium: Kluwer Academic 2, pp. 19–44.

Eicher, C.K., K. Maredia and I. Sithole-Niang. 2006. Crop Biotechnology and the African Farmer. *Food Policy* 31(6): 483–594.

Füssel, H.-M., and R.J.T. Klein. 2006. Climate Change Vulnerability Assessments: An Evolution of Conceptual Thinking. *Climate Change* 75: 301–329.

Gbetibouo, G.A. 2009. Understanding Farmers' Perceptions and Adaptations to Climate Change and Variability: The Case of the Limpopo Basin, South Africa. IFPRI Discussion Paper No. 00849. www.ifpri.org/sites/default/files/publications/rb15_08.pdf (accessed November 3, 2014).

Goulden, M., L.O. Naess, K. Vincent and W.N. Adger. 2009. Accessing Diversification, Networks and Traditional Resource Management as Adaptations to Climate Extremes. In *Adapting to Climate Change: Thresholds, Values, Governance*, edited by W. Adger, I. Lorenzoni and K. O'Brien. Cambridge: University of Cambridge, pp. 448–463.

Greene, W.H. 2000. *Econometric Analysis*, 4th Edition. Upper Saddle River, NJ: Prentice Hall.

Grilli, L., and C. Rampichini. 2007. A Multilevel Multinomial Logit Model for the Analysis of Graduates' Skills. *Statistical Methods and Applications* 16: 381–393.

Gujarati, D.M. 2003. *Basic Econometrics*. New York: McGraw-Hill/Irwin.

Hassan, R., and C. Nhemachena. 2008. Determinants of African Farmers' Strategies for Adapting to Climate Change: Multinomial Choice Analysis. *African Journal of Agricultural and Resource Economics* 2: 83–104.

Heckman, J. 1979. Sample Selection Bias as a Specification Error. *Econometrica* 47: 153–161.

Iglesias, A., S. Quiroga and A. Diz. 2011. Looking into the Future of Agriculture in a Changing Climate. *European Review of Agricultural Economics* 38(3): 427–447.

Ishaya, S., and I.B. Abaje. 2008. Indigenous People's Perception on Climate Change and Adaptation Strategies in Jema'a Local Government Area of Kaduna State in Nigeria. *Journal of Geography and Regional Planning* 1(8): 138–143.

Jain, S. 2006. An Empirical Economic Assessment of the Impacts of Climate Change on Agriculture in Zambia. CEEPA DP27. Pretoria, South Africa: University of Pretoria.

Kabubo-Mariara, J. 2008. Global Warming and Livestock Husbandry in Kenya: Impacts and Adaptations. *Ecological Economics* 68: 1915–1924.

Keil, A., M. Zeller, A. Wida, B. Sanim and R. Birner. 2008. What Determines Farmers' Resilience towards ENSO-Related Drought? An Empirical Assessment in Central Sulawesi, Indonesia. *Climatic Change* 86(3–4): 291–307.

Mano, R., and C. Nhemachena. 2006. Assessment of the Economic Impacts of Climate Change on Agriculture in Zimbabwe: A Ricardian Approach. CEEPA DP11. Pretoria, South Africa: University of Pretoria.

Mathijs, E. 2003. Social Capital and Farmers' Willingness to Adopt Countryside Steward-ship Schemes. *Outlook on Agriculture* 32: 13–16.

Mideksa, T.K. 2009. Economic and Distributional Impacts of Climate Change: The Case of Ethiopia. *Global Environmental Change* 20: 278–286.

Molua, E.L., and C.M. Lambi. 2006. The Economic Impact of Climate Change on Agri-culture in Cameroon. CEEPA DP17. Pretoria, South Africa: University of Pretoria. http://elibrary.worldbank.org/doi/pdf/10.1596/1813-9450-4364 (accessed November 3, 2014).

Montiflor, M.O. 2008. Cluster Farming as a Vegetable Marketing Strategy: The Case of Smallholder Farmers in Southern and Northern Mindanao. *Acta Horticulturae (ISHS)* 7(94): 229–238.

Moussa, K.M., and M. Amadou. 2006. Using the CROPWAT Model to Analyze the Effects of Climate Change on Rainfed Crops in Niger. CEEPA DP32. Pretoria, South Africa: Uni-versity of Pretoria. www.ceepa.co.za/uploads/files/CDP32.pdf (accessed November 3, 2014).

Munasib, A.B.A., and J.L. Jordan. 2011. The Effect of Social Capital on the Choice to Use Agricultural Practices. *Agricultural and Applied Economics* 43: 213–227.

Nhemachena, C., and R. Hassan. 2007. Micro-level Analysis of Farmers' Adaptation to Climate Change in Southern Africa. IFPRI Discussion Paper No. 00714. Washington, DC: International Food Policy Research Institute.

Nzeadibe, T.C., N.A. Chukwuone, C.L. Egbule and V.C. Agu. 2011. Farmers' Perception of Climate Change Governance and Adaptation Constraints in Niger Delta Region of Nigeria. The African Technology Policy Studies Network: ISBN: 978-9966-030-02-3. www.atpsnet.org/Files/rps7.pdf (accessed November 3, 2014).

Ogalleh, S.A., C.R. Vogl, J. Eitzinger and M. Hauser. 2012. Local Perceptions and Responses to Climate Change and Variability: The Case of Laikipia District, Kenya. *Sustainability* 4(12): 3302.

Schofield, N., A.D. Martin, K.M. Quinn and A.B. Whitford. 1998. Multiparty Electoral Competition in the Netherlands and Germany: A Model Based on Multinomial Probit. *Public Choice* 97(3): 257–293.

Sun, F., M.L. Roderick and G.D. Farquhar. 2012. Changes in the Variability of Global Land Precipitation. *Geophysical Research Letters* 39(19): L19402.

United Republic of Tanzania. 2003. Agricultural Sector Development Programme (ASDP): Support through Basket Fund. Government Programme Document. www.kilimo.go.tz/publications/english%20docs/ASDP%20FINAL%2025%2005%2006%20(2).pdf (accessed October 27, 2014).

United Republic of Tanzania. 2005. National Strategy for Growth and Reduction of Pov-erty (NSGRP). Vice President's Office. www.povertymonitoring.go.tz/Mkukuta/MKU-KUTA_MAIN_ENGLISH.pdf (accessed October 28, 2014).

United Republic of Tanzania. 2007. National Adaptation Programme of Action (NAPA). Vice President's Office: Division of Environment. www.preventionweb.net/files/8576_tza01.pdf (accessed October 27, 2014).

USDA. 2003. Production Estimates and Crop Assessment Division Foreign Agricultural Service: Tanzania Regions with Two Agricultural Seasons (Short and Long Rains). www.fas.usda.gov/pecad2/highlights/2003/03/tanzania/images/bimodal.htm (accessed July 2, 2013).

Watson, R.T., M.C. Zinyoera and R.H. Moss. 1998. The Regional Impacts of Climate Change: An Assessment of Vulnerability. A Special Report of IPPC Working Group II. Cambridge: Cambridge University Press.

West, C., C. Roncoli and F. Ouattara. 2008. Local Perceptions and Regional Climate Trends on the Central Plateau of Burkina Faso. *Land Degradation Development* 19: 289–304.

Wooldridge, J.M. 2001. *Econometrics Analysis of Cross Section and Panel Data*. London, Massachusetts: The MIT Press.

Ziervogel, G., S. Bharwani and T.E. Downing. 2006. Adapting to Climate Variability: Pumpkins, People and Policy. *Natural Resources Forum* 30(4): 294–305.

8 Risk preferences and the poverty trap

A look at farm technology uptake among smallholder farmers in the Matzikama Municipality*

Hafsah Jumare, Martine Visser and Kerri Brick

Introduction

The diffusion of new farm technology has been slow in developing countries (Feder et al., 1985; Engle-Warnick et al., 2007; Duflo et al., 2011; Simtowe, 2006; Fernandez-Cornejo et al., 2002; Brick and Visser, 2015), despite the benefits of recent innovations in technology. These include new, extensive ranges of crops that are more nutrient-rich and at the same time more resistant than traditional varieties to insects, disease and drought (Liu and Huang, 2010). These poor technology diffusion rates have been cited as one of the reasons for the persistence of poverty in developing countries. Behavioral economic studies on farmers' decisions to take up technology have identified risk preference and financial constraints as factors that are crucial to the technology adoption process.

In this chapter, we look at the determinants of farm technology uptake in the form of drought-resistant and improved seeds, paying special attention to farmers' risk preference and income. We use a field experiment to elicit measures of risk aversion, loss aversion and nonlinear weights of probability. We then relate these three elicited measures to respondents' reported uptake of drought-resistant and improved seeds. In light of the poverty trap theory, which suggests that low-income people make choices that keep them in poverty, we also consider the role that income plays in risk preference.

Technology uptake has been found to depend not only on financial constraints but also on behavioral traits, such as risk preference, that arise as a result of these constraints. A standard feature of poverty trap models is that they recognize a divergence of behavior between low-income and higher-income groups, where pathways out of persistent poverty are obstructed for low-income groups (Carter and Barrett, 2006, 2007). As a result, the persistence of poverty hinges on the

* This work was supported by the Environmental Research Policy Unit (EPRU) at the University of Cape Town, in conjunction with the Swedish International Development Cooperation Agency (Sida) through the Environment for Development (EfD) initiative. In addition, the financial assistance of the National Research Foundation (NRF) for this research is hereby acknowledged. Opinions expressed and conclusions arrived at are those of the authors and are not necessarily to be attributed to the NRF.

degree to which households are restricted in intertemporal exchange and the extent to which poverty affects farmers' investments in technology, by means of both tangible constraints and behavioral traits, such as risk preferences (Carter and Barrett, 2006).

Many studies have looked at the role of risk preferences in technology uptake. If we assume the expected utility model generally adopted in studies on farmers' choices, farmers will, given their level of risk preference, select the technology that offers the maximum expected utility (Feder et al., 1985; Sunding and Zilberman, 2001; Isik and Khanna, 2003; Marra et al., 2003; Foster and Rosenzweig, 2010; Barham et al., 2014). One downside of the expected utility model is that it considers risk aversion only as a risk preference. Risk aversion describes observed behavior that demonstrates a fear of variance in outcomes (Kahneman and Tversky, 1979; Binswanger and Sillers, 1983; Hill, 2005; Yesuf and Bluffstone, 2009; Tanaka et al., 2010; Brick and Visser, 2010, 2015; Di Falco, 2014).

Measures of risk aversion in expected utility models assume that farmers perceive the impact of losses and gains to be the same in absolute terms and can effectively perceive the probability of outcomes. However, empirical evidence from behavioral literature has established that this is not the case. Individuals do not accurately perceive probabilities, in that they overestimate the probabilities of unlikely events and underestimate the probabilities of likely events. This phenomenon, called nonlinear probability weighting, is another key violation of expected utility (Tversky and Kahneman, 1992; Humphrey and Verschoor, 2004; Ranjan and Shogren, 2006; Brick and Visser, 2010; Tanaka et al., 2010). They also display a greater sensitivity to losses compared to gains, which is described as loss aversion (Tversky and Kahneman, 1992, Thaler et al., 1997; Brick and Visser, 2010; Tanaka et al., 2010; Ward and Singh, 2013; Liu, 2013). Even though loss aversion has been under-represented in the literature, there is now a growing list of studies exploring the role it plays in decision making. We therefore consider probability weighting, along with risk and loss aversion.

We use the method described in Tanaka et al. (2010) (TCN) to measure the three parameters – that is, risk aversion, loss aversion and nonlinear weighting of probabilities – under the assumption of prospect theory. Few studies have considered the role of these three risk preference parameters, collectively, on technology uptake. Among these studies is Liu (2013), who uses the TCN method and a Weibull hazard model to test the effects of risk preferences on the timing of Bt cotton uptake by smallholder farmers in China. Liu (2013) finds that risk aversion and loss aversion are correlated with later adoption of Bt cotton, while farmers who overweight small probabilities are found to adopt Bt cotton earlier. Other studies consider at most two of the parameters. These include Hill (2005), who finds that risk aversion is correlated with replanting of coffee trees, and Ward and Singh (2013), who find loss aversion to be correlated with a switch from traditional rice seeds to a new variety.

Besides the prospect theory parameters, numerous studies have identified drivers of uptake. Some of these studies find household size to be positively related to farm technology uptake, which is attributed to the increased labor available to engage in physically challenging strategies, as well as the ability to spread risk and pool financial resources (Croppenstedt et al., 2003; Nhemachena and Hassan, 2007; Deressa et al., 2009). While some studies have found female-headed households to be more likely to adopt a farm strategy or technology (Nhemachena and Hassan, 2007; Deressa et al., 2009), the majority of studies have found opposite gender effects, indicating that females are less likely to take up farm technologies or strategies. This is attributed to constraints that are usually related to differences in land access, assets, education, health care, markets, extension services, and social/cultural exclusion, which typically pose obstacles for female farmers (Quisumbing, 1995; Doss and Morris, 2001; World Bank, 2001; Githinji et al., 2011; Odame et al., 2002; Meinzen-Dick et al., 2010; Quisumbing and Pandolfelli, 2010; Croppenstedt et al., 2013 in Ndiritu et al., 2014).

Looking at studies that consider crop type, Panda et al. (2013) find total income and total farming income to be linked to a change of crop variety. Etwire et al. (2016) find that uptake of improved seeds, for a sample of smallholder farmers in Northern Ghana, depends on the seed delivery system. Monela (2014) finds that uptake of improved seeds, among a sample of smallholder farmers in Tanzania, depends on land ownership and awareness of improved seeds. So, despite the assumptions that we make about income and risk preferences, we consider a range of other factors – for example, land and gender.

We consider the use of both genetically improved seeds and drought-resistant crops as alternative technologies available to farmers. Drought-resistant crops are natural substitutes to improved seeds. We use experimental and survey data obtained from 125 small-scale farmers from farming communities in the Matzikama Municipality of Western Cape, South Africa.

This chapter proceeds as follows. The next section shows the summary statistics of the farmers' characteristics. The following section describes the experiment design and methodology for eliciting the prospect theory parameters. The fourth section explains the results for the determinants of risk preferences. The fifth section presents the results of the logit regressions on uptake of improved seeds, and draws a comparison with the choice of naturally occurring drought-resistant crops. The concluding section includes policy recommendations.

Background and summary statistics

The data used for the analysis in this study was obtained via survey collection and risk experiments carried out with small-scale farmers in the Matzikama Municipality of the Western Cape, South Africa.[1] Agriculture in Matzikama is supported by the Clanwilliam Dam and Olifants River. The area is dominated by viniculture, vegetables, citrus fruits and livestock production, and is characterized by arid terrain and cool temperatures (Matzikama IDP, 2009–2010).

The sample consists of 125 farmers from the towns of Vanrhynsdorp, Lutzville, Klawer, Clanwilliam and Wupperthal, who were recruited through the Matzikama Emerging Farmers Forum. Table 8.1 presents the summary statistics of farmer characteristics. The average age of farmers in the sample is 43 years, with approximately 44%

Table 8.1 Summary Statistics

Variable	Mean	Standard deviation
Age	42.951	16.337
Female	44%	50%
Male	56%	50%
Primary bread winner	55%	50%
No schooling	3%	18%
Some primary school	34%	48%
Complete primary school	16%	37%
Some secondary school	36%	48%
Matric certificate	9%	28.1%
Higher education	2%	13%
Household size	4.699	1.916
Farm experience in years	6.512	7.023
Commercial farm land	4.8%	21.46%
Employed	30%	46%

Sample Size: 125.

Table 8.2 Farm Uptake

Variable	Obs.	Mean	Std. dev.
Drought-resistant crops	125	8.8%	28.4
Improved seeds (excl. drought-resistant seeds)	125	8.8%	28.4
Intercropping	125	14.4%	35.3
Mulching	125	10.4%	30.6
Fertilizer	125	9.6%	29.6
Organic manure	125	20.8%	40.8
Changing planting date	125	4.8%	21.5
Wind breaks	125	10.4%	30.6
Irrigation	125	19.2%	39.5

Sample size: 125.

of the farmers female and 56% male. Average monthly household income is R2365,[2] with about 60% of the sample below the average relative household income level. Only 30% of the farmers have an alternative source of employment outside of farming.

In the survey, respondents were asked to indicate whether they had started using any new farming practices. The summary statistics for these farming practices are presented in Table 8.2. Just under 9% said they use drought-resistant crops (i.e., naturally occurring crops that are more resilient to poor rainfall), while 8.8% use improved seeds, showing a significantly low rate of uptake of more resilient crops (either natural or genetically modified) among farmers in the sample.

Risk preference elicitation and estimation

This section illustrates the risk preference elicitation and estimation methodology.

Methodology

The experiments in this study were modelled after the design of Tanaka et al. (2010) (TCN), who assume cumulative prospect theory. TCN use a series of gain-only and gain-and-loss pair-wise lotteries with both a risky and a safe option (similar to Holt and Laury, 2002). They assumed the following utility function:

$$U(x,p;y,q) = \begin{cases} v(y) + \pi(p)(v(x) - v(y)) & x > y > 0; x < y < 0 \\ \pi(p)v(x) + \pi(q)v(y) & x < 0 < y \end{cases} \quad (1)$$

U(x,p;y,q) denotes the expected value linked to prospects (x,p;y,q); p and q are the probabilities of receiving outcomes x and y, respectively. The power function $v(x) = x^\sigma$ for gains (x>0) and $v(x) = -\lambda(-x^\sigma)$ for losses (x<0) is assumed, with σ being the risk aversion parameter (i.e., measure of the concavity of the value function) and λ the parameter for loss aversion. The risk aversion parameter (σ) is presumed to be identical in both gains and losses; the inequality $\sigma > 1$ implies risk-seeking preference and $\sigma < 1$ implies risk-averse preference. For λ, $\lambda > 1 (\lambda < 1)$ implies greater sensitivity to losses (gains) compared to gains (losses).

TCN use the nonlinear probability weighting function of Prelec (1998), where $\pi(p) = \exp[-(-\ln p)^a]$, with the function being linear if $\alpha = 1$. If $\alpha = 1$ and $\lambda = 1$, the model reduces to expected utility. If $\alpha < 1$, the function is an inverted S-shape. The inverted S-shape indicates that small probabilities are overweighted and large probabilities are underweighted. The function is S shaped if $\alpha > 1$, indicating that small probabilities are underweighted while large probabilities are overweighted.

Elicitation

Similarly to TCN, the Matzikama farmers were given three sets of multiple price lists (MPLs) with pair-wise lottery sheets.[3] The first two lists (i.e., Series 1 and 2) had a series of 14 decision rows each, both of which were gain-only lotteries. The third sheet (Series 3) had both gain and loss lotteries, with seven decision rows. Subjects had a choice between Lottery A or Lottery B in each row. The lotteries were framed to represent farming seasons, with Lottery A representing the outcome if farmers chose to use traditional seeds and Lottery B representing the outcome if farmers chose to use improved seeds. The payoffs are dependent on whether there is sufficient rainfall for yields to be good. The premise of this framing is that improved seeds require more rain relative to traditional seeds. The probabilities in the lotteries represented the probabilities of good rainfall for the high payoffs and probabilities of bad rain for the low payoffs. The payoffs represent the yields in a farming season.

Subjects were asked to select the row where they wanted to switch from Lottery A (traditional seeds) to Lottery B (improved seeds). Participants could select only one option (A or B) in each decision row. The probabilities of outcomes in the first two series were fixed all through the row. The first row of Series 1 had Lottery A, offering a 30% chance of receiving a high payoff and a 70% chance of receiving a low payoff. The first row of lottery B offered a 10% chance of receiving a higher payoff than the high payoff in Lottery A and a 90% chance of receiving a lower payoff than the low payoff in Lottery B. In Series 2, the first row of Lottery A offered a 90% chance of receiving a high payoff and a 10% chance of receiving a low payoff. The first row of Lottery B offered a 70% chance of receiving a higher payoff than the high payoff in Lottery A and a 30% chance of receiving a lower payoff than the low payoff in Lottery B.

In Series 1, the outcome in Lottery A was also fixed, but in Lottery B the payoffs change as one goes down the rows, until the expected payoff of Lottery B ultimately surpasses that of Lottery A. In both Series 1 and 2, the more risk-averse a participant is, the farther down the row he or she switches to Lottery B. In Series 3, the subjects had a 50% probability of a positive payoff (gain) and a 50% probability of a negative payoff (loss) in both lotteries. The expected value of Lottery A decreases and Lottery B increases as we move down the rows. The more risk-averse a participant is, the farther down the row he or she switches to Lottery B. Subsequent to the completion of each MPL, a subject would draw a numbered ball from numbered balls that were placed in a bag. The balls were numbered 1 to 14 for Series 1 and Series 2 and 1 to 7 for Series 3. The chosen ball then determined which decision row was to be played for money. Rainfall probabilities were also denoted by ten numbered balls. For example, for traditional seeds in Series 1, three balls represented good rainfall levels, while seven balls represented poor rainfall levels. The rainfall level is determined by one of the subjects selecting a ball from the bag.

The MPL lotteries in TCN were structured so that the switching points of the three series produce a permutation of the prospect theory parameters: risk aversion, nonlinear probability weighting and loss aversion. Series 1 and 2 estimate the parameters sigma (the measure of risk aversion) and alpha (the measure for probability weighting). In Series 1, a set of sigma and alpha (σ, α) combinations that rationalize the switching points are estimated. Another combination of sets that justifies the switching point is found for Series 2. For example, if a subject switched in Row 6 of Series 1, the values of sigma and alpha that can rationalize the switch are (0.5, 0.4) (0.6, 0.5), (0.7, 0.6), (0.8, 0.7), (0.9, 0.8), (1.0, 0.9). This implies the following inequalities:

$$5^\sigma + \exp\left[-(-\ln0.3)^\alpha\right]\left(20^\sigma - 5^\sigma\right) < 2.5^\sigma + \exp\left[-(-\ln0.1)^\alpha\right]\left(55^\sigma - 2.5^\sigma\right)$$

$$5^\sigma + \exp\left[-(-\ln0.3)^\alpha\right]\left(20^\sigma - 5^\sigma\right) > 2.5^\sigma + \exp\left[-(-\ln0.1)^\alpha\right]\left(62.5^\sigma - 2.5^\sigma\right)$$

If a subject switched in Row 6 in Series 2, this implies the following inequalities:

$$15^\sigma + \exp\left[-(-\ln 0.9)^\alpha\right](20^\sigma - 15^\sigma) < 2.5^\sigma + \exp\left[-(-\ln 0.7)^\alpha\right](31^\sigma - 2.5^\sigma)$$

$$15^\sigma + \exp\left[-(-\ln 0.9)^\alpha\right](20^\sigma - 15^\sigma) > 2.5^\sigma + \exp\left[-(-\ln 0.7)^\alpha\right](32.5^\sigma - 2.5^\sigma)$$

The combination of sigma and alpha that can rationalize the switch is (0.5, 1), (0.6, 0.9), (0.7, 0.8), (0.8, 0.7), (0.9, 0.6), (1, 0, and 0.5). The crossing point is thus (0.8, 0.7).

In TCN, the coefficient of loss aversion (λ) is derived from Series 3: conditional on the value of sigma derived from Series 1 and Series 2, the switching point in Series 3 implies a range of values for λ. The TCN method produces interval values for the loss aversion parameter. For the value of sigma 0.8 and the subject switching in the 6th Row of Series 3, this implies the following inequalities:

(-λ)(-(-2^0.8))(0.5) + (12.5^0.8) (0.5)> (-λ) (-(-10.5^0.8)) (0.5) + (15^0.8) (0.5),

(-λ)(-(-2^0.8))(0.5) + (2^0.8) (0.5)> (-λ) (-(10.5^0.8)) (0.5) + (15^0.8) (0.5),

(-λ)(-(-2^0.8))(0.5) + (0.5^0.8) (0.5)> (-λ) (-(-10.5^0.8)) (0.5) + (15^0.8) (0.5),

(-λ)(-(-2^0.8))(0.5) + (0.5^0.8) (0.5)> (-λ) (-(-8^0.8)) (0.5) + (15^0.8) (0.5),

(-λ)(-(-4^0.8))(0.5) + (0.5^0.8) (0.5)> (-λ) (-(-8^0.8)) (0.5) + (15^0.8) (0.5),

(-λ)(-(-4^0.8))(0.5) + (0.5^0.8) (0.5) < (-λ) (-(-7^0.8)) (0.5) + (15^0.8) (0.5),

(-λ)(-(-4^0.8))(0.5)+ (0.5^0.8) (0.5) < (-λ) (-(-5.5^0.8)) (0.5) + (15^0.8) (0.5)

The implied interval of lambda is $3.62896 < \lambda < 4.76259$. Note that, if subjects switch in Row 1 or never switch, the intervals are censored. The summary of the payoffs implied by the switching points is presented in Appendix 8.3. Figure 8.1 shows a sample lottery task.

	Rainfall	Traditional seeds	Improved seeds
1	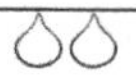 🌢🌢	R20 if ❶	R38.5 if ❶
	🌢	R16 if ❷❸❹❺❻❼❽❾❿	R1 if ❷❸❹❺❻❼❽❾❿

Figure 8.1 Sample Lottery Task

Determinants of risk preferences

In this section, we assess the determinants of risk aversion, nonlinear probability weighting and loss aversion. The results presented in Table 8.3 are, *ceteris paribus*, outcomes of the ordinary least square regressions on normalized σ (risk aversion) (Column 1) and α (probability weighting) (Column 2), respectively, and interval regression on λ (loss aversion) (Column 3). We reframe the parameters so that positive values on σ and interval λ regressions denote an increase in risk aversion and loss aversion, respectively, while positive values on α indicate greater weighting of low-probability events. Thus, for ease of interpretation, we use $\sigma = -\sigma$ and $\alpha = -\alpha$ in order to present a measure that shows a higher σ and α, denoting greater risk aversion and increase in the overweighting of small probabilities. Brick and Visser (2015), using the same method and data, look at the determinants of risk and loss aversion; however, unlike Brick and Visser (2015), our analysis considers the determinants of probability weighting. In addition, we also control for additional variables – namely, farm experience and type of land farmed – that is, ownership status.

The explanatory variables are the farmer's age, gender (dummy variable equals 1 if the farmer is female and 0 if the farmer is male), log of education level of farmer, a dummy variable equal to 1 if the farmer is the primary bread winner of the household and 0 if not, normalized monthly household income, and a dummy variable for land type (1 if the farmer farms on commercial land through an equity share or as an employee and 0 if the farmer owns the land or uses communal land, state land as part of a land reform initiative, or land in proximity to the farmer's residence). The premise is that farmers are more likely to receive greater support in terms of technology, resources or training if they work on commercial land with an external and centralized decision maker. Only three farmers in the sample own the land they farm, so using farmer ownership as the major indicator of land type would not provide meaningful information. The other variables include household size and log of farmer's farm experience in years. We use the log form of education level and farm experience because their distributions are positively skewed.

As can be seen Table 8.3, higher age and education levels are related to lower loss aversion (Table 8.3, Column 3, P-value = 0.045; P-value= 0.009). We find that primary breadwinners are less risk-averse and loss-averse (Table 8.3, Column 1, P-value= 0.031). Primary breadwinners have greater command over the resources of their households and a greater burden in terms of bringing in income to satisfy household consumption (Rouse and Kitching, 2006; Jayawarna et al., 2013). They may be willing to take more risk to bring in income. This is also reflected in their loss aversion because it indicates that they have a preference for upside risk, which shows a greater willingness to take on potential losses as long as potential gains exist.

Lastly we find that greater household monthly income is related to lower loss aversion (Table 8.3, Column 3, P-value = 0.028). This negative relationship is expected, given that the disutility from loss of income decreases as income

Table 8.3 Determinants of Risk Preferences

	Risk aversion	Prob. weighting	Loss aversion
Age	−0.003	0.001	−0.055
	(0.004)	(0.003)	(0.028)**
Female	0.063	0.050	0.561
	(0.084)	(0.065)	(0.901)
Log education level	−0.063	0.048	−3.996
	(0.103)	(0.111)	(1.526)***
Primary bread winner	−0.165	−0.070	−1.982
	(0.097)*	(0.079)	(0.921)**
Normalized HH income	0.078	−0.074	−3.383
	(0.159)	(0.142)	(1.543)**
Land type	−0.074	0.141	0.039
	(0.095)	(0.102)	(1.499)
Log farm experience in yrs.	0.011	0.009	0.425
	(0.044)	(0.033)	(0.439)
Household size	−0.053	0.004	−0.371
	(0.018)***	(0.016)	(0.204)*
Constant	0.072	−0.930	16.380
	(0.354)	(0.364)**	(4.884)***
lnsigna_cons	–	–	1.189
			(0.150)***
R2	0.20	0.05	–
Prob > f	0.0252	0.7384	–
Prob >chi2	–	–	0.0797
N	71	71	71

* $p<0.1$;
** $p<0.05$;
*** $p<0.01$.

increases. It is necessary to mention that, unlike this study, Brick and Visser (2015) find no statistically significant coefficients on any of the variables regressed on the loss aversion parameter.

Similarly to Brick and Visser (2015), we find household size to be related to less risk aversion; we find similar results for loss aversion (Table 8.3, Columns 1 & 3; P-value = 0.005, P-value = 0.068). Wik et al. (2004) found similar results and suggest that household size represents a wealth factor. Bigger households imply a larger household labor force or wealth-generating capacity. Furthermore, household size may be correlated with lower risk aversion because of the greater opportunities for risk sharing/pooling – that is, an implicit form of insurance for members of the household (Barr, 2003; Wik et al., 2004).

Determinants of technology uptake: drought-resistant crops and improved seeds

In this section, we consider the uptake of two farm adaptive mechanisms: drought-resistant crops and improved seeds. The determinants of uptake are obtained using logit regressions and the results are presented in Table 8.4. The coefficients are the expected changes in the probability of taking up an option due to a unit change in the explanatory variable.

Table 8.4 Determinants of Uptake

	Drought-resistant crop			Improved seeds		
	(1)	*(2)*	*(3)*	*(4)*	*(5)*	*(6)*
Risk aversion	−1.139	−0.957	−0.858	0.520	0.590	2.784
	(0.99)	(0.67)	(0.55)	(0.53)	(0.65)	(1.34)
Probability weighting	3.207	4.237	4.172	2.799	2.962	2.574
	(1.84)*	(2.15)**	(2.04)**	(1.08)	(1.18)	(1.10)
Loss aversion	−0.080	−0.010	0.005	0.050	0.067	0.253
	(0.67)	(0.07)	(0.04)	(0.49)	(0.77)	(1.54)
Age	−0.077	−0.104	−0.102	0.014	0.008	0.041
	(1.38)	(1.48)	(1.45)	(0.27)	(0.13)	(0.98)
Female	−3.258	−2.855	−2.911	−4.122	−4.010	−6.707
	(3.48)***	(3.36)***	(3.48)***	(3.10)***	(2.90)***	(2.01)**
Log education level	−0.329	−0.126	0.028	3.759	3.647	6.487
	(0.22)	(0.07)	(0.01)	(1.91)*	(1.76)*	(1.92)*
Primary bread winner	–	2.782	2.786	–	0.532	1.027
		(2.47)**	(2.42)**		(0.44)	(0.68)
Normalized Hh income	4.309	7.310	7.212	8.164	8.611	7.969
	(2.67)***	(3.29)***	(3.15)***	(2.64)***	(3.33)***	(3.27)***
Land type	4.033	4.563	4.606	4.794	4.770	7.375
	(3.15)***	(3.25)***	(3.27)***	(3.32)***	(3.41)***	(2.13)**
Log farm experience in yrs.	0.265	0.157	0.150	−0.207	−0.235	−0.649
	(0.65)	(0.34)	(0.32)	(0.86)	(0.88)	(1.32)
Household size	–	–	0.052	–	–	0.749
			(0.25)			(1.40)
Constant	3.073	2.122	1.448	−9.907	−9.609	−20.983
	(0.61)	(0.38)	(0.24)	(1.96)*	(1.83)*	(2.14)**
Prob > Chi2	0.0066	0.0026	0.0009	0.0339	0.0076	0.1074
Df	9	10	11	9	10	11
N	71	71	71	71	71	71

* p<0.1;
** p<0.05;
*** p<0.01.

The farmers in our sample gave responses about a range of agricultural strate-gies they recently adopted.[4] We select two strategies that represent the cropping choices of these farmers. These choices tell us whether farmers are likely to abandon certain crops for others or to take up an improved version of the crops they already farm. The first strategy we consider is a switch to drought-resistant crops. These are traditional means of mitigating risk. They are not technological innovations, in that they exist naturally. The second is improved seeds, which are considered new technology. They are more costly for farmers but have higher expected yields. They also have a high variance – that is, the risk is higher because improved seeds are more expensive; therefore, in the event of a failed harvest, a farmer will face a greater loss compared to the case when he/she uses traditional seeds. Using these two alternatives to crops that the farmers already cultivate, we explore how risk preference and income affect the uptake of tech-nology in the form of improved seeds and compare these effects with the effects on uptake of drought-resistant crop types.

Looking at the risk preference parameters in Table 8.4, we find no evidence to suggest that risk or loss aversion influences the uptake of either option. How-ever, we find that an increased weighting of small probability events increases the likelihood of uptake of drought-resistant crop varieties (Table 8.4, Col-umns 1, 2 & 3; P-values =0.066; 0.031; 0.042). This result is explained if fear of a drought causes farmers to overestimate its likelihood and thus switch to more drought-resistant crops.

When we consider the other explanatory variables, we find that females are less likely to take up either drought-resistant crop varieties or improved seeds (Table 8.4, Columns 1, 2 & 3; P-values = 0.001; 0.001; 0.001 and Columns 4, 5 & 6; P-values= 0.002; 0.004; 0.044). This gender effect is consistent with the most common findings in the existing literature (Ndiritu et al., 2014). A higher education level is found to increase the likelihood of improved seeds uptake (Table 8.4, Columns 4, 5 & 6; P-values=0.056; 0.078; 0.055). This in line with the common assumption that the education level of farmers has a positive effect on technology uptake because of the connection between education and knowl-edge of technology (Knowler and Bradshaw, 2007).

We find that household income increases the likelihood of both drought-resis-tant seeds and improved seeds uptake (Table 8.4, Columns 1, 2 & 3; P-values= 0.008; 0.001; 0.002 and Columns 4, 5 & 6; P-values =0.008; 0.001; 0.001). One would expect that lower-income farmers would opt for drought-resistant crops, while higher-income farmers would opt for improved seeds; nonetheless, we see that poorer farmers are sticking to non-resilient crops. This to some degree sup-ports the poverty trap hypothesis in so far as households that are poor at the outset do not improve their farm prospects by taking on more resilient crops, while those that are initially more affluent can progress to higher levels of wealth (Adato et al., 2006; Barrett and Carter, 2013). The relatively wealthy farmers take on more advanced crops, in the form of both naturally drought-resistant crops and improved seeds.

We also find that primary breadwinners are more likely to take up drought-resistant crop varieties (Table 8.4, Columns 2 & 3; P-values = 0.014; 0.016). Those who farm on commercial land through equity sharing or as an employee are more likely to use both drought-resistant crop varieties and improved seeds (Table 8.4, Columns 1, 2 & 3; P-values = 0.002; 0.001; 0.001 and Columns 4, 5 & 6; P-values = 0.001; 0.001; 0.033). This again points to the possible influences of commercialization through a centralized body, which are likely to improve the delivery mechanism of improved seeds and improve the knowledge of both improved seeds and drought-resistant crops.

Conclusion

Advancements in farm technology have provided farmers with the means to improve yields while safeguarding farm productivity against harsh climatic and weather events. Nonetheless, the diffusion of these technologies has been slow in developing countries. Evidence from economic and behavioral literature suggests that there are tangible and behavioral attributes of farmers that contribute to the slow uptake of farm technology. The tangible factors include factors such as income, which may determine the willingness and ability to invest in new technology, while behavioral factors, such as risk preference, may play a further role in determining investments in such technology.

In this study, we carry out an analysis to determine the extent to which these income and risk preferences contribute to the farm technology uptake process. Our findings suggest that farm risk management policies need to take into account the role of risk and loss preferences in uptake decisions. We find that farmers do not effectively weigh probabilities and that the weighting of probabilities in turn affects the uptake of adaptive mechanisms. Therefore, farmers should be involved in understanding weather and climate risk and probabilities of loss, so that they can participate in developing responsive, adaptive or mitigation strategies (Patt and Schröter, 2008). An institutional factor that should be considered is access to extension services, which provide information on climate and on the nature of technologies and other adaptive strategies. Farmers can construct viable and efficient farm strategies only if they have comprehensive and accurate information on impending occurrences or outcomes (Smit and Skinner, 2002).

Our findings on the effect of income are to some degree consistent with the poverty trap hypothesis. We find that low incomes have dampening effects on the uptake of resilient crop types, in the form of both naturally drought-resistant crops and technologically modified improved seeds. This signals the need for proactive measures to guarantee access to a minimum package of assets to poor farmers which, as explained by Carter and Barret (2006), is required for their successful rise out of poverty through investment in strategies such as the uptake of more advanced crops.

Appendix 8.1

Questionnaire
Experiment number: _____

Background information

1 Age: _____

2 Gender: [put a tick in the relevant box]

   ☐ Male
   ☐ Female

Education

3 How well can you read in your home language?

   ☐ I cannot read
   ☐ Not well
   ☐ Fair
   ☐ Very well
   ☐ Prefer not to answer

4 How well can you write in your home language?

   ☐ I cannot write
   ☐ Not well
   ☐ Fair
   ☐ Very well
   ☐ Prefer not to answer

5 What is the highest level of education that you have completed?

   ☐ No schooling
   ☐ Sub A

- ☐ Sub B
- ☐ Standard 1
- ☐ Standard 2
- ☐ Standard 3
- ☐ Standard 4
- ☐ Standard 5
- ☐ Standard 6
- ☐ Standard 7
- ☐ Standard 8
- ☐ Standard 9
- ☐ Diploma/certificate with less than a Standard 10/Matric certificate
- ☐ Standard 10/ Matric
- ☐ Diploma or certificate (with a Standard 10/Matric certificate)
- ☐ Degree
- ☐ Postgraduate degree or diploma

Income

6 How many people (including you) live in your household? _____ (here, you should include all those people who sleep in the same household as you on a regular basis)

7 How many people aged less than 18 live in your household? _____ (here, you should include all those people who sleep in the same household as you on a regular basis)

8 Are you the main breadwinner in your household?

- ☐ Yes
- ☐ No

9 Thinking about your *own household's financial situation*, would you describe yourself as:

- ☐ Poor
- ☐ Lower income
- ☐ Middle income
- ☐ Upper income
- ☐ Rich

10 What is your household's monthly income? R _____
Do you have a sufficient amount of food in your household?

- ☐ We always have enough food in our household
- ☐ Most of the time we enough food in our household
- ☐ We often do not have enough food in our household
- ☐ We never have enough food in our household

Employment

12 Besides your own farming activities, do you have a job?

☐ Yes
☐ No

13 If yes, what job do you do? _____
What is your monthly income from this job? R _____
Is the job full-time or part-time?

☐ Full-time
☐ Part-time
☐ I do not have a job

16 If you are not working, do you have any other form of income?

☐ Pension: if so, how much do you receive each month? R _____
☐ Child Care Grant: if so, how much do you receive each month?
R _____
☐ Disability Grant: if so, how much do you receive each month?
R _____
☐ Remittances: if so, how much do you receive each month?
R _____

17 In addition to your farming activities and any job that you have already told us about, do you have a part-time job or do you do any activity to earn money for yourself?

☐ Yes: if so, tell us what you do: _____
☐ No

18 How much do you earn each month from this job or activity?
R _____

Farming Activities

19 How many years have you been involved in farming? _____
20 What kind of crops do you grow? _____
21 How much do you earn during a farming season from farming activities?
R _____
22 On what type of land do you grow crops or rear animals?

☐ Land which you or a household member owns
☐ Land which you or a household member has access to as an employee on a commercial farm
☐ A land reform project on state land

☐   An equity share scheme on a commercial farm
☐   Communal land
☐   Land in/near an informal or urban settlement in which the household lives

23   How many hectares is the land that you farm? _____

Climate Change

24   Have you noticed any of the following changes?

Changes in the frequency and timing of rainfall? ☐ Yes ☐ No
Changes in the rainfall level? ☐ Yes ☐ No
Changes in the rainfall intensity? ☐ Yes ☐ No
An increase in temperature? ☐ Yes ☐ No
An increase in the number of pests? ☐ Yes ☐ No

25   Which of the changes have affected your crop yield?

☐   These changes have not affected my crop yield
☐   Changes in the frequency and timing of rainfall
☐   Changes in the level of rainfall
☐   Changes in the rainfall intensity
☐   An increase in the temperature
☐   An increase in the number of pests

26   How has your crop yield been affected?

☐   My yield has increased
☐   My yield has decreased
☐   My yield has been affected

New Farming Practices

27   Please indicate whether you have adopted any of the farming strategies listed here:

[Please tick all the options that apply to you]

☐   I have not adopted any new farming practices
☐   Growing more drought-resistant crops → when: Year: ___ Month: ___
☐   Using improved seeds   → when: Year: ___ Month: ___
☐   Intercropping   → when: Year: ___ Month: ___
☐   Mulching   → when: Year: ___ Month: ___
☐   Applying fertilizer   → when: Year: ___ Month: ___
☐   Applying organic manure   → when: Year: ___ Month: ___
☐   Changing planting dates   → when: Year: ___ Month: ___
☐   Planting wind breaks   → when: Year: ___ Month: ___
☐   Using irrigation   → when: Year: ___ Month: ___
☐   Other: _____   → when: Year: ___ Month: ___

28 If you have adopted new farming practices, how have they affected your yield?

☐ My yield has increased
☐ My yield has decreased
☐ My yield has stayed the same

29 If you have not adopted new farming practices, why have you not?

☐ I do not know what measures to take (or what methods to use)
☐ I do not have the money to adopt these measures
☐ The risk of crop failure is too great
☐ Other: _____

Credit And Insurance

30 Are you a member of a savings group?

☐ Yes
☐ No
☐ I used to

31 If YES, have you contributed this year?

☐ Yes; if so: how much did you contribute this year? R _____
☐ I have not yet contributed
☐ Will not contribute this year
☐ I prefer not to answer

32 If you want to invest in farming equipment or other farming inputs, where do you obtain the money for this?

☐ From my savings
☐ I borrow money from my savings group
☐ I request a loan from the bank
☐ I request a loan from a financial institution
☐ I borrow money from friends and/or relatives
☐ Other; please specify: _____

33 Have you ever applied for a loan from a bank or other formal institution for farming activities?

☐ Yes
☐ No

34 Did you take any bank loans for farming this year?

☐ Yes; if so: how much was requested: R _____

If so: was the loan granted? ☐ Yes ☐ No

☐ No

35 If you have never attempted to borrow money, why have you not?

☐   There are no formal lending institutions
☐   I did not need credit
☐   I dislike any borrowing
☐   The loans are too expensive
☐   I would have like to apply for a loan but did not apply because I felt that the loan would not be granted
☐   Other; if so, please specify: _____

36 Have you heard of insurance?

☐   Yes
☐   No

37 Would you consider purchasing insurance?

☐   Yes
☐   No

Social

38 Which of the following statements describes you the best?

☐   I often take risks
☐   I sometimes take risks
☐   I never take risks

Appendix 8.2

Series 1

Once again please assume that it is planting season. You must decide whether you would like to plant traditional seeds or improved seeds.

This game consists of 14 rows. For each row, you must decide between planting traditional seeds or improved seeds.

Let's do an example [turn to the poster]. Look at row 1:

> Let's start with traditional seeds. The level of rainfall will be enough for a high yield if you draw ball number 1, 2 or 3 out of this bag. If you draw ball number 4, 5, 6, 7, 8, 9 or 10 out of the bag, there is drought. If you planted traditional seeds and there is enough rain for a high yield, your harvest will be worth R20. If you planted traditional seeds and there is a drought, your harvest will be worth R5.
>
> With improved seeds: there will be enough rain for a good harvest if you draw ball number 1 out of this bag. If you draw ball number 2, 3, 4, 5, 6, 7, 8, 9 or 10 out of the bag, there is a drought. If you had planted improved seeds and there is enough rain for a good harvest, your harvest will be worth R34. If you had planted improved seeds and there is a drought, your harvest will be worth R2.50.

Now let's move to row 2:

> Let's start with traditional seeds. There will be enough rain for a good harvest if you draw ball number 1, 2 or 3 out of this bag. If you draw ball number 4, 5, 6, 7, 8, 9 or 10 out of the bag, there is a drought. If you planted traditional seeds and there is enough rain for a good harvest, your harvest will be worth R20. If you planted traditional seeds and there is a drought, your harvest will be worth R5.
>
> With improved seeds: there will be enough rain for a good harvest if you draw ball number 1 out of this bag. If you draw ball number 2, 3, 4, 5, 6, 7, 8, 9 or 10 out of the bag, there is a drought. If you had planted improved seeds and there is enough rain for a good harvest, your harvest will be worth R34.

If you had planted improved seeds and there is a drought, your harvest will be worth R2.50.

Notice that the balls showing whether there is enough rain for a good harvest or whether there is drought stay the same throughout the game. The value of the harvest for planting traditional seeds also stays the same throughout the game. The only thing that changes is the value of the harvest for planting improved seeds when there is enough rain for a good harvest.

In the first row, if you plant improved seeds and there is enough rain for a good harvest, your harvest is worth R34. In the very last row, if you plant improved seeds and there is enough rain for a good harvest, you harvest is worth R850.

Remember, because the payoffs are so high for this game, if this game is chosen to be played for real money, two of you will randomly be chosen to play the game for money. We don't know who those two of you will be, so it is important to play this game as if you are playing for real money.

Just like before, we won't play all the rows for money. Once you have made your decisions, one of you will draw a ball from this bag which has 14 balls inside it.

		Traditional seeds	Improved seeds
1	🌢🌢	R20 if ❶❷❸	R34 if ❶
	🌢	R5 if ❹❺❻❼❽❾❿	R2.5 if ❷❸❹❺❻❼❽❾❿
2	🌢🌢	R20 if ❶❷❸	R37.5 if ❶
	🌢	R5 if ❹❺❻❼❽❾❿	R2.5 if ❷❸❹❺❻❼❽❾❿
3	🌢🌢	R20 if ❶❷❸	R41.5 if ❶
	🌢	R5 if ❹❺❻❼❽❾❿	R2.5 if ❷❸❹❺❻❼❽❾❿
4	🌢🌢	R20 if ❶❷❸	R46.5 if ❶
	🌢	R5 if ❹❺❻❼❽❾❿	R2.5 if ❷❸❹❺❻❼❽❾❿
5	🌢🌢	R20 if ❶❷❸	R53 if ❶
	🌢	R5 if ❹❺❻❼❽❾❿	R2.5 if ❷❸❹❺❻❼❽❾❿
6	🌢🌢	R20 if ❶❷❸	R62.5 if ❶
	🌢	R5 if ❹❺❻❼❽❾❿	R2.5 if ❷❸❹❺❻❼❽❾❿
7	🌢🌢	R20 if ❶❷❸	R75 if ❶
	🌢	R5 if ❹❺❻❼❽❾❿	R2.5 if ❷❸❹❺❻❼❽❾❿
8	🌢🌢	R20 if ❶❷❸	R92.5 if ❶
	🌢	R5 if ❹❺❻❼❽❾❿	R2.5 if ❷❸❹❺❻❼❽❾❿
9	🌢🌢	R20 if ❶❷❸	R110 if ❶
	🌢	R5 if ❹❺❻❼❽❾❿	R2.5 if ❷❸❹❺❻❼❽❾❿
10	🌢🌢	R20 if ❶❷❸	R150 if ❶
	🌢	R5 if ❹❺❻❼❽❾❿	R2.5 if ❷❸❹❺❻❼❽❾❿
11	🌢🌢	R20 if ❶❷❸	R200 if ❶
	🌢	R5 if ❹❺❻❼❽❾❿	R2.5 if ❷❸❹❺❻❼❽❾❿
12	🌢🌢	R20 if ❶❷❸	R300 if ❶
	🌢	R5 if ❹❺❻❼❽❾❿	R2.5 if ❷❸❹❺❻❼❽❾❿
13	🌢🌢	R20 if ❶❷❸	R500 if ❶
	🌢	R5 if ❹❺❻❼❽❾❿	R2.5 if ❷❸❹❺❻❼❽❾❿
14	🌢🌢	R20 if ❶❷❸	R850 if ❶
	🌢	R5 if ❹❺❻❼❽❾❿	R2.5 if ❷❸❹❺❻❼❽❾❿

This will tell us which row you are playing for money. If ball number 1 is drawn from the bag, you will play row 1 for money. If ball number 2 is drawn from the bag, you will play row 2 for money. If ball number 14 is drawn from the bag, you will play row 14 for money.

Does anyone have any questions before we start?

Let's start. Please write the number we gave you at the start of the experiment on the top left-hand side of the sheet where it says experiment number [gesture to where they must put their number].

For each row in the sheet in front of you, indicate whether you would like to plant traditional seeds or improved seeds.

Answer

I choose traditional seeds for rows 1-

I choose improved seeds for rows – 14

Series 2

This game works exactly the same as the previous game. Once again please assume that it is planting season. You must decide whether you would like to plant traditional seeds or improved seeds. This game also consists of 14 rows. For each row, you must decide between planting traditional seeds or improved seeds.

Let's do an example [turn to the poster]. Look at row 1:

> Let's start with traditional seeds. There will be enough rain for a good yield if you draw ball number 1, 2, 3, 4, 5, 6, 7, 8 or 9 out of this bag. If you draw ball number 10 out of the bag, there is a drought. If you planted traditional seeds and there is enough rain for a good yield, your harvest will be worth R20. If you planted traditional seeds and there is a drought, your harvest will be worth R15.
>
> With improved seeds: there will be enough rain for a good yield if you draw ball number 1, 2, 3, 4, 5, 6 or 7 out of this bag. If you draw ball number 8, 9 or 10 out of the bag, there will be a drought. If you had planted improved seeds and there is enough rain for a good yield, your harvest will be worth R27. If you had planted improved seeds and there is a drought, your harvest will be worth R2.50.

Now let's move to row 2:

> Let's start with traditional seeds. There will be enough rain for a good yield if you draw ball number 1, 2, 3, 4, 5, 6, 7, 8 or 9 out of this bag. If you draw ball number 10 out of the bag, there is a drought. If you planted traditional seeds and there is enough rain for a good yield, your harvest will be worth

R20. If you planted traditional seeds and there is a drought, your harvest will be worth R15.

With improved seeds: there will be enough rain for a good yield if you draw ball number 1, 2, 3, 4, 5, 6 or 7 out of this bag. If you draw ball number 8, 9 or 10 out of the bag, there is a drought. If you had planted improved seeds and there is enough rain for a good yield, your harvest will be worth R28. If you had planted improved seeds and there is a drought, your harvest will be worth R2.50.

Notice that the balls showing whether there is enough rain for a good harvest or whether there is drought stay the same throughout the game. The value of the harvest for planting traditional seeds also stays the same throughout the game. The only thing that changes is the value of the harvest for planting improved seeds when there is enough rain for a good yield.

In the first row, if you plant improved seeds and there is enough rain for a high yield, your harvest is worth R27. In the very last row, if you plant improved seeds and there is enough rain for a high yield, your harvest is worth R65.

		Traditional seeds	Improved seeds
1	◌◌ / ◌	R20 if ❶❷❸❹❺❻❼❽❾ R15 if ⑩	R27 if ❶❷❸❹❺❻❼ R2.5 if ❽❾⑩
2	◌◌ / ◌	R20 if ❶❷❸❹❺❻❼❽❾ R15 if ⑩	R28 if ❶❷❸❹❺❻❼ R2.5 if ❽❾⑩
3	◌◌ / ◌	R20 if ❶❷❸❹❺❻❼❽❾ R15 if ⑩	R29 if ❶❷❸❹❺❻❼ R2.5 if ❽❾⑩
4	◌◌ / ◌	R20 if ❶❷❸❹❺❻❼❽❾ R15 if ⑩	R30 if ❶❷❸❹❺❻❼ R2.5 if ❽❾⑩
5	◌◌ / ◌	R20 if ❶❷❸❹❺❻❼❽❾ R15 if ⑩	R31 if ❶❷❸❹❺❻❼ R2.5 if ❽❾⑩
6	◌◌ / ◌	R20 if ❶❷❸❹❺❻❼❽❾ R15 if ⑩	R32.5 if ❶❷❸❹❺❻❼ R2.5 if ❽❾⑩
7	◌◌ / ◌	R20 if ❶❷❸❹❺❻❼❽❾ R15 if ⑩	R34 if ❶❷❸❹❺❻❼ R2.5 if ❽❾⑩
8	◌◌ / ◌	R20 if ❶❷❸❹❺❻❼❽❾ R15 if ⑩	R36 if ❶❷❸❹❺❻❼ R2.5 if ❽❾⑩
9	◌◌ / ◌	R20 if ❶❷❸❹❺❻❼❽❾ R15 if ⑩	R38.5 if ❶❷❸❹❺❻❼ R2.5 if ❽❾⑩
10	◌◌ / ◌	R20 if ❶❷❸❹❺❻❼❽❾ R15 if ⑩	R41.5 if ❶❷❸❹❺❻❼ R2.5 if ❽❾⑩
11	◌◌ / ◌	R20 if ❶❷❸❹❺❻❼❽❾ R15 if ⑩	R45 if ❶❷❸❹❺❻❼ R2.5 if ❽❾⑩
12	◌◌ / ◌	R20 if ❶❷❸❹❺❻❼❽❾ R15 if ⑩	R50 if ❶❷❸❹❺❻❼ R2.5 if ❽❾⑩
13	◌◌ / ◌	R20 if ❶❷❸❹❺❻❼❽❾ R15 if ⑩	R55 if ❶❷❸❹❺❻❼ R2.5 if ❽❾⑩
14	◌◌ / ◌	R20 if ❶❷❸❹❺❻❼❽❾ R15 if ⑩	R65 if ❶❷❸❹❺❻❼ R2.5 if ❽❾⑩

Just like before, we won't play all the rows for money. Once you have made your decisions, one of you will draw a ball from this bag which has 14 balls inside it. This will tell us which row you are playing for money. If ball number 1 is drawn from the bag, you will play row 1 for money. If ball number 2 is drawn from the bag, you will play row 2 for money. If ball number 14 is drawn from the bag, you will play row 14 for money.

Does anyone have any questions before we start?

Let's start. Please write the number we gave you at the start of the experiment on the top left-hand side of the sheet where it says experiment number [gesture to where they must put their number].

For each row in the sheet in front of you, indicate whether you would like to plant traditional seeds or improved seeds.

Answer

I choose traditional seeds for rows 1 –

I choose improved seeds for rows – 14

Series 3

This game works exactly the same as the previous game.

Once again please assume that it is planting season. You must decide whether you would like to plant traditional seeds or improved seeds.

This game consists of 7 rows. For each row, you must decide between planting traditional seeds or improved seeds.

The difference in this game is that, now, you can lose money. Any money you lose will be taken from your earnings for this session.

Let's do an example [turn to the poster]. Look at row 1:

Let's start with traditional seeds. There is enough rain for a good yield if you draw ball number 1, 2, 3, 4 or 5 out of this bag. If you draw ball number 6, 7, 8, 9 or 10 out of the bag, there is a drought. If you planted traditional seeds and there is enough rain for a good yield, your harvest will be worth R12.50. If you planted traditional seeds and there is a drought, you will lose R2.

With improved seeds: there is enough rain for a good yield if you draw ball number 1, 2, 3, 4 or 5 out of this bag. If you draw ball number 6, 7, 8, 9 or 10 out of the bag, there is a drought. If you planted improved seeds and there is enough rain for a good yield, your harvest will be worth R15. If you planted improved seeds and there is a drought, you will lose R10.50.

Now let's move to row 2:

Let's start with traditional seeds. There is enough rain for a good yield if you draw ball number 1, 2, 3, 4 or 5 out of this bag. If you draw ball number 6, 7,

8, 9 or 10 out of the bag, there is a drought. If you planted traditional seeds and there is enough rain for a good yield, your harvest will be worth R12.50. If you planted traditional seeds and there is a drought, you will lose R2.

With improved seeds: There is enough rain for a good yield if you draw ball number 1, 2, 3, 4 or 5 out of this bag. If you draw ball number 6, 7, 8, 9 or 10 out of the bag, there is a drought. If you planted improved seeds and there is enough rain for a good yield, your harvest will be worth R15. If you planted improved seeds and there is a drought, you will lose R10.50.

Just like before, we won't play all the rows for money. Once you have made your decisions, one of you will draw a ball from this bag, which has 7 balls inside it. This will tell us which row you are playing for money. If ball number 1 is drawn from the bag, you will play row 1 for money. If ball number 2 is drawn from the bag, you will play row 2 for money. If ball number 7 is drawn from the bag, you will play row 7 for money.

Does anyone have any questions before we start?

Let's start. Please write the number we gave you at the start of the experiment on the top left-hand side of the sheet where it says experiment number [gesture to where they must put their number].

For each row in the sheet in front of you, indicate whether you would like to plant traditional seeds or improved seeds.

	Rainfall	Traditional seeds	Improved seeds
1	△△	R12.5 if **1 2 3 4 5**	R15 if **1 2 3 4 5**
	△	-R2 if **6 7 8 9 10**	-R10 if **6 7 8 9 10**
2	△△	R2 if **1 2 3 4 5**	R15 if **1 2 3 4 5**
	△	-R2 if **6 7 8 9 10**	-R10 if **6 7 8 9 10**
3	△△	R0.5 if **1 2 3 4 5**	R15 if **1 2 3 4 5**
	△	-R2 if **6 7 8 9 10**	-R10 if **6 7 8 9 10**
4	△△	R0.5 if **1 2 3 4 5**	R15 if **1 2 3 4 5**
	△	-R2 if **6 7 8 9 10**	-R8 if **6 7 8 9 10**
5	△△	R0.5 if **1 2 3 4 5**	R15 if **1 2 3 4 5**
	△	-R4 if **6 7 8 9 10**	-R8 if **6 7 8 9 10**
6	△△	R0.5 if **1 2 3 4 5**	R15 if **1 2 3 4 5**
	△	-R4 if **6 7 8 9 10**	-R7 if **6 7 8 9 10**
7	△△	R0.5 if **1 2 3 4 5**	R15 if **1 2 3 4 5**
	△	-R4 if **6 7 8 9 10**	-R5.5 if **6 7 8 9 10**

Answer

I choose traditional seeds for rows 1 – [   ]

I choose improved seeds for rows – 7 [   ]

Appendix 8.3

Table 8.A1 Summary of Payoffs

Switching rows	Rainfall level	Series 1				Series 2				Series 3			
		Traditional seed	Probability	Improved seed	Probability	Traditional seeds	Probability	Improved seeds	Probability	Traditional seeds	Probability	Improved seeds	Probability
1	Good	R 20.00	0.3	R 34.00	0.1	R 20.00	0.9	R 27.00	0.7	R 12.50	0.5	R 15.00	0.5
	Bad	R 5.00	0.7	R 2.50	0.9	R 15.00	0.1	R 2.50	0.3	-R 2.00	0.5	-R 10.00	0.5
2	Good	R 20.00	0.3	R 37.50	0.1	R 20.00	0.9	R 28.00	0.7	R 2.00	0.5	R 15.00	0.5
	Bad	R 5.00	0.7	R 2.50	0.9	R 15.00	0.1	R 2.50	0.3	-R 2.00	0.5	-R 10.00	0.5
3	Good	R 20.00	0.3	R 41.50	0.1	R 20.00	0.9	R 29.00	0.7	R 0.50	0.5	R 15.00	0.5
	Bad	R 5.00	0.7	R 2.50	0.9	R 15.00	0.1	R 2.50	0.3	-R 2.00	0.5	-R 10.00	0.5
4	Good	R 20.00	0.3	R 46.50	0.1	R 20.00	0.9	R 30.00	0.7	R 0.50	0.5	R 15.00	0.5
	Bad	R 5.00	0.7	R 2.50	0.9	R 15.00	0.1	R 2.50	0.3	-R 2.00	0.5	-R 8.00	0.5
5	Good	R 20.00	0.3	R 53.00	0.1	R 20.00	0.9	R 31.00	0.7	R 0.50	0.5	R 15.00	0.5
	Bad	R 5.00	0.7	R 2.50	0.9	R 15.00	0.1	R 2.50	0.3	-R 4.00	0.5	-R 8.00	0.5
6	Good	R 20.00	0.3	R 62.50	0.1	R 20.00	0.9	R 32.50	0.7	R 0.50	0.5	R 15.00	0.5
	Bad	R 5.00	0.7	R 2.50	0.9	R 15.00	0.1	R 2.50	0.3	-R 4.00	0.5	-R 7.00	0.5
7	Good	R 20.00	0.3	R 75.00	0.1	R 20.00	0.9	R 34.00	0.7	R 0.50	0.5	R 15.00	0.5
	Bad	R 5.00	0.7	R 2.50	0.9	R 15.00	0.1	R 2.50	0.3	-R 4.00	0.5	-R 5.50	0.5
8	Good	R 20.00	0.3	R 92.50	0.1	R 20.00	0.9	R 36.00	0.7	—	—	—	—
	Bad	R 5.00	0.7	R 2.50	0.9	R 15.00	0.1	R 2.50	0.3	—	—	—	—

(Continued)

Table 8.A1 (Continued)

Switching rows	Rainfall level	Series 1				Series 2				Series 3			
		Traditional seed	Probability	Improved seed	Probability	Traditional seeds	Probability	Improved seeds	Probability	Traditional seeds	Probability	Improved seeds	Probability
9	Good	R 20.00	0.3	R 110.00	0.1	R 20.00	0.9	R 38.00	0.7	–	–	–	–
	Bad	R 5.00	0.7	R 2.50	0.9	R 15.00	0.1	R 2.50	0.3	–	–	–	–
10	Good	R 20.00	0.3	R 150.00	0.1	R 20.00	0.9	R 41.00	0.7	–	–	–	–
	Bad	R 5.00	0.7	R 2.50	0.9	R 15.00	0.1	R 2.50	0.3	–	–	–	–
11	Good	R 20.00	0.3	R 200.00	0.1	R 20.00	0.9	R 45.00	0.7	–	–	–	–
	Bad	R 5.00	0.7	R 2.50	0.9	R 15.00	0.1	R 2.50	0.3	–	–	–	–
12	Good	R 20.00	0.3	R 300.00	0.1	R 20.00	0.9	R 50.00	0.7	–	–	–	–
	Bad	R 5.00	0.7	R 2.50	0.9	R 15.00	0.1	R 2.50	0.3	–	–	–	–
13	Good	R 20.00	0.3	R 500.00	0.1	R 20.00	0.9	R 55.00	0.7	–	–	–	–
	Bad	R 5.00	0.7	R 2.50	0.9	R 15.00	0.1	R 2.50	0.3	–	–	–	–
14	Good	R 20.00	0.3	R 850.00	0.1	R 20.00	0.9	R 60.00	0.7	–	–	–	–
	Bad	R 5.00	0.7	R 2.50	0.9	R 15.00	0.1	R 2.50	0.3	–	–	–	–

Source: Brick and Visser (2010).

Notes

1 See Appendix 8.1 for sample questionnaire.
2 In August 2017, 1 South African Rand = USD .077.
3 A sample of the multiple price lists is presented in Appendix 8.2.
4 See Table 8.2 for the summary statistics of all the uptake options in the survey.

References

Adato, M., M.R. Carter and J. May. 2006. Exploring Poverty Traps and Social Exclusion in South Africa Using Qualitative and Quantitative Data. *Journal of Development Studies* 42(2): 226–247.

Barham, B.L., J. Chavas, D. Fitz, V.R. Salas and L. Schechter. 2014. The Roles of Risk and Ambiguity in Technology Adoption. *Journal of Economic Behaviour and Organization* 97: 204–218.

Barr, A. 2003. *Risk Pooling, Commitment, and Information: An Experimental Test of Two Fundamental Assumptions.* Oxford: University of Oxford: Centre for the Study of African Economies.

Barrett, C.B., and M.R. Carter. 2013. The Economics of Poverty Traps and Persistent Poverty: Empirical and Policy Implications. *Journal of Development Studies* 49(7): 976–990.

Binswanger, H.P., and D.A. Sillers. 1983. Risk Aversion and Credit Constraints in Farmers' Decision Making: A Reinterpretation. *Journal of Development Studies* 20(1): 5–21.

Brick, K., and M. Visser. 2010. Risk Preferences in South African Farmers: Experimental Evidence. Environment Economics Policy Research Unit. Working Paper.

Brick, K., and M. Visser. 2015. Risk Preferences, Technology Adoption and Insurance Uptake: A Framed Experiment. *Journal of Economic Behaviour and Organization* 118: 383–396.

Carter, M.R., and C.B. Barrett. 2006. The Economics of Poverty Traps and Persistent Poverty: An Asset-Based Approach. *Journal of Development Studies* 42(2): 178–199.

Carter, M.R., and C.B. Barrett. 2007. Asset Thresholds and Social Protection: A 'Think-Piece'. *IDS Bulletin* 38(3): 34–38.

Croppenstedt, A., M. Demeke and M.M. Meschi. 2003. Technology Adoption in the Presence of Constraints: The Case of Fertilizer Demand in Ethiopia. *Review of Development Economics* 7(1): 58–70.

Croppenstedt, A., M. Goldstein and N. Rosas. 2013. Gender and Agriculture: Inefficiencies, Segregation and Low Productivity Traps. Policy Working Paper No. 6370. World Bank, Development Economics Vice-Presidency, Partnerships, Capacity Building Unit.

Deressa, T.T., R.M. Hassan, C. Ringler, T. Alemu and M. Yesuf. 2009. Determinants of Farmers' Choice of Adaptation Methods to Climate Change in the Nile Basin of Ethiopia. *Global Environmental Change* 19(2): 248–255.

Di Falco, S. 2014. Adaptation to Climate Change in Sub-Saharan Agriculture: Assessing the Evidence and Rethinking the Drivers. *European Review of Agricultural Economics* 41(3): 405–430.

Doss, C.R., and M.L. Morris. 2001. How Does Gender Affect the Adoption of Agricultural Innovations? The Case of Improved Maize Technology in Ghana. *Agricultural Economics* 25(1): 27–39.

Duflo, E., M. Kremer and J. Robinson. 2011. Nudging Farmers to Use Fertilizer: Theory and Experimental Evidence from Kenya. *American Economic Review* 101(6): 2350–2390.

Engle-Warnick, J., J. Escobal and S. Laszlo. 2007. Ambiguity Aversion as a Predictor of Technology Choice: Experimental Evidence from Peru. CIRANO-Scientific.

Etwire, E., A. Ariyawardana and M.Y. Mortlock. 2016. Seed Delivery Systems and Farm Characteristics Influencing the Improved Seed Uptake by Smallholders in Northern Ghana. *Sustainable Agriculture Research* 5(2): 27.

Feder, G., R.E. Just and D. Zilberman. 1985. Adoption of Agricultural Innovations in Developing Countries: A Survey. *Economic Development and Cultural Change* 33: 255–298.

Fernandez-Cornejo, J., S. Daberkow and W.D. McBride. 2002. Decomposing the Size Effect on the Adoption of Innovations. *Agbioforum* 4(2): 2.

Foster, A.D., and M.R. Rosenzweig. 2010. Microeconomics of Technology Adoption. *Annual Review of Economics* 2(1): 395–424.

Githinji, M., C. Konstantinidis and A. Barenberg. 2011. Small and as Productive: Female-Headed Households and the Inverse Relationship between Land Size and Output in Kenya. Department of Economics Working Paper No. 2011–31. Amherst: University of Massachusetts.

Hill, R.E. 2005. Risk, Production and Poverty: A Study of Coffee in Uganda. Doctoral Dissertation. University of Oxford, Oxford, UK.

Holt, C.A., and S.K. Laury. 2002. Risk Aversion and Incentive Effects. *American Economic Review* 92(5): 1644–1655.

Humphrey, S.J., and A. Verschoor. 2004. The Probability Weighting Function: Experimental Evidence from Uganda, India and Ethiopia. *Economics Letters* 84(3): 419–425.

Isik, M., and M. Khanna. 2003. Stochastic Technology, Risk Preferences, and Adoption of Site-Specific Technologies. *American Journal of Agricultural Economics* 85(2): 305–317.

Jayawarna, D., J. Rouse and J. Kitching. 2013. Entrepreneur Motivations and Life Course. *International Small Business Journal* 31(1): 34–56.

Kahneman, D., and A. Tversky. 1979. Prospect Theory: An Analysis of Decision under Risk. *Econometrica: Journal of the Econometric Society* 47(2): 263–291.

Knowler, D., and B. Bradshaw. 2007. Farmers' Adoption of Conservation Agriculture: A Review and Synthesis of Recent Research. *Food Policy* 32(1): 25–48.

Liu, E., and J. Huang. 2010. Risk Preferences and Pesticide Use by Cotton Farmers in China. *Journal of Development Economics* 103: 202–215.

Liu, E.M. 2013. Time to Change What to Sow: Risk Preferences and Technology Adoption Decisions of Cotton Farmers in China. *Review of Economics and Statistics* 95(4): 1386–1403.

Marra, M., D.J. Pannell and A. Abadi Ghadim. 2003. The Economics of Risk, Uncertainty and Learning in the Adoption of New Agricultural Technologies: Where Are We on the Learning Curve? *Agricultural Systems* 75(2): 215–234.

Meinzen-Dick, R., A. Quisumbing, J. Behrman, P. Biermayr-Jenzano, V. Wilde, M. Noordeloos, C. Ragasa and N. Beintema. 2010. Engendering Agricultural Research. IFPRI Discussion Paper No. 973. Washington, DC: International Food Policy Research Institute.

Monela, A.G. 2014. Access to and Adoption of Improved Seeds by Smallholder Farmers in Tanzania: Cases of Maize and Rice Seeds in Mbeya and Morogoro Regions. Doctoral Dissertation, Sokoine University of Agriculture, Morogoro, Tanzania.

Ndiritu, S.W., M. Kassie and B. Shiferaw. 2014. Are There Systematic Gender Differences in the Adoption of Sustainable Agricultural Intensification Practices? Evidence from Kenya. *Food Policy* 49: 117–127.

Nhemachena, C., and R. Hassan. 2007. *Micro-Level Analysis of Farmers' Adaption to Climate Change in Southern Africa.* Washington, DC: International Food Policy Research Institute.

Odame, H.H., N. Hafkin, G. Wesseler and I. Boto. 2002. Gender and Agriculture in the Information Society. ISNAR Briefing Paper No. 55. The Hague: International Service for National Agricultural Research.

Panda, A., U. Sharma, K.N. Ninan and A. Patt. 2013. Adaptive Capacity Contributing to Improved Agricultural Productivity at the Household Level: Empirical Findings Highlighting the Importance of Crop Insurance. *Global Environmental Change* 23(4): 782–790.

Patt, A.G., and D. Schröter. 2008. Perceptions of Climate Risk in Mozambique: Implications for the Success of Adaptation Strategies. *Global Environmental Change* 18(3): 458–467.

Prelec, D. 1998. The Probability Weighting Function. *Econometrica* 66(3): 497–527.

Quisumbing, A.R. 1995. Gender Differences in Agricultural Productivity: Survey of Empirical Evidence. FCND Discussion Paper No. 5. Washington, DC: International Food Policy Research Institute, Food Consumption and Nutrition Division.

Quisumbing, A.R., and L. Pandolfelli. 2010. Promising Approaches to Address the Needs of Poor Female Farmers: Resources, Constraints, and Interventions. *World Development* 38(4): 581–592.

Ranjan, R., and J.F. Shogren. 2006. How Probability Weighting Affects Participation in Water Markets. *Water Resources Research* 42(8).

Rouse, J., and J. Kitching. 2006. Do Enterprise Support Programmes Leave Women Holding the Baby? *Environment and Planning C* 24(1): 5.

Simtowe, F. 2006. Can Risk Aversion towards Fertilizer Explain Part of the Non-Adoption Puzzle for Hybrid Maize? Empirical Evidence from Malawi. *Journal of Applied Sciences* 6(7): 1490–1498.

Smit, B., and M.W. Skinner. 2002. Adaptation Options in Agriculture to Climate Change: A Typology. *Mitigation and Adaptation Strategies for Global Change* 7(1): 85–114.

Sunding, D., and D. Zilberman. 2001. The Agricultural Innovation Process: Research and Technology Adoption in a Changing Agricultural Sector. In *Handbook of Agricultural Economics*, Volume 1, edited by R. Evenson and P. Pingali. pp. 207–261. Amsterdam, Netherlands: Elsevier Science and Technology.

Tanaka, T., C. Camerer and Q. Nguyen. 2010. Risk and Time Preferences: Linking Experimental and Household Survey Data from Vietnam. *American Economic Review* 100(1): 557–571.

Thaler, R.H., A. Tversky, D. Kahneman and A. Schwartz. 1997. The Effect of Myopia and Loss Aversion on Risk Taking: An Experimental Test. *The Quarterly Journal of Economics*: 647–661.

Tversky, A., and D. Kahneman. 1992. Advances in Prospect Theory: Cumulative Representation of Uncertainty. *Journal of Risk and Uncertainty* 5(4): 297–323.

Ward, P.S., and V. Singh. 2013. Risk and Ambiguity Preferences and the Adoption of New Agricultural Technologies: Evidence from Field Experiments in Rural India. Paper Selected for Presentation at the Agricultural and Applied Economics Association 2013 AAEA & CAES Joint Annual Meeting, Washington, DC.

Wik, M., T.A. Kebede, O. Bergland and S.T. Holden. 2004. On the Measurement of Risk Aversion from Experimental Data. *Applied Economics* 36: 2443–2451.

World Bank. 2001. Engendering Development. A World Bank Policy Research Report. Washington, DC: World Bank.

Yesuf, M., and A. Bluffstone. 2009. Poverty, Risk Aversion and Path Dependence in Low-Income Countries: Experimental Evidence from Ethiopia. *American Journal of Agricultural Economics* 91(4): 1022–1037.

9   Good things come in packages

Sustainable intensification systems in smallholder agriculture

Cyndi Spindell Berck and Hailemariam Teklewold

Introduction

Agricultural production has increased in sub-Saharan Africa in recent decades. Despite the fact that all land in that region that is classified as very suitable for cultivation is already under cultivation, farmers are still increasing their agricultural production by utilizing marginal lands, converting communal grazing lands to arable lands, or reducing or eliminating fallow periods (Pretty et al., 2010). Agricultural extensification (clearing more land for agriculture) compounds the problem of degrading the natural resource base of production and the associated loss of carbon sequestration potential, while decreasing fallow periods degrades the land's productivity (Teklewold, Kassie, and Shiferaw, 2013). By contrast, sustainable agricultural intensification has been defined as producing more outputs with more efficient use of all inputs on a durable basis – that is, while reducing environmental damage and building resilience, natural capital, and the flow of environmental services (Pretty et al., 2010; Kassie, Teklewold, Marenya, et al., 2015). With the population growing, agricultural intensification is needed to improve food security in Africa, even without consideration of climate change. There has been much effort in sub-Saharan Africa to intensify farming systems to increase the productivity of land already under cultivation. With rising average temperatures and increasing uncertainty about rainfall patterns, sustainable agricultural intensification becomes even more important to maximize the productivity of agricultural lands.

Agricultural production can be intensified by practices and inputs that are neutral, negative, or positive in their environmental impacts. These impacts include both the external effects on the environment and concerns about sustainability of agricultural production. For instance, regarding the environmental effects of productivity-enhancing technologies, Kassie, Teklewold, Jaleta, et al. (2015) summarize the trade-offs experienced during the "Green Revolution" in Asia: high-yielding seeds, fertilizer, and irrigation may lead to groundwater depletion, soil fertility degradation, and chemical runoff. Such technologies will not increase farm income and food security, at least not for very long, if the result is depletion of nutrients and groundwater (Kassie, Teklewold, Jaleta, et al., 2015). However, sustainable intensification of agricultural production is central to achieving sustainable agricultural development by halting environmental degradation, conserving the natural resource base while increasing agricultural productivity, and increasing resilience and adaptation of farmers to climate variability and change (Pretty et al., 2010).

Sustainable intensification practices are commonly used in combination under smallholder farming conditions. In many cases, adoption of a system, or package, improves yields more than adoption of individual practices (Nyangena and Ogada, 2014, for Kenya; Kassie, Teklewold, Marenya, et al., 2015, for Malawi; Teklewold, Kassie, and Shiferaw, 2013, for Ethiopia). "Packages" means a combination of practices. Under some circumstances, introducing a combination of sustainable intensification practices into a farming system can reduce the quantity of external inputs (inorganic fertilizer and/or pesticides), which are costly and affect the environment, while maintaining or increasing yield (Teklewold, Kassie, and Shiferaw, 2013).

This chapter, therefore, will summarize some recent research on smallholder farmers' adoption of sustainable intensification systems that have the potential to increase productivity while taking into account environmental effects. The chapter focuses on research in Ethiopia, with some attention to Kenya, Malawi, and Tanzania.

This chapter emphasizes on-farm decisions made at the farm household level, for two reasons. First, the suitability of practices varies greatly depending on agroecological conditions and the mix of agricultural products (including the type of crop and the role of mixed crop-livestock systems), all the way down to farm-level conditions (soil quality and steep versus flat plots) and household characteristics (e.g., education and gender of the head of household). Second, these adoption decisions are within the control of individual farm households, with a large role for public policy to reduce barriers to farmers' choices.

This chapter is organized as follows. The next section presents evidence that adoption of a sustainable intensification system can improve productivity, manage production risk, and address the implications for the environment (e.g., through reduction of the quantity of agrochemicals used). The third section discusses farmers' decisions to adopt interdependent practices (i.e., practices that are complements or substitutes) as part of sustainable intensification packages. The chapter concludes with a discussion of some policy implications.

Intensification practices and packages

Examples of intensification practices

The inputs and practices that can intensify productivity include improved seed varieties; inorganic fertilizer; manure; pesticides, including insecticides and herbicides/weed-killers; diversification of agricultural production, including rotation of maize and legumes and intercropping of maize and legumes at the same time; soil conservation practices, including conservation tillage; and other soil and water conservation practices. Soil and water conservation practices include terracing and bunds (barriers against runoff and erosion that can be made of soil or stone). This is not an exhaustive list.

We begin by discussing three commercially purchased inputs: improved seed varieties, inorganic fertilizer, and pesticides. Improved seeds and inorganic fertilizers have been promoted in a number of sub-Saharan African countries in various ways: through subsidies in Malawi and Tanzania, through a deregulated private sector in Kenya, and through low-interest credit in Ethiopia (Kassie, Teklewold, Jaleta, et al., 2015). While inorganic fertilizer and pesticides are not thought of as "sustainable" inputs, their use in optimal quantities may be part of a sustainable intensification system.

The arrival of climate change has spurred interest in water-efficient or drought- and heat-tolerant crop varieties better suited to a warmer and drier climate. Improved varieties of maize (and other crops) may be higher in yield, more tolerant of heat and drought, and/or more resistant to pests. Maize is the main staple in Kenya, Tanzania, and Malawi, and is important in Ethiopia (Kassie, Teklewold, Jaleta, et al., 2015). In Malawi, the area of land farmed with improved maize varieties was positively correlated with own maize consumption, income, and asset holdings (Bezu et al., 2014). Not surprisingly, different results have been found for different types of improved seeds, and the results also depend on the context in which they are used (Nyangena and Ogada, 2014). In Kenya between 1992 and 2002, the use of improved maize seeds did not increase yields, suggesting that some local varieties could perform as well as the improved varieties in some circumstances (De Groote et al., 2005; Nyangena and Ogada, 2014).

Cereal crops, such as maize, deplete nitrogen, phosphorus, and/or potassium in the soil. Inorganic fertilizer and organic fertilizer (e.g., manure and compost) provide these nutrients. One or the other or both are essential for cereal farming, although, as discussed ahead, lower quantities may be needed if crop diversification and soil conservation practices are used.

Inorganic fertilizer has been shown to improve yield in many circumstances, but it does increase risk. While average returns from inorganic fertilizer are good, there is variation in yield across different agroecologies. For instance, chemical fertilizers provided a greater crop return per hectare in wetter areas (Alem et al., 2010; Teklewold, Kassie, and Shiferaw, 2013; Teklewold et al., 2017). Contrasting findings in Kenya suggest socioeconomic differences, although it is not certain whether wealthier or poorer farmers benefit more. Marenya and Barrett (2009) found that inorganic fertilizer application was beneficial to farmers with high soil organic matter, suggesting to Nyangena and Ogada (2014) that degraded soils, and the poorer farmers who farm poor soil, are less able to benefit from fertilizer. This indicates the relationship between inorganic and organic fertilizer for replenishing the soil and maintaining the soil organic matter level (Place et al., 2003). Conversely, Nyangena and Ogada's study in Kenya found a positive effect of "partial adoption" of fertilizer-seed packages only in the lower quartile of yield and suggested that the lowest-yielding/poorest farmers should be targeted to promote new technology adoption, especially of packages. However, the results of

Nyangena and Ogada (2014) are not directly comparable to those of Marenya and Barrett (2009) because "partial adoption" was not specific to fertilizer.

An advantage of manure relative to inorganic fertilizer is its ready availability in a mixed crop-livestock system, without the need for cash or credit. It also adds organic matter to the soil. A disadvantage compared to commercially prepared fertilizer is that manure doesn't necessarily provide the right mix of nutrients for optimal crop growth. While manure has historically been considered to be an important resource, its competing use is for household fuel (Erkossa and Teklewold, 2009). This illustrates the relationship between agricultural intensification and overall development, in that many subsistence farmers still rely on biomass instead of alternatives, such as natural gas. The availability of manure (as well as the availability of other products from livestock) can be increased through high-yielding livestock breeds and improved forage legumes (Teklewold, Kassie, and Shiferaw, 2013).

Pesticides (insecticides and herbicides/weed-killers) may be used at various times in the cropping cycle, including postharvest. Some of the environmentally friendly practices discussed ahead may reduce the need for pesticides by controlling weeds and insects in other ways.

We now turn to practices that are generally considered sustainable. Diversification, soil conservation (including conservation tillage), and water management are sometimes considered together as "integrated soil fertility management."

The diversification strategies considered in this chapter are diversification over time (i.e., crop rotation between maize and legumes) and diversification at the same time (intercropping). Because legumes are nitrogen-fixing plants, maize-legume rotation or intercropping restores nitrogen to the soil (again, not necessarily in the same quantities as provided by purchased fertilizer). Diversification can improve the water-holding capacity of the soil and its organic content (Kassie, Teklewold, Jaleta, et al., 2015). Crop rotation also can disrupt pests and disease, which are likely to become more of a problem due to the warming climate. Crop diversification can contribute to control of pests, diseases, and weeds (Kassie, Teklewold, Jaleta, et al., 2015).

As a form of diversification, crop rotation can be a risk management mechanism, in that one crop may provide food or income, even if another crop fails or there is a fluctuation in the market price for a crop. One way in which diversification manages risk is through growing crops with different environmental needs and stress responses (Kassie, Teklewold, Jaleta, et al., 2015). For instance, if weather or rainfall for one crop is unfavorable in a given year, the intercropped crop might do better.

In some cases, crop rotation can take advantage of different seasonal needs for different crops, so that a farmer can harvest more than one type of crop in a year. In such cases, crop rotation can help smooth out the demands on a household's resources, such as labor (planting, harvesting, preparing for storage, etc.), over the course of a year. Similarly, having different crops mature at different times can smooth out the availability of food or income for household consumption (Kassie, Teklewold, Jaleta, et al., 2015). It is a constant challenge for subsistence households to meet daily consumption needs when crops or income become available only infrequently.

Traditionally, farmers in Africa have tilled (plowed) the farm to loosen the soil, control weeds, and allow moisture to penetrate. In Ethiopia, for example, fields planted with wheat and teff (a small grain) are plowed with oxen three to five times in a growing season (Teklewold and Mekonnen, 2017). However, excessive tillage causes surface runoff and soil erosion, contributing to losses of soil and water, plant nutrients, and organic matter, as well as accelerated breakdown of organic matter, with the resulting release of CO_2. By contrast, conservation tillage eliminates plowing, or reduces its frequency to only one pass per growing season, and lets crop residue remain on the ground. This practice promotes soil aeration, reduces erosion and loss of nutrients, reduces the loss of water through evaporation, and promotes sequestration of carbon in the soil. Another benefit is less need for draft power (usually oxen). However, when a farmer does not plow repeatedly for weed control, weeds must be controlled in some other way, either by weeding by hand or using weed-killing chemicals, such as glyphosate. These trade-offs are discussed ahead.

Agricultural water management is critical because the farmer's challenge is to get the right amount of water to crops at the right time; therefore, both drought and flooding are risks, and both irrigation and drainage are issues. Most small-holder farms in Africa depend on rainfall rather than irrigation systems. Moreover, long-term soil fertility requires minimizing soil erosion by managing runoff; managing runoff also reduces the siltation of water bodies. Managing water so that it seeps into the soil (infiltration) promotes vegetation cover, which in turn retains water. Smallholder farmers in Africa have traditionally used a variety of practices, including water-holding structures and terracing, to address these needs.

Adoption of agricultural water management practices improves water balance and availability; improves infiltration and retention by the soil; reduces water loss due to runoff and evaporation; and improves the quality and availability of ground and surface water. Water management (irrigation, drainage, and water conservation and control) stabilizes crop production by maintaining soil conditions close to the optimum for crop growth. Water management practices that change slope profile (e.g., terracing) can reduce runoff speed – especially on erosion-prone highlands – thus reducing soil erosion. It also allows some water to seep into the soil (infiltration), improving the soil to allow more vegetation cover. This practice also increases groundwater recharge and protects the topsoil (FAO, 2014).

Adoption of a package of practices often improves productivity

Adoption of various types of practices in combination often improves productivity more than adoption of a single intensification practice. For instance, a combination of farmyard manure and chemical fertilizer optimized soil fertility and yield in a cereal-legume rotation farming system in the Ethiopian Highlands (Erkossa and Teklewold, 2009).

In a study of maize farming in Kenya, Nyangena and Ogada (2014) showed that adopting improved seeds and fertilizer as a complete package (applying fertilizer at planting time and again as a top dressing) improved yield more than adopting only part of the package. This is an example of inorganic fertilizer and modern seeds as complementary inputs.

In a study of Ethiopia, Kenya, Tanzania, and Malawi, there was a positive association between the number of sustainable intensification practices used and the average maize yield (Kassie, Teklewold, Jaleta, et al., 2015). That study analyzed the following sustainable intensification practices: conservation tillage; improved maize seeds; crop diversification (both maize-legume rotation and maize-legume intercropping); soil and water conservation practices; inorganic fertilizers; and manure. The survey data for the study covered household and plot-level data gathered in the four countries in 2010–11, for a total of 5,779 households and 9,837 plots on which maize was grown.

Adding water management to a system can boost the productivity-enhancing effects of modern seeds and inorganic fertilizer. In 2015, farm household and plot-level data was collected on 929 households and 4,702 plots in five regional states in the Blue Nile Basin of Ethiopia (Teklewold et al., 2017). Ethiopian smallholder households typically farm multiple plots, which are not necessarily contiguous. The average farm size in the study area is 1.8 hectares and the main crops grown are maize, wheat, teff, barley, and legumes. A mix of crops and livestock is typical. The agroecological zones represent cool, humid highlands; temperate, cool sub-humid highlands; warm, semiarid lowlands; and hot and very arid areas. Farmers mainly depend on rainfall rather than irrigation, with a main rainy season from June to October and a short rainy season from January to April.

Teklewold et al. (2017) found that agricultural water management was used on 41% of the plots studied, improved crop seeds on 24%, and inorganic fertilizer on 53%. All three were used on 9% of the plots. On 28% of the plots, none of the practices was used. One or two of these practices were used on the remaining plots. In further analysis, Teklewold et al. (2017) considered eight mutually exclusive combinations of practices obtained from combining the three practices (modern seeds, inorganic fertilizers, and agricultural water management practices). The outcome variable was net farm income. The analysis evaluated the average effect of a combination of practices, using a counterfactual approach to ask how much income would have been obtained from a plot had a different practice been used. However, a given farmer would not necessarily get identical results from a change in practices, because of the unobservable effects driving farm income. Therefore, it was necessary to control for the possibility that unobservable characteristics drove both the choice of farming practices and the outcome variable (net farm income). Such self-selection bias was controlled for using a multinomial endogenous switching treatment effects approach.

According to Teklewold et al. (2017), adoption of any one of the three practices studied, whether in isolation or combination, provided higher net crop income compared with non-adoption. Adoption of inorganic fertilizers in isolation provided higher net income than adoption of other practices in isolation. Adoption of inorganic fertilizers in combination with agricultural water management practices or in combination with modern crop seeds also provided higher net income than a combination of water management and modern crop seeds. The highest net income was obtained from adoption of agricultural water management practices jointly with inorganic fertilizers and modern crop seeds. In other words, the three practices are complements.

Teklewold et al. (2017) also showed that the net income of adopting a combination of agricultural water management with modern seeds or inorganic fertilizer was higher than adopting modern seeds or inorganic fertilizer alone. The net farm income from modern seeds and fertilizer increased by about 45% if agricultural water management was combined with modern seeds and fertilizers.

Teklewold et al. (2017) also simulated crop income response to future climate change. The simulations suggest that, in a warming climate with less predictable rainfall, adoption of water management practices will be essential in order to increase farm income. By 2060, the two purchased inputs will no longer provide increased income unless accompanied by water management. The highest income is projected for farmers who adopt the three practices as a package.

Sustainable agricultural intensification can reduce the quantity of agrochemicals used

Teklewold, Kassie, Shiferaw, et al. (2013) looked at the role of sustainable intensification practices in reducing the environmental impacts of agricultural inputs. That analysis focuses on 900 households and 1,644 plots on which maize is grown in Ethiopia. Eight possible combinations of three practices (maize-legume rotation, conservation tillage, and improved maize seeds) were analyzed. The outcomes variables are net income from maize, use of agrochemicals (inorganic fertilizer, insecticides, and herbicides), and agricultural labor. The use of inorganic fertilizer was measured in terms of intensity, or quantity used, of nitrogen fertilizer. The quantities of pesticides include both insecticides and weed-killing herbicides. A multinomial endogenous switching treatment effects approach is used to control for possible selection bias between the choice of farming practices and the outcome variables.

About one-quarter of the plots examined in Teklewold, Kassie, Shiferaw, et al. (2013) used none of these practices, while 5.4% used all three. The unconditional probabilities of adopting rotation, improved seeds, and conservation tillage are described ahead in the discussion of Teklewold, Kassie, and Shiferaw (2013). The conditional probabilities also show complementarity in the adoption decisions. For instance, the probability of adopting conservation tillage increased from about 36% to 50% when farmers rotate maize with legumes, and the probability of adopting improved seeds increased from 53% to 58% when farmers adopt crop rotation.

Once again, adoption of a package of practices resulted in higher income than adoption of individual practices, and adoption of all three parts of the package resulted in the highest income. In Teklewold, Kassie, Shiferaw, et al. (2013), the additional maize income per hectare from adopting these three practices was 5,580 *birr* per hectare (USD 242).

Teklewold, Kassie, Shiferaw, et al. (2013) also found that crop rotation is associated with a reduced quantity of inorganic fertilizer. This means that farmers "credit" the nitrogen that is fixed by legumes. This implies an environmental benefit without an economic trade-off, in that the source of the nitrogen isn't what matters for purposes of yield. The reduced need for expenditures on a purchased input also can benefit farmers who are cash- and credit-constrained.

However, the use of improved seeds was significantly associated with increased use of inorganic fertilizer and pesticides (Teklewold, Kassie, Shiferaw, et al., 2013), regardless of whether conservation tillage and crop rotation were practiced as well. This may be because these purchased inputs often are used as complements. In addition, high-yielding maize varieties are susceptible to pests, which may induce farmers to manage risk by adding pesticides to the package.

Pesticide use also increased with conservation tillage. One reason that farmers plow is to destroy weeds. Conservation tillage calls for alternative methods of controlling weeds, usually the application of an herbicide. Over the longer term, there is likely to be less need to use herbicides in conjunction with conservation tillage, for two reasons. One, crop rotation is another way to control both weeds and insects. Two, leaving residue on the ground as part of conservation tillage also controls weeds (Teklewold, Kassie, Shiferaw, et al., 2013).

The problem of weeds leads to the last undesirable effect of conservation tillage noted in Teklewold, Kassie, Shiferaw, et al. (2013): increased need for farm labor, particularly female labor. In fact, adoption of all the packages studied in Teklewold, Kassie, Shiferaw, et al. (2013) resulted in more hours of female labor.

In sum, Teklewold, Kassie, Shiferaw, et al. (2013) showed that adoption of a package of improved seeds, conservation tillage, and crop rotation offers both environmental and income-increasing benefit, but not without some trade-offs.

Farmers' decisions to adopt sustainable intensification systems

A broad body of research has looked at the factors influencing adoption decisions in developing countries, including sub-Saharan Africa. One set of factors includes everything that affects a household's ability to purchase inputs, including the household's assets, income (including income from off-farm employment), liquidity, and access to credit. Another set of factors is related to risk, and includes farmers' risk aversion, as well as risk-coping mechanisms, such as formal insurance, informal social structures that help people cope with risk, public assistance in case of crop failure, and diversification of income through a household member's off-farm employment. Yet another category includes knowledge and information: general education, quality and quantity of agricultural extension services, age and experience of farmers, and knowledge that may come from social networks. Other factors include the costs of obtaining inputs and transporting inputs to plots, such as access to roads and transportation and distance to farm plots. A wide range of agroecological conditions, including those at the plot level, such as steepness or soil quality, and a number of household characteristics, such as household size, enter into adoption decisions.

Teklewold, Kassie, and Shiferaw (2013) looked at factors influencing Ethiopian farmers' adoption of multiple practices, comprising the following five intensification practices: improved seeds, maize-legume rotation, conservation tillage, inorganic fertilizer, and manure. This research is based on surveys conducted in 2010 with 898 farming households growing maize on 1,616 plots in three Ethiopian regional states.

The education of both spouses can be a factor in adoption decisions (Teklewold, Kassie, and Shiferaw, 2013). Both household heads (92% male) and spouses had low levels of primary education. However, each additional year of education of the spouse increases the probability of adopting more than two intensification practices by 12%.

Access to input and output markets influences adoption decisions (Teklewold, Kassie, and Shiferaw, 2013). Close to half of households (44%) rely on walking to markets, which takes an hour on average. Somewhat more than half (56%) use a bicycle, cart, horse, donkey, or public transportation. Farmers whose only alternative is walking are 9% less likely to adopt more than two sustainable intensification practices.

Resource constraints are a major barrier to adoption (Teklewold, Kassie, and Shiferaw, 2013). In Teklewold, Kassie, and Shiferaw (2013), wealth was measured in ownership of land and non-land assets and livestock. Off-farm employment and remittances also were considered. Credit-constrained households were found to be more likely to adopt soil and water conservation, which can rely on household labor (Kassie, Teklewold, Jaleta, et al., 2015). However, at least in Malawi and Ethiopia, credit-constrained households were less likely to adopt conservation tillage, because of the need to purchase herbicides.

Ethiopia has a Productive Safety Net Program that allows farmers to work for food or cash in case their own crops fail. Respondents who believed they could rely on such help were 20% more likely to adopt more than two practices (Teklewold, Kassie, and Shiferaw, 2013). The result shows the insurance effect of the social protection program.

Three measures of social capital and social networks were assessed (Teklewold, Kassie, and Shiferaw, 2013): membership in a rural association, kinship network, and the number of trusted traders the household head knows. These had positive effects on adoption of a system of intensification practices. Similarly, in Kassie, Teklewold, Jaleta, et al. (2015), membership in groups made it more likely that farmers would adopt improved varieties and inorganic fertilizer in Kenya; soil and water conservation in Malawi; crop diversification, minimum tillage, and fertilizer in Ethiopia; and soil and water conservation, manure use, improved maize varieties, and fertilizer in Tanzania.

The ratio of extension agents to farm households differs greatly among countries. For instance, the ratio is 1:2,500 in Tanzania, 1:476 in Ethiopia, 1:1,000 in Kenya, and 1:1,603 in Malawi (Kassie, Teklewold, Jaleta, et al., 2015).The mere existence of agricultural extension services may not be enough to promote adoption. Extension agents in Ethiopia, for example, play multiple roles. As well as providing information, they deliver inputs, administer credit, and collect payments (Teklewold, Kassie, and Shiferaw, 2013). Therefore, trust in these agents is important. Farmers who had confidence in extension agents were more likely to adopt intensification practices (Teklewold, Kassie, and Shiferaw, 2013). This result shows that it is not the quantity of information per se that affects adoption, but rather the quality of the information.

Plot-level characteristics (fertility, depth, slope, tenure, and plot distance to the farmer's home) affect adoption decisions. For example, inorganic fertilizers (rather than manure) and improved seeds are more likely on steep slopes (Teklewold, Kassie, and Shiferaw, 2013). Distance from home to plot also affects adoption, often negatively (probably due to the need to transport inputs to distant plots) but sometimes positively (Teklewold, Kassie, and Shiferaw, 2013). For instance, a shorter distance to the plot makes it more likely a farmer will use manure, which is bulky (Teklewold, Kassie, and Shiferaw, 2013).

Teklewold et al. (2017) also found that slope of the farm and depth of the soil are particularly associated with the likelihood of climate-smart practices. The type of crop grown is also important in the choice of adaptation practices. While adoption of modern seeds and/or inorganic fertilizer is more likely for cereal than vegetable crops, adoption of water management is more likely for vegetable crops than for cereal and legume crops. As expected, farmers are less likely to apply nitrogen fertilizer to legume crops.

Both heat and cold affect adoption decisions. Higher temperature affected the adoption of agricultural water management in combination with modern seeds or inorganic fertilizer in Ethiopia in a hill-shaped pattern. Higher temperature up to a point promoted adoption, but very high temperatures beyond that discouraged adoption (Teklewold and Mekonnen, 2017). In the Ethiopian Highlands, waterlogging and frost discourage the adoption of crop rotation (Teklewold, Kassie, and Shiferaw, 2013).

Livestock ownership points in two different directions: as a source of wealth that allows the purchase of commercial inputs, and as a source of manure. While Teklewold, Kassie, and Shiferaw (2013) and Kassie, Teklewold, Jaleta, et al. (2015) found that households that owned more livestock were more likely to use manure (as well as soil and water conservation), Kassie, Teklewold, Jaleta, et al. (2015) also found that credit-constrained households were more likely to use manure. Kassie, Teklewold, Jaleta, et al. (2015) found a similar effect on the adoption of soil and water conservation and crop diversification in Malawi and inorganic fertilizer in Tanzania. Further, the 2015 survey discussed earlier found that the extent of livestock holdings in Ethiopia influenced the adoption of a combination of modern seeds and inorganic fertilizer. Apparently, livestock ownership operates through two channels: an income effect (i.e., wealthier households are more able to purchase inputs) and a substitution effect (in which livestock-owning households have their own supply of nitrogen fertilizer), with the dominant effect depending on specific circumstances.

Complementarity and substitution in adoption decisions

When a farmer is deciding whether to adopt a system involving multiple sustainable practices, the practices can be either substitutes or complements. Practices are complements when the likelihood of using one practice is positively associated with the use of another practice. Practices are substitutes when the likelihood of using one practice is negatively associated with the use of another practice.

Among the plots studied in Teklewold, Kassie, and Shiferaw (2013), all five practices under consideration – improved seeds, maize-legume rotation, conservation tillage, inorganic fertilizer, and manure – were used in combination on only 10% of the plots. At least one practice was used on 1,509 plots. Inorganic fertilizer was applied to 67.3% of plots. On the plots where fertilizer was used, the quantities were considerably lower than the quantity recommended by local agricultural extension services. As percentages of all plots, 11% of plots were treated with inorganic fertilizer but no other intensification practices, 16% of plots used both inorganic fertilizer and improved seed, and 10% of plots combined inorganic fertilizer, improved seed, and conservation tillage.

Manure was used on 27.3% of the sampled plots in Teklewold, Kassie, and Shiferaw (2013). As a percentage of all plots, manure was used as the sole intensification practice on 4.9% of plots. Manure and inorganic fertilizer were used (with no other intensification practices) on 3.5% of plots and in combination with improved seeds on 4.3% of the plots. Despite the apparent complementarity of manure and inorganic fertilizer, the analysis of conditional probabilities of adoption shows that manure and inorganic fertilizer are generally substitutes rather than complements in this context. The likelihood of adopting inorganic fertilizer is reduced by more than 19% when manure is applied to a plot. This finding differs from that in Marenya and Barrett (2007), who found that inorganic fertilizer and manure were complements, in a study in western Kenya. These findings illustrate that adoption decisions are greatly affected by farm-specific conditions. For instance, Teklewold, Kassie, and Shiferaw (2013) found that inorganic fertilizer was more likely on plots with poor soil quality and manure was more likely on plots with good soil quality. Teklewold, Kassie, and Shiferaw (2013) also found that manure was a substitute for crop rotation and conservation tillage; in other words, the likelihood of crop rotation and conservation tillage was lower when manure was used on a plot.

Maize-legume rotation, for instance, was practiced on 23.2% of the plots studied in Teklewold, Kassie, and Shiferaw (2013). In most cases, if crop rotation was practiced, it was done as part of a package; only 1.6% of all plots had maize-legume rotation as the sole intensification practice. Improved varieties of maize were planted on 52.5% of all the plots sampled in Teklewold, Kassie, and Shiferaw (2013) and conservation tillage was practiced on 36.3% of the total. As percentages of all plots, 4.3% combined crop rotation, improved seed, and inorganic fertilizer, while 3.5% of plots combined four practices: maize-legume rotation, inorganic fertilizer, improved seeds, and conservation tillage. These four practices are complements with each other.

Teklewold, Kassie, and Shiferaw (2013) show that adoption decisions are interrelated, whether practices function as substitutes or complements. For instance, the unconditional adoption rate of inorganic fertilizer is about 67.3%. The probability of adoption of inorganic fertilizer increases to 78.1% conditional on adoption of one practice (improved seed), 73.2% conditional on adoption of two practices (improved seed and rotation), and 76.4 % conditional on adoption of three practices (improved seed, rotation, and conservation tillage).

Kassie, Teklewold, Jaleta, et al. (2015) looked at complementarity over a wider range of practices and countries, considering Malawi, Tanzania, and Kenya as well as Ethiopia, and considering water management as well as the five practices evaluated in Teklewold, Kassie, and Shiferaw (2013). Analyzing the probabilities of adoption of one practice conditional on adoption of another, Kassie, Teklewold, Jaleta, et al. (2015) found that inorganic fertilizer and improved seeds were complements in all four countries. Inorganic fertilizer and manure were substitutes in three countries but not in Kenya. In Tanzania, adoption of inorganic fertilizer increased with the adoption of improved seeds, soil and water conservation, and conservation tillage, indicating complementarity of the components of that package. In general in the study areas, the conditional probabilities of adopting one sustainable intensification practice were higher than the unconditional probabilities (with the exception of the manure-inorganic fertilizer substitution), indicating complementarity of the components of these systems.

Adoption of a package of practices can be part of a risk management strategy

Risk is ever-present in farming. For smallholder farmers, the downside risks (crop failure) can be extreme. The costs of mitigating risks place farmers in a poverty trap; for instance, in the event of crop failure, a farmer might liquidate assets in order to meet immediate household needs, thus reducing her ability to invest in the next production cycle.

Typical risks include erratic rainfall, crop diseases and pests, and fluctuations in commodity prices. Small farmers in sub-Saharan Africa are particularly averse to downside risk, and seek to avoid variance in yields, as Kassie, Teklewold, Marenya, et al. (2015) discussed in a study of maize farmers in Malawi.

Some intensification techniques increase risk to smallholder farmers, while others reduce risk. For instance, fertilizer and high-yielding seeds increase average yield but also increase the variance in yield. Thus, fertilizer and improved seeds are risk-increasing practices. Conversely, water management can be a risk-reducing strategy, because inadequate or unpredictable rainfall is a major risk, which is increasing with a changing climate. This subsection presents evidence of the role of risk in adoption decisions.

Teklewold, Kassie, and Shiferaw (2013) found a higher probability of adopting inorganic fertilizer and crop rotation in times and places where rainfall was more reliable. If the climate becomes wetter, the probability of choosing agricultural water management practices in isolation or in combination with modern seeds and inorganic fertilizer increases (Teklewold et al., 2017). Conversely, farmers in the Nile Basin of Ethiopia do not apply fertilizer if application to a crop is perceived as too risky (Teklewold et al., 2017). Similarly, Teklewold, Kassie, Shiferaw, et al. (2013) found that high-yielding maize varieties are susceptible to pests, which may induce farmers to manage risk by adding pesticides to the package.

Agricultural water management can minimize the risk of a yield shortfall arising from application of fertilizer and new seeds in the event of unfavorable rain. In Ethiopia, the amount of rainfall in the growing season is important for the decision to use fertilizer and for the choice of a combination of water management practices in combination with improved seed or fertilizer; however, adoption of modern seeds and fertilizer might be less affected by changes in precipitation when they are combined with water management practices (Teklewold et al., 2017).

However, when modern seeds and inorganic fertilizer are not supported by water management, farmers are less likely to adopt these two inputs when annual rainfall is higher or lower than an optimal level of 800 mm (Teklewold et al., 2017). Similarly, increasing rainfall variability significantly decreases the likelihood of adoption of fertilizer, either in isolation or in combination with modern seeds, thus reflecting the adverse effects of rainfall variability on adoption of risk-increasing inputs (Teklewold et al., 2017). However, the likelihood of adoption of agricultural water management increases with high rainfall variability, whether it is adopted in isolation or in combination with modern seeds and inorganic fertilizer (Teklewold et al., 2017). In low-rainfall areas, adoption of improved crop seeds and/or fertilizer in combination with agricultural water management is more likely under highly variable rainfall conditions (Teklewold et al., 2017). These findings suggest that water management acts as a buffer against rainfall variability.

In addition to water management, there are other sustainable intensification systems that can reduce risk. In Malawi, a study of maize farmers found that most of the cost of risk comes from downside risk, and that simultaneous adoption of diversification (rotation and intercropping) and conservation tillage (combined with leaving crop residue on the ground) reduced both downside risk and the cost of risk, while improving food security (Kassie, Teklewold, Marenya, et al., 2015). These findings are noteworthy because Malawi has been identified as especially vulnerable to climate change, with heavy dependence on rain-fed, smallholder maize farming for its food security (Kassie, Teklewold, Marenya, et al., 2015). In a nationally representative sample, crop diversification was adopted on 58% of plots, and conservation tillage on 35% of plots, while both adaptations were in use on 21% of plots and neither was in use on 28%. When adopted simultaneously, conservation tillage and diversification conserve soil moisture and limit disease and pest outbreaks. In the analysis, downside risk is measured using maize yield skewness distribution. After correcting for endogeneity, and using a counterfactual analysis, the greatest improvement in yield is obtained from adoption of both practices. While yield improves when farmers adopt either practice, there is greater improvement with joint adoption. Similarly, adoption of either practice decreases exposure to downside risk, and adoption of both practices has a greater effect in reducing risk. Further, the cost of risk (which is mostly explained by downside risk) is lowest for package adoption.

Package adoption decisions weigh alternative uses of limited cash, inputs, and labor

One theme running through the adoption literature is that lack of access to cash or credit is a barrier to adoption. A related issue is that very poor households cannot take the risk of a new technology. Thus, household wealth is associated with greater adoption of improved seed, inorganic fertilizer, and conservation tillage (Teklewold, Kassie, and Shiferaw, 2013). Therefore, there is some evidence that rising income – in other words, a reduction of poverty overall – will remove some of the constraints to adoption. However, Teklewold, Kassie, and Shiferaw (2013) demonstrated two results that show that there is no simple correlation between adoption of specific practices and typical measures of prosperity: (1) distance to input markets had a positive effect on adoption of conservation tillage, presumably because it is comparatively easy for farmers to adopt this practice despite living far from a market; (2) larger holdings of livestock (one typical measure of wealth in rural developing areas) were associated with greater reliance on manure, for the obvious reason that this asset produces manure.

Input subsidies are one response to the cost barrier. Malawi implemented fertilizer subsidies in the late 1990s in response to declining maize production and resulting food and political crises and later added subsidies for improved seeds for maize and legumes (Kassie, Teklewold, Jaleta, et al., 2015). Tanzania's fertilizer subsidies have tended not to benefit the poorest farmers (Kassie, Teklewold, Jaleta, et al., 2015), because local officials obtained a large share of the vouchers (Pan and Christiaensen, 2011) and because the partial subsidy did not cover the full market price (Hepelwa et al., 2013). Kenya has an active private sector selling inputs and has generally deregulated input markets, but, following the food crisis of 2007–08, targeted subsidies to smallholders (Kassie, Teklewold, Jaleta, et al., 2015). Ethiopia has offered low-interest credit guaranteed by the government for seed and fertilizer purposes, and its national extension service is a major distributor of fertilizer and improved seeds (Kassie, Teklewold, Jaleta, et al., 2015). Kassie, Teklewold, Jaleta, et al. (2015) argue that agricultural policies need to look beyond one or two purchased inputs in isolation and, instead, to look at sustainable intensification practices in an integrated manner.

Farm by-products are not available in unlimited quantities. As mentioned earlier, farm households have to choose whether to use manure for fertilizer or for fuel. Similarly, there are competing uses for crop residue. For instance, conservation agriculture offers the greatest benefits to farmers when it is practiced as part of a system, in particular in a system of mulching and crop rotation, as Thierfelder et al. (2013) showed in a study in Zambia. However, in a mixed crop-livestock system, farmers have to decide whether to use crop residue for soil mulching or as livestock feed (Teklewold and Mekonnen, 2017). Further, low tillage permits a greater accumulation of weeds, which increases labor demand for weeding (or reliance on chemical weed control) (Teklewold, Kassie, Shiferaw, et al., 2013; Teklewold and Mekonnen, 2017).

A long time lag in realizing the benefits of investment can affect adoption decisions

The benefits from conservation practices may not be evident for several years – up to seven years in the case of soil conservation (Schmidt and Tadesse, 2013). Soil conservation practices actually decreased net revenue per hectare in a recent study in Ethiopia (Beyene et al., 2017).

In the case of low tillage, grain yield increases with an increasing number of plowings, in the short run and all else equal (Institute of Agricultural Research, 1998). In the meantime, farmers not only incur the cost of investment but also may experience short-term decreases in yield. It's possible that the longer-term negative effects of erosion due to extensive tillage are less salient to a farmer than the short-term benefits of increased yield.

Because of the delay in realizing a return on some investments, a greater level of tenure security may encourage such investments. Farmers are more likely to adopt water management on plots that they own (Teklewold et al., 2017). This suggests that tenure security encourages investment, which is especially pertinent to the construction of water management structures, which require a large initial investment that will generate returns over time, as opposed to the annual investment in purchased inputs that are intended to generate returns in the short run. While an investment in improved seeds, for example, might be lost in a single bad rainfall year, an investment in a water management structure will still be useful in later years.

Conclusion/policy responses

The overarching conclusions of this chapter are that package adoption usually results in a greater increase in productivity than does adoption of individual practices; the appropriate adaptation response varies greatly depending on farm-level and agroecological conditions; purchased and/or polluting inputs, such as agrochemicals, have proven benefits in improving yield, while their use can be optimized as part of a sustainable system of intensification; farmers' limited resources make it difficult for them to adopt the adaptation strategies that are best for their specific needs; risk is a barrier to adoption; strengthening local institutions, social and human capital, and tenure security are important to help farmers better adapt; and agricultural adaptation is closely related to overall economic development.

In most cases, adoption of a package or system improves farm productivity, compared to adoption of single practices. However, we note that adopting multiple practices at the same time is not always the right adaptation choice for every farmer. In a study in Ethiopia by Beyene et al. (2017), tree planting alone produced a slightly higher payoff (602 Ethiopian birr/USD 25.86 per hectare) than the combination of intercropping, tree planting, and soil conservation (589 birr/25.30 per hectare), possibly because of the long lag time in realizing the benefits of soil conservation. Their finding highlights the importance of specific conditions of the farm household, agroecological setting, and so forth.

Regional agroecological conditions, farm-level and plot-level characteristics, and household characteristics influence which intensification practices are most suitable for a particular region or farm from a purely agronomic point of view, which seeks yield-maximizing "recipes" that might or might not be practical for all smallholders. While farmers' individual choices are influenced by the local characteristics that are taken into account by an extension agent who is recommending best practices, the farmer also faces financial constraints and risk aversion that cause suboptimal productivity.

Farmers who own their plots are more likely to adopt sustainable intensification practices. This indicates the role of tenure security.

The lack of cash and credit is an overarching issue. Purchased inputs such as chemical fertilizer and improved seeds have proven benefits in increasing yield and income. However, it is costly for poor farmers to purchase such inputs, and many farmers lack the necessary capital or access to credit (Teklewold, Kassie, and Shiferaw, 2013). By contrast, sustainable intensification systems may include inputs that do not have to be purchased (e.g., manure, crop residue, locally available building materials, and household labor). Such intensification practices can be adopted by households that are short of cash and can't get credit. However, it would be too much of a generalization to state that richer households are more likely than poorer households to adopt any given practice or package. For instance, households that are rich in livestock holdings might use manure even if they can afford commercially purchased fertilizer.

Regarding human capital, both general education and agriculture-specific education play a role in adaptation decisions. Agricultural extension is the major source through which many climate-related adaptation practices and information are channeled (Teklewold et al., 2017). The quality of agricultural extension services is as important as the quantity of contacts (Teklewold et al., 2017). However, quantity matters as well. As noted earlier, the ratio of extension agents to farmers differs among countries. Access to extension services, through roads and communication, is also important. This illustrates that agricultural development depends in large part on overall development.

The spouse's level of education can increase the probability that a household will adopt a productivity-enhancing package. In the households surveyed earlier, most "heads of household" were male and therefore most of the spouses were female. This illustrates the importance of general education for girls and women.

As well as human capital, social capital is important in adaptation. Local institutions can play a critical role in providing farmers with timely information, inputs (e.g., labor, credit, and insurance), and technical assistance.

Finally, overall economic development can help farmers obtain the cash and credit, information, and other resources needed to adopt the intensification practices that are right for their individual farms.

References

Alem, Y., M. Bezabih, M. Kassie and P. Zikhali. 2010. Does Fertilizer Use Respond to Rainfall Variability? Panel Data Evidence from Ethiopia. *Agricultural Economics* 41(2): 165–175.

Beyene, A.D., A. Mekonnen, M. Kassie, S. Di Falco and M. Bezabih. 2017. Determinants of Adoption and Impacts of Sustainable Land Management and Climate-Smart Agricultural Practices (SLM-CSA): Panel Data Evidence from the Ethiopian Highlands. Environment for Development Discussion Paper Series No. 17–10. Addis Ababa, Ethiopia: Environment and Climate Research Center; Gothenburg, Sweden: Environment for Development; Washington, DC: Resources for the Future.

Bezu, S., G.T. Kassie, B. Shiferaw and J. Ricker-Gilbert. 2014. Impact of Improved Maize Adoption on Welfare of Farm Households in Malawi: A Panel Data Analysis. *World Development* 59: 120–131.

De Groote, H., G. Owuor, C. Doss, J. Ouma, L. Muhammad and K. Danda. 2005. The Maize Green Revolution in Kenya Revisited. *E-Journal of Agricultural and Development Economics* 2(1): 32–49.

Erkossa, T., and H. Teklewold. 2009. Agronomic and Economic Efficiency of Manure and Urea Fertilizers Use on Vertisols in Ethiopian Highlands. *Agricultural Sciences in China* 8(3): 352–360.

FAO. 2014. *Adapting to Climate Change through Land and Water Management in Eastern Africa: Results of Pilot Projects in Ethiopia, Kenya and Tanzania.* Rome: Food and Agriculture Organization of the United Nations.

Hepelwa, A., O. Selejio and J.K. Mduma. 2013. The Voucher System and the Agricultural Production in Tanzania: Is the Model Adopted Effective? Evidence from the Panel Data Analysis. Paper Presented by Environment Development Initiative-Tanzania. https://docs.google.com/viewer?url=www.efdinitiative.org/sites/default/files/fertilizer_voucher_system_and_the_agricultural_production_in_tanzania.pdf (accessed August 24, 2017).

Institute of Agricultural Research. 1998. Holleta Agricultural Research Center. Progress Report for the Period April 1997 to March 1998. Addis Ababa, Ethiopia: IAR.

Kassie, M., H. Teklewold, M. Jaleta, P. Marenya and O. Erenstein. 2015. Understanding the Adoption of a Portfolio of Sustainable Intensification Practices in Eastern and Southern Africa. *Land Use Policy* 42: 400–411.

Kassie, M., H. Teklewold, P. Marenya, M. Jaleta and O. Erenstein. 2015. Production Risks and Food Security under Alternative Technology Choices in Malawi: Application of a Multinomial Endogenous Switching Regression. *Journal of Agricultural Economics* 66(3): 640–659.

Marenya, P.P., and C.B. Barrett. 2007. Household-Level Determinants of Adoption of Improved Natural Resources Management Practices among Smallholder Farmers in Western Kenya. *Food Policy* 32: 515–536.

Marenya, P.P., and C.B. Barrett. 2009. State-Conditional Fertilizer Yield Response on Western Kenyan Farms. *American Journal of Agricultural Economics* 91(4): 991–1006.

Nyangena, W., and M.J. Ogada. 2014. Impact of Improved Farm Technologies on Yields: The Case of Improved Maize Varieties and Inorganic Fertilizer in Kenya. Environment for Development Discussion Paper Series No. 14–02. Gothenburg, Sweden: Environment for Development; Washington, DC: Resources for the Future.

Pan, L., and L. Christiaensen. 2011. Who Is Vouching for the Input Voucher? Decentralized Targeting and Elite Capture in Tanzania. The World Bank East Asia and Pacific Region Social, Environment and Rural Sustainable Development Unit Policy Research Working Paper No. 5651.

Place, F., C.B. Barrett, H.A. Freeman, J.J. Ramisch and B. Vanlauwe. 2003. Prospects for Integrated Soil Fertility Management Using Organic and Inorganic Inputs: Evidence from Smallholder African Agricultural Systems. *Food Policy* 28(4): 365–378.

Pretty, J., W.J. Sutherland, J. Ashby, J. Auburn, D. Baulcombe, M. Bell, J. Bentley, S. Bickersteth, K. Brown, J. Burke, H. Campbell, K. Chen, E. Crowley, I. Crute, D. Dobbelaere, G. Edwards-Jones, F. Funes Monzote, H.C.J. Godfray, M. Griffon, P. Gypmantisiri, L. Haddad, S. Halavatau, H. Herren, M. Holderness, A.-M. Izac, M. Jones, P. Koohafkan, R. Lal, T. Lang, J. McNeely, A. Mueller, N. Nisbett, A. Noble, P. Pingali, Y. Pinto, R. Rabbinge, N.H. Ravindranath, A. Rola, N. Roling, C. Sage, W. Settle, J.M. Sha, L. Shiming, T. Simons, P. Smith, K. Strzepeck, H. Swaine, E. Terry, T.P. Tomich, C. Toulmin, E. Trigo, S. Twomlow, J.K. Vis, J. Wilson and S. Pilgrim. 2010. The Top 100 Questions of Importance to the Future of Global Agriculture. *International Journal of Agricultural Sustainability* 8(4): 219–236.

Schmidt, E., and F. Tadesse. 2013. Sustainable Agriculture in the Blue Nile Basin: Land and Watershed Management Practices in Ethiopia. *Environment and Development Economics* 19(5): 648–667.

Teklewold, H., M. Kassie and B. Shiferaw. 2013. Adoption of Multiple Sustainable Agricultural Practices in Rural Ethiopia. *Journal of Agricultural Economics* 64(3): 597–623.

Teklewold, H., M. Kassie, B. Shiferaw and G. Köhlin. 2013. Cropping System Diversification, Conservation Tillage and Modern Seed Adoption in Ethiopia: Impacts on Household Income, Agrochemical Use and Demand for Labor. *Ecological Economics* 93: 85–93.

Teklewold, H., and A. Mekonnen. 2017. The Tilling of Land in a Changing Climate: Panel Data Evidence from the Nile Basin of Ethiopia. *Land Use Policy* 67: 449–459.

Teklewold, H., A. Mekonnen, G. Köhlin and S. Di Falco. 2017. Does Adoption of Multiple Climate-Smart Practices Improve Farmers' Climate Resilience? Empirical Evidence from the Nile Basin of Ethiopia. *Climate Change Economics* 8(1). www.worldscientific.com/doi/abs/10.1142/S2010007817500014

Thierfelder, C., M. Mwila and L. Rusinamhodzi. 2013. Conservation Agriculture in Eastern and Southern Provinces of Zambia: Long-Term Effects on Soil Quality and Maize Productivity. *Soil and Tillage Research* 126(0): 246–258.

Part III

On-farm practices other than those related to food crop productivity

10 Climate change adaptation and livestock activity choice in the Nile Basin of Ethiopia

An economic analysis

Tsegazeab Gebremariam and Zenebe Gebreegziabher

Introduction

Ethiopia has a larger livestock resource than most countries in Africa. It is estimated that more than 80 million people live in rural areas and depend mainly on agriculture for their livelihoods; the sector contributes about 40% of the country's gross domestic product (World Bank, 2016). Developing countries are more vulnerable to climate change relative to developed countries due to the dominance of the agriculture sectors in their economies, as well as the scarcity of capital for adaptation measures (Fischer et al., 2005). This study investigates the drivers of climate change adaptation and livestock activity choice, using a cross-sectional household survey dataset from a sample of 1,000 households in the Nile Basin of Ethiopia that is enriched with long-term averages of climate change data. Specifically, the chapter investigates the impact of climate change on the farmer's decision about whether to engage in livestock activity and also on the choice of the different livestock species. Findings suggest that farmers adapt their livestock-holding decisions in light of changing climate. Findings also call for adaptation measures that minimize the impact of climate change on livestock production, such as ownership systems that promote pasture security, among other measures.

An increase in temperature and decrease in precipitation is expected to cause substantial loss in Africa because of its reliance on rain-fed traditional agriculture and absence of irrigation facilities. This loss is predicted to negatively influence the overall economy of African countries because about 30% of these countries' gross national product comes from agriculture. The livestock subsector is a major component of agriculture and is a source of cash income, meat, milk and wool for smallholder keepers in different farming systems and agroecological zones (Getahun, 2008; FAO, 2009). Livestock also serves as a source of foreign currency (Gebremedhin et al., 2006). At the farm level, small ruminants serve as an investment and as insurance due to their high fertility, short generation interval, ability to produce under limited feed resources, and adaptation capacity in harsh environments (Tsedeke, 2007). Apart from their economic importance, livestock are socially and culturally important in Africa in general and particularly in Ethiopia's Nile Basin, for payment of dowry, celebrations and gifts to family members, and also as a source of savings. Despite diseases and drought, livestock are safer than the banking system and easier to manage for remote farmers.

However, as several studies prove, climate change has had significant and alarming negative impacts on the livestock activity of farmers in different parts of the world (Kabubo-Mariara, 2008). Therefore, adaptation is important. For example, previous studies on climate change and adaptation of livestock farmers have shown that climate change affects livestock farming directly and indirectly (Hoffmann, 2008; Kabubo-Mariara, 2008). The direct effects include retardation of animal growth and low quality of animal products, including hides, skins and overall animal production. In small-scale, mixed crop-livestock farming, livestock provides animal power, transportation and manure for fertilizing crop lands. Indirect effects occur through influences on the quantity and quality of feedstuffs, such as pasture, forage and grain, and the severity and distribution of livestock diseases and parasites (Seo and Mendelsohn, 2008a). For instance, a higher temperature and changes in rainfall patterns may translate into an increased spread of vector-borne diseases and macro-parasites that currently affect farm animals, as well as the emergence and spread of new diseases.

Hence, efforts to mitigate greenhouse gases are essential. Mitigation tackles the cause of climate change while adaptation tackles its effects (MERET, 2009). Mitigation efforts, though, take time and require international cooperation. Adaptation, on the other hand, can reduce climate-related risks in human-managed systems on regional as well as local scales, and often with a short lead time, although its scope is generally limited to specific systems and risk types (Fussel and Klein, 2004). Therefore, adaptation measures that take into account local conditions are critical in developing countries (Hassan and Nhemachena, 2008).

Livestock production has historically been one mechanism through which rural producers have coped with changes in climate. Examples include the accumulation of small stock to buffer against newly arising climate stress during the past century (in the Nile Basin of Ethiopia and elsewhere) and the small, year-to-year modifications in livestock types that enable livestock keepers to continuously manage the shifting resources of Ethiopia, particularly in the Nile Basin. In fact, mobile pastoralism arose in response to long-term climate change several millennia ago.

However, most of the existing literature concentrates on crop agriculture (Deressa, 2007; Deressa and Hassan, 2009; Kassahun, 2009), and there is a dearth of literature about climate change and livestock production. So far, there has been no study conducted in the Ethiopian context to analyze the determinant factors affecting climate change adaptation and livestock activity choice. Understanding these factors is important in order to provide policy guidance on ways to promote adaptation mechanisms in the livestock subsector.

The specific objectives of this study are to analyze the factors that determine farmers' decisions about whether to engage in livestock activities; to investigate climate factors and household socioeconomic characteristics underlying farmers' adaptation decisions about primary animal livestock activity choices; and to examine the impact of climate variables on the choice of different combinations of livestock species (the optimal portfolio) under variable climatic conditions.

The rest of the chapter is organized as follows. In the next section, we review related literature. The following section presents the theoretical model. The fourth section presents the empirical model and data. The fifth section discusses the results. The concluding section makes some policy recommendations.

Review of related literature

Generally, developing countries are more vulnerable to climate change because of their location in temperature-sensitive areas and also due to their dependency on rain-fed traditional agriculture. In fact, studies assert that, even though they contribute very little to causing climate change, developing countries, especially the very poor, are more vulnerable and will be worse affected by climate change than developed countries, largely because of their limited capacity to adapt (Thomas and Twyman, 2005; Todaro and Smith, 2009). For example, Apata et al. (2009) and Deressa et al. (2009) show that Africa is most vulnerable to climate change, with agriculture identified as the sector most sensitive to the negative effects of climate change due to the erratic and unreliable weather, which calls for farmers to be aware of the effects of climate change in terms of both immediate weather patterns and long-term production decisions. The latest report of the Intergovernmental Panel on Climate Change (IPCC, 2007) also confirmed that, in the twenty-first century, global warming will be more intense in Africa than in the rest of the world.

Obviously, such changes will affect natural resources and all related production systems, including livestock. Hence, in the immediate run, adaptation measures are suggested in order to control the negative effects, including those related to livestock production. In this context, 'adaptation' refers to whether a farmer does something different in response to climate change with the goal of increasing her income, production, household welfare and so forth. Adaptation related to livestock could include deciding whether to keep livestock, to change the quantity of animals or mix of species, to use a different animal husbandry technology and so forth.

In general, livestock production plays an important role both globally and in Africa. According to Blummel et al. (2010), livestock contributes around 40% to agricultural GDP and employs more than a billion people. It also creates a source of livelihoods for more than 1 billion poor people. Livelihood enhancement provided by the extra income from livestock production allows farmers to invest in adaptive farming technology, which is an important aspect of climate change adaptation (Smith et al., 2001). Studies also confirm that livestock production is the world's dominant land use, accounting for about 45% of the earth's land surface, much of which is in harsh and variable environments that are unsuitable for other uses (Sejian et al., 2015; Baumgard et al., 2012). From a nutritional point of view, livestock contributes about 30% of protein in human diets globally and more than 50% in developed countries. Livestock has been considered to be the backbone of agriculture in many developing countries, as it provided draught power/traction and farmyard manure before the promotion of modern agriculture in the middle of the twentieth century, and still provides these inputs to farming in many parts of the developing world. One-third of the world's people depend in part on farm animals. Of the 1.3 billion people living in absolute poverty, 80% live in rural areas and, out of these, 678 million keep livestock (Emboga and Rowlinson, 2008). Poor livestock keepers are among those whose livelihoods are most vulnerable to climate change (Garforth, 2008) and loss of livestock might lead to chronic poverty, with long-term effects on the livelihoods of rural communities (Ainslie, 2002).

Climate change affects the amount and quality of production, reliability of production and the natural resource base on which agriculture, including livestock production, depends. For example, Sunil et al. (2011) show that a rise in temperature affects health and performance along with productive capabilities of livestock in the tropics. That is, an increment in heat stress decreases the feed intake of animals, which in turn reduces milk production, body weight and reproductive capacity (libido, fertility and embryonic survival). Interestingly, the results of this study could be compared with the Seo and Mendelsohn (2008a) study on climate change and livestock. Using a Ricardian approach, Seo and Mendelsohn predict that large livestock farms will be hurt by warming but that small farms will in fact gain, perhaps because small farmers will be the most likely to adapt to climate change. That analysis suggests that farmers will switch species and move away from cattle and toward goats and sheep. Oseni and Bebe (2010) argue that climate change is expected to have maximum impact on vulnerable pastoral communities engaged in extensive livestock production systems, especially in dry-lands. They also argue that design interventions for adaptation to climate change should be based on comprehensive knowledge of the overall structure and dynamics of livestock production systems, including information about indigenous breeding strategies relating to animal adaptation and management in climate-sensitive dry-lands. Hoffmann (2008) also argues that the portfolio of breeds demanded by society is likely to change, which will depend upon the ecosystem changes brought about by climate change and other pressures, as well as on the trade-offs among the public goods considered.

Generally, the choice of animals in Africa today is very temperature-sensitive. Regional studies have also shown that the choice of livestock species is sensitive to climate changes (Seo and Mendelsohn, 2007). However, there has been very little quantitative research on animal husbandry in Africa, and thus there are few empirical studies to compare to these results. Seo and Mendelsohn (2007) analyze the probability of choosing a species in relationship to annual temperature. Seo and Mendelsohn's paper took the mean temperature in sub-Saharan Africa as 22°C. The authors find that a rise in temperature increases the probability of choosing beef cattle compared to milk cattle. However, beef cattle are no longer chosen when farms reach the mean temperature; as temperature increases, the probability of choosing goats and sheep goes up. The estimated probability of keeping chickens is hill-shaped, with a maximum at the mean temperature of Africa of 22°C.

Results from studies on the impact of climate change and adaptation on cattle and sheep farming indicate that long distances decrease the likelihood of adaptation. Distance to weather station had a negative impact on livestock adaptation by farmers (Mandleni and Anim, 2011). A shorter distance to input markets was positively related to livestock adaptation choices. Access to input markets allows farmers to acquire inputs needed for adaptation choices, such as planting of supplementary feed, windbreaks, purchase of new livestock species, and vaccination (Mandleni and Anim, 2011). According to Mandleni and Anim, distances to input markets positively and significantly influence the likelihood of livestock adaptation.

Access to information, particularly weather forecasts, has been found to be an important factor in determining technology adoption choices among livestock producers or pastoralists (Luseno et al., 2003).

Considering demographic variables, the influence of age on a farmer's decision to use livestock as an adaptation strategy appears to be mixed (Galvin et al., 2001). Some researchers have found a negative relationship between age and farmers' livestock adaptation decisions, with older farmers less likely to adapt (Sherlund et al., 2002), while others have found a positive relationship (Imai, 2003).

In addition, some studies have looked into covariates of adaptation in general. Characteristics of the head of household, such as gender, nonfarm income, access to credit, access to information, access to market, and distance to infrastructure, such as a weather station, are variables regarded as crucial in adaptation decision-making among farmers (Gbetibouo and Hassan, 2005). The impact of gender is uncertain. Bayard et al. (2006) find that female farmers are more likely to adopt natural resource management and conservation practices than are their male counterparts. However, Bekele and Drake (2003) find that this variable has no significant value in the adaptation decision-making process.

The impact of climate change on agriculture indicates that climate attributes significantly affect net farm income and reduce adaptation (Mano and Nhemachena, 2006). Households with windy and higher temperatures over the specific areas are less likely to adapt to climate change through adoption of different practices.

Access to credit has a positive impact on adaptation. The existence of access to credit increases the likelihood of adaptation by farmers (Nhemachena and Hassan, 2007). Those results showed that institutional support in terms of the provision of credit was an important factor in promoting adaptation options to reduce the negative effects of climate change (Deressa et al., 2009). Various studies have shown that access to credit by farmers is an important determinant of the adoption of various technologies (Kandlinkar and Risbey, 2000). Studies also indicate that years of education had negative impacts on farmers' likelihood of adapting to climate change. Education has been found to be negatively correlated with farmers' decisions about climate change and adaptation measures (Gould et al., 1989), while access to information has been found to have mixed impacts on farmers' decision making (Dolisca et al., 2006).

Empirical studies on climate change and adaptation have shown that climate characteristics in different agricultural zones significantly affect adaptation decisions (Kurukulasuriya and Mendelsohn, 2006). Therefore, the evidence shows that high levels of warming lead farmers to employ different adaptation measures to offset the climate-related shocks (Deressa et al., 2009).

Theoretical model

Here, we outline the random utility theory underlying our analysis. First, we consider whether the farmer will engage in livestock activity at all. Then, we will consider the choice of various livestock species.

Suppose that the decision of the i_{th} farmer to engage in livestock activity depends on unobservable utility index/latent variable Π_i^*, which is determined by some explanatory variables in such a way that the larger the value of index Π_i^* above a certain threshold, the greater the probability of farmer i engaging in livestock activity. Note that, although the underlying unobservable latent variable is a continuous random variable, all we observe is a binary response variable that takes a value of 1 or 0 according to whether the unobservable latent variable crosses a threshold (Cameron and Trivedi, 2005). Therefore, the choice model in the first step decision is a dichotomous choice model of whether to engage in livestock production activity. We assume that a livestock farmer chooses the optimal combination of outputs and inputs that maximizes the net revenue subject to the prices, climate, soils and other external variables that he or she faces. Thus, the farmer must determine whether engaging in livestock activity is profitable. Next, the farmer also must choose which of the different livestock species, such as cattle for meat, cattle for milk, oxen for draught power, goats, sheep or chicken, to manage. Now, suppose that the regressors are composed of climate, soils and various socioeconomic factors. Suppose the disturbance/stochastic term is known to the farm households and unknown to the econometrician (the farmer is more likely to select an animal species that is highly profitable; Seo and Mendelsohn, 2008a), but the cumulative distribution function (CDF) is a function $F(\varepsilon)$ that is known up to a finite parameter vector. Thus, the profit-maximizing farmer chooses to have livestock if $\Pi_i^* > 0$ or $\varepsilon < X\beta$. Then, the probability that this happens, given X, is $P(\varepsilon < X\beta) = F(X\beta)$. In other words, given the different species available, the farmer compares the expected profit in order to choose which type of livestock species to adopt/manage.

Because the farmer chooses among several livestock species, it is necessary to model the choice of the livestock species by applying different approaches (Seo and Mendelsohn, 2008a). In particular, analyzing such a polychotomous choice using a binary response model is impossible. Rather, a multinomial logit model which can take more than two categorical values would be appropriate (Verbeek, 2008).

However, multinomial logit dictates that the probability that the particular farmer can choose a particular livestock species is given by the probability that the utility or profit of that alternative to that farmer is greater than the utility or profit of any of the other alternatives. Then, the farmer chooses the livestock species that maximizes his or her utility or profit. Hence, we apply a multinomial logit model to analyze all the polychotomous choices among livestock types/species. The primary animal choice of farmers is modeled using a multinomial logit 'primary animal' model; the portfolio (the selected combination based on the entire set of choices) is modeled using a multinomial logit 'portfolio animal' model; and the demand system is modeled using a multivariate probit 'demand system' model. Applying the three models allows us to compare the results of the three different models for the choices of livestock type/ species. Basically, the *primary animal* model assumes that the only choice of importance to the farmer is the primary animal – that is, the species that earns him or her the highest net revenue on the farm. In that model, the farmer chooses a single primary animal from the list of available species; this decision will be addressed via multinomial logit. The second portfolio model shows all the possible combinations of species that a farmer can choose, which will be addressed through the multinomial logit portfolio animal selection model. Furthermore, this model treats specific combinations of species as distinct choices. To continue with these models, we assume that choices are mutually exclusive and, as a result, the farmer can select only one choice of the species. Now, we assume that farmer *i*'s profit in choosing the *j*th animal, where ($j=1, 2,\ldots, J$), is given as:

$$\prod_{ij} = V(k_j, s_j) + \varepsilon(k_j, s_j) \tag{1}$$

where k shows the vector of the exogenous characteristics of the farm and S shows the vector of characteristics of farmer *i*. Note that, for the *portfolio model*, the *j*th choice is a combination of animals instead of a single animal. Here, the vector K includes climate, soils and access variables and S includes the farmer's age, education and family size. The profit function of the farmer is composed of two components: these are the deterministic component V and a stochastic component, ε, where the stochastic component is unknown to the investigator, but is known to the farmer. Then, the farmer chooses the livestock type that gives him or her the highest profit/utility. Defining Z = (K, S), with regard to this model, the farmer chooses the *j*th animal over all the other animals if:

$$\pi^*(Z_{ji}) > \pi^*(Z_{ki}) \text{ for all k} \neq \text{j. [or if } \varepsilon(Z_{ki}) - \varepsilon(Z_{ji}) < V(Z_{ji}) - V(Z_{ki}) \text{ for k} \neq \text{j]} \tag{2}$$

More compactly, farmer *i*'s problem is as follows:

$$\text{Argmax}[\pi^*(Z_{1i}), \pi^*(Z_{2i}),\ldots, \pi^*(Z_{ji})] \tag{3}$$

The probability of P*ji* – that is, the probability for the j_{th} animal to be chosen by farmer *i* – is as follows:

$$
\begin{aligned}
P_{ji} &= \Pr\left[\varepsilon(Z_{ki}) - \varepsilon(Z_{ji}) < V_j - V_k\right] \\
&= \Pr\left[\varepsilon(Z_{ki}) < \varepsilon(Z_{ji}) + V_j - V_k\right] \text{ for all k} \neq \text{j where } V_j = V(Z_{ji})
\end{aligned} \tag{4}
$$

Empirical model and data

Empirical models

In what follows, we outline the empirical framework – that is, the discrete choice models employed in our analysis.

The probit model

If F (ε) is a standard logistic CDF, then, after integration, the probability of the adaptation decision can be stated as follows:

$$P_i = \frac{e^{X\beta}}{e^{X\beta} + 1} \tag{5}$$

Instead, if we assume that ε is IN(0, δ^2), then the decision to hold livestock would be estimated using a probit model (Maddala, 1995). Indeed, if we define y_i as a binary response variable, the probit model can be defined as P_i ($y_i \neq 0 \mid X$) = F ($X\beta$).

Here, the likelihood of observing our sample can be constructed and the maximum likelihood estimators can be obtained through a nonlinear optimization technique (McFadden, 1999). The likelihood function can be expressed as:

$$L = \prod_{\pi^*_i < 0} \left[F(-X\beta)\right] \prod_{\pi^*_i > 0} \left[1 - F(-X\beta)\right] \tag{6}$$

Multinomial logit model

This is considered an unordered dependent variables case (there should be no order within the categories of the dependent variable livestock type).

If V is linear in parameters, then this integration reduces to a simple form of:

$$P_{ji} = \frac{e^{Z_{ji}\gamma_j}}{\sum_{k=1}^{J} e^{Z_{ki}\gamma_k}} \tag{7}$$

This gives the probability that farmer *i* will choose alternative *j* among J alternatives (McFadden, 1999; Train, 2003). The parameters can be estimated by applying the maximum likelihood method, using an iterative nonlinear optimization technique, such as the Newton-Raphson method. These estimates are CAN (consistent and asymptotically normal) under standard regularity conditions (McFadden, 1999).

Multivariate probit model

The multivariate probit models accommodate more than two correlated equations. Moreover, this third approach estimates a system of demand equations for each animal. Then, the farmer determines whether production of a species is profitable. The more profitable the species, the more likely that the farmer will select it. Note that the choices in this theoretical model are not mutually exclusive and farmers can select more than one species. Let Y_{ij} denote the binary response of the i_{th} farmer on the j_{th} animal and then let $Y_i=(Y_i1,\ldots,Y_{ij})$ denote the aggregation of responses on all J animals. According to the multivariate probit model (Chib and Greenberg, 1998), the probability that $Y_i=y_i$, in the condition of parameters β, Σ, and a setup of covariates X_{ij}, is given by:

$$P(Y_i = y_i \mid \Sigma) = \int_{A_{ij}} \cdots \int_{A_{ij}} \phi_j(t_1 \ldots, t_2 \mid 0, \Sigma) d_{t1} \ldots d_{t2} \tag{8}$$

where

$\varphi_j(t \mid 0, \Sigma)$ is J-variate normal distribution with vector mean of 0 and correlation matrix $\Sigma = \{\delta_{jk}\}$, and A_{ij} is the interval

$$A_{ij} = \frac{\left(-\infty, x_{ij}\beta_j\right) \text{ if } y_{ij} = 1}{\left(x_{ij}\beta_j, \infty\right) \text{ if } y_{ij} = 0} \tag{9}$$

Equation (9) shows a system of probit equations for livestock-holding decisions. The estimation accounts for possible correlation across errors in the regressions. Surprisingly, the alternatives are not mutually exclusive and the sum of probabilities is greater than 1.

If we explain z as a vector of climate variables, then the marginal effect of climate change on the probability of holding livestock can be defined as:

$$\frac{\partial P_j}{\partial z} = (1 - P_j) * P_j * [\beta_1 + 2\beta_2 z] \tag{10}$$

Generally, all three approaches to selecting species are theoretically sound. However, each approach is best suited to particular circumstances. The primary animal approach is well suited when the secondary animals are of little economic importance. The portfolio approach is well suited when there are few choices but specific combinations of species are unique and important. Finally, the demand system approach is best suited to the case when the choice of each species is independent of the choice of others (Seo and Mendelsohn, 2008a). For these reasons, the researcher often cannot decide ahead of time which method is to be preferred.

Data and study area description

The data for this study comes from a cross-sectional household-level survey dataset of 1,000 farm households in the Nile Basin of Ethiopia conducted during the 2004/05 production year by the International Food Policy Research Institution (IFPRI) in collaboration with the Ethiopian Development Research Institution (EDRI).

The household survey covered five regional states of Ethiopia and 20 districts. Sample districts were purposely selected to include different attributes of the basin, such as traditional agroecological zones, the degree of irrigation activity (percent of cultivated land), average annual rainfall, rainfall variability, and food vulnerability (aid-dependent population) and included the livestock activity (the decision to engage or not to engage in livestock production and the practice of owning different livestock species). Then, using a purposive sampling technique which was targeted to incorporate all the attributes of the farm households, one peasant association (PA) was selected from each district, making a total of 20 PAs. Once the PAs were chosen, 50 sample farmers were selected from each peasant association using a simple random sampling technique. Thus, a total of 1,000 farmers were selected and interviewed from the entire basin. Furthermore, to enrich the household survey data, long-term data on temperature and rainfall was obtained from the NMA (National Meteorological Agency).

The Nile Basin of Ethiopia covers a total area of about 358,889 km^2, which is equivalent to 34% of Ethiopia's total geographic area, and contains about 40% of the country's population. Portions of six different regional states of Ethiopia are contained within the basin. Specifically, the basin encompasses 38% of the total land area of Amhara, 24% of Oromia, 15% of Benishangul-Gumuz, 11% of Tigrai, 7% of Gambella and 5% of Southern Nations Nationalities and Peoples (SNNP). The basin contains three major rivers: the Abbay River, which originates from the central highlands; the Tekeze River, which originates from the northwestern parts of the country; and the Baro-Akobo River, which originates from the south-western part of the country. The total annual surface runoff of the three rivers is estimated at 80.83 billion cubic meters per year, which amounts to nearly 74% of the total runoff from Ethiopia's 12 river basins (Deressa et al., 2009).

In the Nile Basin of Ethiopia, most farm households engage in livestock farming as a source of livelihood or savings. In the pastoral and agro-pastoral systems, livestock serve as key assets for the poor, providing a cushion against economic, social and environmental shocks. In short, they perform risk management functions. Livestock are kept for various reasons, such as income, manure, plowing, status and savings (Notenbaert et al., 2010). Thus, livestock owners may sell their livestock, move their livestock or provide inputs into their livestock management system as demanded. These management strategies are institutionalized into the culture. Adaptive management of livelihoods may ameliorate effects of climate change (incorporating increased variation in climatic factors) over time, so that households adapt to climatic uncertainty in an incremental way, gradually diverging from initial conditions. However, such adaptations may be less effective or absent when change in environmental conditions is rapid.

Results and discussion

Descriptive statistics

As Table 10.1 shows, the most common response to climate change in the Nile Basin of Ethiopia is to make no changes (50.3%). The surveyed farmers who changed their livestock production activities to cope with the effects of climate

Table 10.1 Livestock Activity and Adjustment to Climate Change

Livestock activity	No. of farmers	% of farmers
Did nothing	503	50.3
Sold livestock	263	26.3
Sold livestock and borrowed from relatives	106	10.6
Sold livestock and ate less	35	3.5
Sold livestock and engaged in food-for-work program	34	3.4
Depend on food aid and liquidate other assets	21	2.1
Sought off-farm opportunities	18	1.8
Other	20	2.0
Total number of farmers	1000	100

Source: Authors' own computation.

Table 10.2 Average Livestock Holding of Households in the Nile Basin of Ethiopia

Livestock type	Obs	Mean	Std. dev.	Min	Max
Cattlemeat	12	1.14	.71	0	3
Cattlemilk	162	2.74	1.48	0	30
Oxenfatt	8	1.56	.87	0	4
Breedbulls	11	2.41	3.03	1	17
Goat	101	4.16	3.32	0	6
Sheep	180	4.49	3.84	1	13
Lamb	10	2.10	2.96	0	21
Beehives	13	2.38	8.99	0	51
Donkeys	21	1.33	.65	0	17
Horse	13	1.81	1.38	1	13
Mules	6	1.28	.47	1	2
Pig	2	2.17	.75	1	3
Oxen	205	2.83	.86	0	6
Chicken	200	4.90	4.15	0	40
Fishownpond	2	1.5	.71	5	6
Other	2	1.75	1.5	1	4
Calf	8	1.64	.93	1	5
Heifer	6	1.28	.45	1	2

Source: Authors' own computation.

change pursued a variety of strategies. They sold livestock; sold livestock and borrowed food or money from relatives; sold livestock and ate less; or sold livestock and engaged in food-for-work programs.

In the study area, over 95% of the sample households engage in livestock production. The average livestock holding of the households (Table 10.2) shows that households hold a diversified portfolio of animal species, with oxen, cattle (for milk), chickens, sheep and goats forming the main livestock types. Based on the foregoing results, the highly specialized livestock types are oxen, cattle (for milk), chickens, sheep and goats. Other types include cattle (for meat), pigs, breeding bulls, oxen (for fattening), lamb, donkey, calf, heifer, mule, horse, bee-hives, fish (own pond) and others. We focus in this analysis on the five major animals because they serve as a store of value and also provide traction (especially oxen) and manure required for soil fertility maintenance (Deressa, 2010). In addition, the key household variables of interest for this study include diversified livestock species held by farmers, costs associated with livestock inputs (including labor) and household characteristics. As depicted in Table 10.2, oxen, chickens, sheep, cattle for milk, and goats were reared by the largest number of households (205, 200, 180, 162, 101 respectively). Consequently, the main livestock products produced were milk and eggs. Fish (own pond), pigs, heifers and other animals are reared by a small number of households in the Nile Basin of Ethiopia.

The average summer temperature experienced by the surveyed households is 18.25 °C and average winter temperature, spring temperature and fall temperature are 18.9, 20.69 and 18.16 (°C) respectively. The average summer, winter, spring and fall rainfalls are 265.36, 7.68, 89.66 and 100.91 mm respectively. In addition, as Table 10.3 shows, the average rainfall in the long-rains season (December–May) and short-rains season (June–November) is 39.64 and 36.41 (mm/mo) respectively. In this case, the winter temperature and spring temperature affect the temperature during the long-rains season, whereas the summer temperature and fall temperature affect the temperature during the short-rains season.

Table 10.3 Summary Statistics for Temperatures (°C) and Precipitation (mm/mo) by Season

Season	Mean	Std. dev.
Stemp	18.25	2.16
Wtemp	18.94	3.22
Ptemp	20.69	2.67
Ftemp	18.16	2.44
Srain	265.35	72.21
Wrain	7.68	12.64
Prain	89.66	48.70
Frain	100.91	49.29
Longtemp	39.64	5.84
Shorttemp	36.41	4.47
Longrain	366.27	115.90
Shortrain	97.34	58.91
Annavtemp	6.34	.83
Annavrain	38.63	11.82

Source: Authors' own computation.

On the other hand, summer and fall rainfalls affect the long-rains season, while winter and spring rainfalls affect the short rain season. In this chapter, the long rainfall (June–November) and short rainfall (December–May) refer to the extended part of the wet and dry climate conditions respectively.

The average age of the household head is 44.92 years. This variable serves as a proxy for the experience of the farm households. This average age shows that almost all of the respondents are economically active in the labor force, which helps them to engage in livestock production ownership activity choice. Moreover, this young age of the household heads can mean that they need training in order to engage in livestock production and ownership activity choice. Furthermore, as Table 10.4a shows, the average household size is 6.16, with a standard deviation of 2.23. That is, on average, the size of the household is large. Those households with large family size can perform different activities within a short period of time. With large family size, a household can divert part of its labor to nonfarm activities and can adopt labor-intensive technology.

Furthermore, the linear and quadratic means of rain during the short-rains season are 97.34 and 12941.64 mm/mo, respectively.

Table 10.4a Summary Statistics of Key Household and Climate Variables (Using Continuous Variables)

Variable	Mean	Std. dev.	Min	Max
Household characteristics				
Age	44.92	13.79	5	92
Education	2.09	2.71	0	13
Hhsize	6.17	2.23	1	15
age_sq	2207.74	1353.16	25	8464
Wet/dry climate conditions				
short_temp	36.41	4.47	26	48.47
short_tempsq	1345.55	334.88	676.26	2348.95
long_temp	39.64	5.84	27.85	52.3
long_tempsq	1605.24	474.6	775.62	2734.99
short_rain	97.34	58.91	−7.499	301.57
short_rainsq	12941.64	13409.6	56.24	90947.1
long_rain	366.27	115.9	70.06	537.68
long_rainsq	147572.6	82742.5	4908.68	
Annual climate conditions				
ann_temp	76.04642	9.949	56.02	99
atemp_sq	5883.847	1569.77	3138.72	9801.37
ann_rain	463.6076	141.79	145.74	734.41
arain_sq	234991.8	132328.6	21239.96	539359.3

Source: Authors' own computation.

Note: **,*** significant at 5%, 1% probability level of significance.

The local agroecological zone variables, such as Dega, Weynadega and Kolla agroecological zones, have a significant impact on the probability of holding live-stock. Farmers living in these agroecological zones are likely to hold livestock. Table 10.4b shows the probability of holding livestock by zone and by access to a livestock facility.

Table 10.4b Percent Livestock Ownership by Access to Livestock Facility and Ecological Zone

All zones

	Livestock ownership		p-value
	No	Yes	
Access to facility			
No	6.67	93.33	
Yes	8.24	91.76	0.374

Dega

	Livestock ownership		p-value
	No	Yes	
Access to facility			
No	8.86	91.14	
Yes	5.00	95.00	0.036**

Weynadega

	Livestock ownership		p-value
	No	Yes	
Access to facility			
No	9.82	90.18	
Yes	5.11	94.89	0.005***

Kolla

	Livestock ownership		p-value
	No	Yes	
Access to facility			
No	5.07	94.93	
Yes	15.60	84.40	0***

Source: Authors' own computation.

Note: **,*** significant at 5%, 1% probability level of significance.

Empirical results

Decision to engage in livestock activities

The first outcome is the choice of whether to engage in livestock production. As noted earlier, more than 95% of the sample households were found to hold livestock. Here, we present two variants of the livestock ownership model: the first one is a climate-only model without household characteristics, and the second one includes some household characteristics. First of all, we model the impact of seasonal climate variables on the decision of the household to hold livestock. In this study, all the results of probit model regressions present the marginal effects (rather than the coefficients) of each variable on the probability of engaging in livestock production because the marginal effects measure the change in the unobservable y^* associated with a unit change in one of the explanatory variables. This is a more useful measure. Moreover, the coefficients of probit models are more difficult to interpret than the marginal effect coefficients of the probit model.

Furthermore, in the climate impact–only model, all the seasonal climate variables except spring rainfall are statistically significant determinants of the decision to hold livestock. As depicted in Table 10.5, higher summer and spring temperatures are positively correlated with the decision to hold livestock, whereas higher winter and fall temperatures are negatively correlated with that decision.

Table 10.5 Marginal Effects of Probit Regression of Livestock Ownership (Dependent Variable =1 If Own, 0 Otherwise) Seasonal Climate Variable Model

Variable	Coefficient	Std. error	$P>|z|$
Stemp	.047*	(.02704)	0.078
wtemp	−.067***	(.016)	0.000
Ptemp	.099***	(.021)	0.000
Ftemp	−.074*	(.041)	0.073
srain	.0012***	(.00047)	0.007
wrain	.0032**	(.0016)	0.046
prain	−.00045.	(.00057)	0.432
frain	−.00086**	(.00044)	0.047
Number of observations	1000		
LR chi2 (8)	66.87***		
Prob> chi2	0.0000		
Pseudo R2	0.1093		
Log likelihood	−241.70922		

Source: Authors' own computation.

Note: Robust standard errors in parentheses. * Significant at 10%;
** significant at 5%;
*** significant at 1%.

Table 10.5 shows the marginal effects of temperature change. A summer temperature increase of 1 °C would increase the probability of holding livestock by 4.76%, holding other things constant. This result of the analysis is consistent with the result found by Seo and Mendelsohn (2008b). A winter temperature increase of 1 °C would decrease the probability of holding livestock by 6.73%, holding other things constant. This result is consistent with the results of Kabubo-Mariara (2008) and Seo and Mendelsohn (2008b). Similarly, summer rainfall is statistically significant at (p<0.01) and positively affects livestock holding. Specifically, a summer rainfall increase of 1 mm/month would increase the probability of holding livestock by 0.13%, holding other things constant. The probable reason is that livestock holding demands availability of enough water, grazing land and pasture for rearing.

Here, the results of the seasonal temperatures are surprising. Spring and fall temperatures have higher marginal effects than temperatures during other seasons. In a similar manner, summer and spring temperatures have statistically significant effects at (p<0.05) and at (p<0.01) respectively and are positively correlated with livestock holding. Similar studies by Apata et al. (2009) confirm that higher temperature positively affects adaptation to climate change. Furthermore, winter and fall temperatures are statistically significant at (p<0.01) and at (p<0.1) respectively and both are negatively correlated with livestock holding. During the long rains, farmers are encouraged to engage in livestock holding or to keep more livestock due to the availability of water, forage and pasture. The marginal effects of rainfall, however, are low in magnitude. Importantly, these marginal effects suggest that the decision to hold livestock is more sensitive to temperature variations than to rainfall variations.

Turning to demographic variables, Table 10.6 shows that both the log of the household size and education are significant at (p<0.01; p<0.05) and are positively correlated with livestock holding. A unit percentage change of the household size increases the probability of holding livestock by 6.55%. The results suggest that households with larger family size are more likely to engage in livestock production than are smaller households. Although livestock production may not be labor-intensive in pastoral regions, it is rather labor-intensive in smallholder farming due to scarcity of pasture or grazing land.

Education level is significant at (p<0.05) and positively affects the decision to engage in livestock holding. In other words, among livestock farmers in the area of study, education appears to be a contributing factor to adaptation. As the farmer's education increases by one year, the probability of holding livestock increases by 0.055%. Previous research by Apata et al. (2009) confirms that education positively influences adaptation.

In addition, farmers living in Weynadega are 79.9% more likely to engage in livestock holding as compared to those living in Kolla. The probable reason is that in Weynadega there is more grazing land and water availability. Though there is little empirical evidence that farmers living in hot climate conditions are more likely to engage in livestock holding to adapt to climate change, ownership of more grazing land and sufficient water are major factors that induce

Table 10.6 Marginal Effects of Explanatory Variables from the Probit Regression of Whether to Own Livestock Model: All-Variables Model

Variable	Coefficient	Std. error	P>|z|
Stemp	.0496**	(.0242)	0.040
Wtemp	−.0564***	(.016)	0.000
Ptemp	.0784***	(.0189)	0.000
ftemp	−.0656*	(.038)	0.085
Srain	.0011**	(.00042)	0.011
Wrain	.0024	(.0015)	0.120
Prain	−.0003	(.0005)	0.557
Frain	−.00075**	(.00038)	0.047
Loghhsize	.0655***	(.015)	0.000
Age	.0019	(.00199)	0.332
age_sq	−.000024	(.00002)	0.230
Education	.00548**	(.0026)	0.033
accessfacility	−.1129	(.1452)	0.437
landinhectare	−.0112	(.0529)	0.833
Dega	.7261**	(.3123)	0.020
Weynadega	.7994***	(.2569)	0.002

Source: Authors' own computation.

Note: Dependent variable: livestock ownership (if own=1, otherwise=0). Number of obs: 1000. LR chi2 (12): 103.29***. Prob> chi2: 0.0000. Pseudo R2: 0.19. Robust standard errors are in parentheses.
*Significant at 10%;
**significant at 5%;
***significant at 1%.

farmers to engage in livestock activity; the result here is consistent with this reasoning.

The results indicate that the significant variables of the wet/dry climate variables are the linear and quadratic terms of the long-rains temperature at (p<0.1), the linear term of the short-rains temperature at (p<0.1) and the linear and quadratic terms of the long-rains temperature at (p<0.05) and at (p<0.1) respectively. Among the household characteristics, the important characteristics are log of household size and education at (p<0.01; p<0.05) respectively. As long-rains temperature increases by 1°C, the probability of holding livestock increases by 8.12% Also, as the quadratic of long-rains temperature increases by 1°C, the probability of holding livestock decreases by 0.14%.

Furthermore, Table 10.7 shows that the amount of rainfall during the short rains is significantly and negatively correlated with the farmer's livestock holding. When short rainfall increases by 1 mm/month, the probability of holding livestock decreases by 0.17%. A similar study by Kabubo-Mariara (2008) indicated that more rain during the short-rains season negatively affected livestock holding. The livestock response to temperature during the short rains is U-shaped, but the

Table 10.7 Marginal Effect of Explanatory Variables from Probit Model Results of Live-stock Adoption: Wet/Dry Climate Conditions

Variable	Coefficients	Std. error	$P>|z|$
shortt~p	−.0537	(.0419)	0.200
shortt~q	.0011	(.0007)	0.107
Longtemp	.0812*	(.0484)	0.093
longte~q	−.0014*)	(.00074)	0.058
shortr~n	−.0017*	(.001)	0.096
sho~n_sq	0.000005)	(.00000)	0.171
lon~n_sq	−0.00000096*	(.00000)	0.076
Loghhsize	.0767***	(.0168)	0.000
Age	.0029413	(.0023)	0.202
age_sq	−.000037	(.00002)	0.113
Education	.0073**	(.0029)	0.011
Accessfacility	−.0157	(.014)	0.257
Landinhectare	−.00063	(.0055)	0.909
Dega	.065778**	(.026)	0.011
Weynadega	.0732***	(.026)	0.005

Source: Authors' own computation.

Note: Dependent variable: livestock ownership (if own=1, otherwise=0) Number of observations: 1000. LR chi2 (12): 75.36***. Prob> chi2: 0.0000. Pseudo R2: 0.137. Robust standard errors in parentheses.
*Significant at 10%;
**significant at 5%;
***significant at 1% probability level of significance.

livestock response to temperature during the long rains is hill-shaped. Similarly, the livestock response to short rainfall and long rainfall is U-shaped and hill-shaped respectively.

As Table 10.7 shows, the overall fit of the model increases by 10.9% as a result of the introduction of household and agroecological zone characteristics. With the introduction of household characteristics, household size and education of the household head are significant at ($p<0.01$; $p<0.05$) respectively and both have a positive correlation with livestock holding.

Table 10.8 shows the last model for the household's decision of whether to keep livestock, with regard to the variables of the annual climate conditions. The significant variables are the quadratic term of annual temperature at ($p<0.1$), the linear and the quadratic terms of annual rainfall at ($p<0.05$), log of the household size at ($p<0.01$) and education of the household head at ($p<0.05$). The quadratic annual temperature is significant at the ($p<0.1$) probability level and is negatively correlated with livestock holding. A 1 °C change in the quadratic term of annual temperature would decrease the probability of holding livestock by 0.011%. Similarly, the linear and quadratic terms for rainfall are positively and negatively correlated with livestock holding, respectively. When the linear term of rainfall

Table 10.8 Marginal Effect of Explanatory Variables from Probit Model Results of Livestock Adoption: Annual Climate Conditions

Variable	Coefficients	Std. error	P>|z|
ann_temp	.0141	(.0106)	0.183
atemp_sq	−.0001*	(.00007)	0.098
ann_rain	.00069**	(.00033)	0.038
arain_sq	−0.0000008**	(.00)	0.018
Age	.0033	(.0025)	0.192
age_sq	−.000042	(.00003)	0.101
Loghhsize	.0685***	(.0175)	0.000
Education	.0061**	(.00295)	0.038
accessfacility	−.0181	(.014)	0.204
landinhectare	−.00202)	(.0056)	0.720
Dega	.0733***	(.0184)	0.000
Weynadega	.0862***	(.0235)	0.000

Source: Authors' own computation.

Note: Dependent variable: livestock ownership (if own=1, otherwise=0). Number of observations: 1000. LR chi2 (8): 70.25***. Prob>chi2: 0.0000. Pseudo R2: 0.14. Log likelihood: −234.6. Robust standard errors are in parentheses. *Significant at 10%;
**significant at 5%;
***significant at 1%.

increases by 1 mm/month, the probability of holding livestock increases by 0.069%, but only up to a certain point. But a similar change in the quadratic value would lead to a negligible effect. With regard to the Dega and Weynadega agroecological zones, there is a statistically significant and economically meaningful correlation between living in these zones and the probability of holding livestock.

In this analysis, the interest is to examine the impact of the combination of the seasonal and annual climate conditions on livestock holding. As the results displayed earlier show, due to the powerful correlation between the short-rains season temperature and the amount of rainfall in the short-rains season, they are excluded from the model. Therefore, the results indicate that the marginal impacts of the wet/dry and annual climate conditions are insignificant when evaluated without household characteristics and the overall fit of the model is quite poor, which is not uncommon in cross-sectional data. But it is a little bit better than a model with only an intercept. The coefficients for the climate variables are insignificant.

The results from the marginal effect of the probit model presented in Table 10.9 indicate that the model has poor overall predictive power. This is indicated by the fact that the overall measure of fit of the model of the Pseudo R2 is a 2.76% prediction for the marginal effect of the wet/dry and annual climate conditions. The likelihood ratio/Wald chi2 test is 13.72 for the marginal effect of the probit model for the wet/dry and annual climate conditions. As a result, the likelihood ratio/

Table 10.9 Marginal Effects of Explanatory Variables from Probit Model of Livestock Adoption

Variable	Coefficients	Std. error	P>|z|
Longtemp	−.0073	(.0088)	0.411
Anntemp	.0018	(.0046)	0.701
longrain	−.00012	(.00021)	0.574
Ann_rain	.000062	(.00014	0.651

Source: Authors' own computation.

Note: Dependent variable: livestock ownership (if own=1, otherwise=0). Number of obs: 1000. LR chi2 (4): 13.72***. Prob> chi2: 0.0082. Pseudo R2: 0.0276. Log likelihood: −263.88. Robust standard errors are in parentheses. *Significant at 10%; **significant at 5%; ***significant at 1%.

Wald chi2 tests were used to test the null hypothesis that all coefficients were zero for each of the models. Given the p-value of 0.01 for both the likelihood ratio and the Wald chi2 tests, the null hypothesis for each model was rejected. The log likelihood of the marginal effect of the probit model on livestock holding is −263.88234, which corresponds to the value of the log likelihood at convergence.

Choice of livestock species

For this study, the estimation of the multinomial logit 'primary animal' selection model is undertaken by normalizing one category, which is normally referred to as the 'reference' or 'base category.' In this technique, the first category (oxen) is the reference category.

The second decision that we investigate is the choice of the livestock species to adopt, separate from making the decision to engage in livestock production. The analysis of the different choices of livestock species is based on an annual climate model because the powerful correlation of the seasonal as well as the wet/dry climate variables is a little bit higher than that of the annual climate conditions. The results concentrate on the decision of the farm household to hold cattle for milk, goats, sheep, oxen and chickens. However, as explained earlier, the parameter estimates of the MNL model merely provide the direction of the impact of the independent variables on the response variables. Therefore, the estimates do not represent the magnitude of change or probabilities. Based on this fact, the marginal effects from the MNL, which measure the expected change in probability of a particular choice being made with respect to a unit change in an independent variable, are reported and discussed. In all cases, the estimated coefficients should be compared with the base category of oxen in response to the primary animal selection model.

Table 10.10a shows the marginal effect of the probability of choosing oxen relative to the base reference of the five major animals chosen in the multinomial logit primary animal model. In this model, the primary animals are selected for

Table 10.10a Marginal Effects of Multinomial Logit 'Primary Animal' Selection Model Results for Choice of Livestock Species

	Cattlemilk			Goats			Sheep		
	Coefficients	Std. error	P>|z|	Coefficients	Std. error	P>|z|	Coefficients	Std. error	P>|z|
ann_temp	-.0392***	(.0114)	0.001	-.0096***	(.0036)	0.008	-.008*	(.0047)	0.085
atemp_sq	.00025***	(.00007)	0.001	.000067***	(.000023)	0.003	.00005*	(.00003)	0.097
ann_rain	-.0015***	(.0004)	0.000	-.000022	(.0002)	0.920	-.0004**	(.0002)	0.024
arain_sq	2E-6***	(4E-7)	0.001	-1E-8	(2E-7)	0.957	4E-7**	(2E-7)	0.023
Loghhsize	.0199	(.0229)	0.383	-.0030	(.0096)	0.754	.0046	(.0082)	0.57
Age	.00059	(.00077)	0.449	.00036*	(.0002)	0.061	.0002	(.0003)	0.503
Education	.00051	(.00389)	0.896	.0014	(.0014)	0.192	-.0003	(.0024)	0.909

	oxen			chicken		
	Coefficients	Std. error	P>|z|	Coefficients	Std. error	P>|z|
ann_temp	.1046***	(.02098)	0.000	-.0347***	(.0083)	0.000
atemp_sq	-.0007***	(.00013)	0.000	.00024***	(.00005)	0.000
ann_rain	.0036***	(.0007)	0.000	-.00088**	(.00034)	0.014
arain_sq	-4E-6***	(1E-6)	0.000	1E-6**	(4E-7)	0.036
Loghhsize	.0024	(.034)	0.485	-.0259	(.0179)	0.149
Age	-.002*	(.0011)	0.096	-.00044	(.00057)	0.444
Education	-.002	(.0054767)	0.736	.0001	(.00249)	0.967

Note: Dependent variable: livestock species choice (if cattlemilk=1, goat=2, sheep=3, oxen=4, chicken=5). Number of obs: 1000. LR chi2 (4): 152.35***. Prob> chi2: 0.0000. Pseudo R2: 0.0757. Log likelihood: −989.856. Robust standard errors in parentheses. *Significant at 10%; **significant at 5%; ***significant at 1%.

the purpose of selling the animal or its products. Taking annual temperature into consideration, both the linear and quadratic terms of annual temperature are significant at (p<0.01) but the direction of the signs are opposite: the linear term for temperature is negatively correlated with the household's choice of cattle for milk but the quadratic term is positively correlated with that decision. Annual temperature has a U-shaped effect on the choice of cattle for milk. This result is in line with the finding of Seo and Mendelsohn (2008b).

The marginal effects of individual variables suggest that temperature changes are the key drivers of the decision to hold cattle for milk. The probability of keeping cattle for milk exhibits a U-shaped relationship with annual rainfall. However, the marginal effect of annual temperature on the choice of cattle for milk is 26 times as large as the marginal effect of annual rainfall on the choice of cattle for milk. In Table 10.10a, annual temperature exhibits a U-shaped relationship with the probability of selecting goats.

Age of the household head is significant at (p<0.1) and it positively affects the probability of choosing goats. A one-year increase in the age of the household head increases the probability of selecting goats by 0.036%. The probability of selecting goats is high when age increases because it is simpler to feed goats compared to cattle or oxen. Thus, more elderly households can keep the less labor-demanding animals, such as goats, without requiring any labor support.

Annual temperature and annual rainfall exhibit a U-shaped effect on the probability of selecting sheep. The results presented in Table 10.10a indicate that both annual temperature and annual rainfall exhibit a hill-shaped relationship with the probability of selecting oxen. The probable reason is that rural farm households use oxen as a basic instrument for traction for plowing their land.

The study also shows that age of the household head is significantly (at p<0.1) and negatively correlated with the probability of keeping oxen. When households become older, their probability of selecting oxen declines because oxen need a large amount of forage and pasture. The study reveals that both annual temperature and annual rainfall have a U-shaped relationship with the probability of selecting chickens. All the marginal impacts of the climate variables significantly affect the probability of keeping goats, sheep, oxen, chickens, and cattle for milk.

Table 10.10b reveals that goats and sheep are less responsive to annual temperature than are oxen or dairy cattle. The marginal impacts for oxen and cattle for milk are higher than the marginal effects for goats and sheep. Goats are chosen in high-temperature areas, because they are capable of resisting high temperature. In heavy rain areas, the probability of selecting goats is low.

As depicted in Table 10.10b, annual temperature negatively affects the probability of holding oxen, implying that the probability of selecting oxen is quite low in areas that are arid and lack forage. Oxen demand more forage, water and pasture than do small ruminants, such as goats and sheep. Besides, the marginal effect of the annual temperature on the probability of choosing chickens is higher than the choice of goats, which is the next-most important choice of species next to oxen. However, the marginal effect of annual rainfall is stronger for the probability of selecting chickens rather than oxen.

Table 10.1ob Marginal Effects of Choice of Livestock Species

	Coefficient	Std. error	P>|z|	Coeff.	Std. error	P>|z|	Coeff.	Std. error	P>|z|
	Cattlemilk			Goat			Sheep		
ann_temp	-.0018	(0013)	0.168	.0015***	(.0003)	0.000	-.0006	(.0007)	0.332
ann_rain	-.0002*	(.0001)	0.053	-.00005*	(.00003)	0.076	-.00002	(.00004)	0.703
	Oxen			Chickens					
ann_temp	-.0046***	(.0017)	0.006	.0029***	(.0009)	0.002			
ann_rain	.0002	(.00011)	0.115	-.0002***	(.0001)	0.004			

Source: Authors' own computation.

Note: Number of obs: 1000. Wald chi2 (10): 55.22**. Prob> chi2: 0.0000. Pseudo R2: 0.0362. Log pseudolikelihood: −1032.1855. Robust standard errors in parentheses.
*Significant at 10%;
**significant at 5%;
***significant at 1%.

The model was run and tested for the validity of the independence of irrelevant alternatives (IIA) assumptions by using both the Hausman test for IIA and the seemingly unrelated post-estimation procedure (SUEST). The results of the two post-estimation tests failed to reject the null hypothesis of independence of the climate change adaptation alternatives and confirmed that the multinomial logit (MNL) model specification is appropriate to model climate change adaptation strategies of farm households (chi2 ranged from 2.03 to 2.10, with the probability values ranging from 1.0000 to 0.8352 in the Hausman test, and chi2 ranging from 5.37 to 5.74, with probability value of 0.1 to 0.3 in the case of SUEST).

This result reveals that the multinomial logit is the optimal portfolio animal selection model. That model simultaneously contains all the combinations of the species that a farm household has chosen. In this analysis, the base category of the farm household is the combination of sheep, oxen and chicken together. In the optimal portfolio model result, educational level of the household head increases the probability of choosing goats and oxen together. More educated households have the means of holding diversified species, either for traction of land (especially oxen) or for selling the small animals during idiosyncratic shocks, for consumption smoothing within the families. These activities benefit from education because some basic skills are required.

Similarly, annual temperature exhibits a U-shaped relationship with the probability of choosing goats and chicken simultaneously. This is apparently because chickens have less capacity than goats to resist high temperature. So, the effect is weighted by the greater effect of annual temperature on chickens than on goats. Size of the farm household has a positive and significant effect at ($p<0.01$) on the probability of selecting goats and chicken simultaneously.

As shown in Table 10.11, the probability of simultaneously keeping goats, chicken and oxen exhibits a U-shaped relationship with annual temperature. Of the household characteristics, educational level negatively and significantly affects the probability of holding goats, chickens and oxen. This result supports some literature that suggests that more educated farmers are likely to keep less livestock, compared to their less educated counterparts, because education diversifies alternative income earning opportunities (Kabubo-Mariara, 2008).

The multivariate probit model was run for the choice of non-mutually exclusive species. The multivariate probit results presented in Table 10.12 show the coefficients rather than the marginal effects presented in the probit as well as the multinomial model. Here, the annual temperature is significantly correlated with the probability of selecting the different species in all cases except for chickens. In addition, the log transformed of the household size is significantly and positively correlated with the probability of selecting the different species in all cases. The Wald chi2 test indicates that the overall model is quite significant and also suggests that the system of equations is completely stable.

The results for the choice of dairy cattle, sheep and oxen show a hill-shaped response to annual temperature, whereas the choice of goats shows a U-shaped response to annual temperature. Contrary to the multinomial logit primary animal

Table 10.11 Marginal Effects of Explanatory Variables from Multinomial Logit 'Optimal Portfolio Animal' Model

	Goats + oxen		Goats + chickens		Goats + chickens + oxen	
	Coefficient	Std. error	Coefficient	Std. error	Coefficient	Std. error
ann_temp	.0019	(.0038)	-.0114*	(.006)	-.02029**	(.0086)
atemp_sq	-.000014	(.00003)	.000075*	(.00004)	.00012**	(.000056)
ann_rain	.0001	(.00024)	-.00036	(.00031)	-.0006*	(.00036)
arain_sq	-0.0000001	(0.0000003)	0.00000044	(0.0000003)	0.000001	(0.0000004)
Loghhsize	.0144	(.0123)	.056***	(.015)	.0043	(.0212963)
Age	-.00039	(.00032)	.00025	(.0005)	.007	(.000597)

	Sheep + goats + chickens		Oxen + chickens		Sheep + goats + oxen	
	Coefficient	Std. error	Coefficient	Std. error	Coefficient	Std. error
ann_temp	-.02167**	(.0085)	.0088	(.0184)	.002	(.009)
atemp_sq	.00012**	(.000054)	-.000045	(.00012)	-.000017	(.00006)
ann_rain	.0199***	(.0005)	-.0012*	(.00062)	-.00012	(.00028)
arain_sq	-0.000002***	(0.0000005)	0.0000008	(0.0000007)	0.00000011	(0.0000003)
Loghhsize	-.00497	(.024)	.0023	(.035)	-.016	(.012)
Age	-.00059	(.00067)	-.00025	(.0011)	.00299	(.0005)
Education	.0014	(.0031896)	-.00008	(.0052)	.0444**	(.002076)

	Sheep + chickens		Sheep + oxen + chickens		Sheep + oxen	
	Coefficient	Std. error	Coefficient	Std. error	Coefficient	Std. error
ann_temp	.026	(.0179)	-.0012	(.0199)	-.0009	(.01)
atemp_sq	-.00016	(.00011)	.00002	(.00013)	-.0000043	(.000066)
ann_rain	-.00019	(.0006)	-.0012*	(.00069)	-.0003239	(.00032)
arain_sq	0.00000021	(0.0000006)	0.0000008	(0.0000007)	0.0000003	(0.00000035)
Loghhsize	.044*	(.0262)	-.00169	(.0377)	-.003	(.0207)
Age	-.00065	(.0008)	.00007	(.0011)	.00008	(.00056)
Education	-.0034	(.004)	.08	(.006)	-.00026	(.0026)

(Continued)

Table 10.11 (Continued)

Band 1

	Sheep + goats Coefficient	Std. error	Cattlemilk + goats + sheep + oxen + chickens Coefficient	Std. error	Cattlemilk + oxen Coefficient	Std. error
ann_temp	.0087	(.0083)	.00012	(.0013)	.0011	(.004)
atemp_sq	-.00006	(.000055)	-0.0000004	(0.000008)	-0.000008	(.000024)
ann_rain	-.000099	(.00028)	.00021*	(.00011)	.00014	(.00016)
arain_sq	-0.000000014	(0.0000003)	-0.00000017*	(0.0000001)	-0.0000001	(0.0000002)
loghhsize	.00651	(.0146)	-.0058105*	(.003527)	.00457	(.0052)
Age	-.0006	(.00044)	.000043	(.0000588)	-.00026	(.00018)
Education	.00375**	(.00176)	-.00047	(.00029)	.00049	(.0006)

Band 2

	Cattlemilk + chickens Coefficient	Std. error	Cattlemilk + goats + sheep Coefficient	Std. error	Cattlemilk + goats Coefficient	Std. error
ann_temp	.00374	(.0043)	.0077**	(.00325)	.016924	(.0029)
atemp_sq	-.000022	(.000026)	-.000046**	(.000019)	-.0000128	(.000018)
ann_rain	-.000066	(.000124)	.000085	(.00016)	.00249**	(.00013)
arain_sq	0.00000001	(0.00000012)	-0.00000008	(0.0000002)	-0.0000023*	(0.00000014)
Loghhsize	-.0057	(.0045)	.0059	(.0062)	.0015	(.0048)
Age	.0000887	(.0001)	-.000081	(.00017)	-.000099	(.00015)
Education	.000074	(.00086)	-.0003	(.00079)	.0097*	(.0005)

Band 3

	Cattlemilk + sheep Coefficient	Std. error	Cattlemilk + goats + oxen Coefficient	Std. error	Cattlemilk + sheep + oxen Coefficient	Std. error
ann_temp	-.000929	(.001465)	-.000097	(.0026)	.00083	(.0032)
atemp_sq	0.00000395	(0.0000092)	-0.00000031	(.000016)	-0.0000054	(.000019)
ann_rain	.000296**	(.00014)	.00022*	(.00012)	.00023	(.000158)
arain_sq	-0.00000025*	(0.00000014)	-0.0000002	(0.00000013)	-0.0000002	(0.0000002)
loghhsize	.0058234	(.0069)	-.0049	(.0034)	-.0016	(.0071)0.817
Age	.000021	(.00019)	.000062	(.000082)	.0001	(.00013)
Education	.000298	(.00073)	.00064	(.00056)	-.00059	(.00086)

	(Model 1)	(Model 2)	(Model 3)
ann_temp	-.0207 (.0137)	.0057* (.003086)	.0020089 (.0029)
atemp_sq	0.0000056 (0.0000074)	-.000034* (.000019)	-.0000139 (.00002)
ann_rain	.00017 (.00015)	0.250 .0003395* (.00018)	.0003119** (.00016)
arain_sq	-0.00000012 (0.00000014)	-0.00000033** (0.0000002)	-0.0000003* (0.0000002)
Loghhsize	-.0042 (.0037)	-.0019 (.00292)	.0044732 (.0065)
Age	.000103 (.00013)	-.000024 (.00012)	-.0023* (.00013)
Education	.0002 (.000498)	.0022 (.00035)	.0003 (.000499)
	Cattlemilk + oxen + chickens	Cattlemilk + goats + sheep + oxen	Goats + sheep + oxen + chickens
ann_temp	00361 (.002497)	.0099 (.00144)	.0036 (.00245)
atemp_sq	.000022 (.000015)	-.000006 (0.000009)	-.000022 (.000015)
ann_rain	.0025* (.00013)	.00064 (.0001)	.000234* (.00014)
arain_sq	-0.00000024** (0.00000012)	-0.0000001 (0.0000001)	-0.00000022* (0.00000013)
loghhsize	.00135 (.00296)0.649	(.000292)	-.0165 (.00177)
Age	.00012 (.000094)	-0.0000008 (0.0000079)	-.0000242 (.00006)
Education	.000066 (.000315)	0.000004 (.00003)	-.00091 (.0006314)

Note: Robust standard errors in parentheses.
Number of observations: 1000. LR chi2 (168): 627.00***. Pseudo R2: 0.0632. Log likelihood: −2521.0632.
*Significant at 10%;
**significant at 5%;
***significant at 1%.

Table 10.12 Multivariate Probit 'Demand System' Regression Results for Choice of Livestock Species

	cattle_milk1			goat_1			sheep_1		
	Coefficient	(se)	P>|z|	Coefficient	(se)	P>|z|	Coefficient	(se)	P>|z|
ann_temp	.2493***	(.053)	0.000	-.2346***	(.062)	0.000	.2038***	(.0546)	0.000
atemp_sq	-.0017***	(.0003)	0.000	.00169***	(.0004)	0.000	-.0014***	(.00035)	0.000
ann_rain	.0018	(.002)	0.372	.0021	(.003)	0.477	.0019	(.002)	0.336
arain_sq	-0.000002	(0.000003)	0.262	-0.000004	(0.0000032)	0.240	-0.000003	(0.000002)	0.193
Loghhsize	.668***	(.1039)	0.000	.536***	(.152)	0.000	.336***	(.1015)	0.001
Age	.00025	(.0031)	0.935	-.0109**	(.0044)	0.013	-.0035	(.0031)	0.249
Education	.0154	(.0156)	0.325	.0389**	(.01985)	0.050	-.00036	(.0154)	0.981
_cons	-10.49	(2.083)	0.000	6.046	(2.479)	0.015	-8.040	(2.118)	0.000

	oxen_1			chicken_1		
	Coefficient	(se)	P>|z|	Coefficient	(se)	P>|z|
ann_temp	.45***	(.0565)	0.000	.0367	(.05188)	0.479
atemp_sq	-.003***	(.0004)	0.000	-.00024	(.00033)	0.469
ann_rain	.0028	(.0021)	0.171	-.0036*	(.002)	0.073
arain_sq	-0.000003	(0.000002)	0.12	0.000002	(0.000002)	0.363
Loghhsize	.6901***	(.106)	0.000	.389***	(.1026)	0.000
Age	-.0038	(.0032)	0.229	-.00356	(.0031)	0.241
Education	-.01064	(.0161)	0.509	.0147	(.0153)	0.337
_cons	-18.325	(2.215)	0.000	-.8777	(2.028)	0.665

Source: Authors' own computation.

Note: Number of obs.: 1000.
Wald chi2 (35) 311.70***
Prob> chi2 0.0000
Log likelihood: -2779.45
Robust standard errors in parentheses.

*Significant at 10%;
**significant at 5%;
***significant at 1%.

model results, in this model the choice of dairy cattle, sheep and chickens exhibits a hill-shaped response to annual temperature.

Likewise, the choice of dairy cattle, goats and sheep shows a hill-shaped response to annual rainfall, whereas, similarly to the multinomial logit primary animal model, the choice of oxen depicts a hill-shaped response to both annual temperature and rainfall. The choice of goats also exhibits a U-shaped response to annual temperature. Similarly, the choice of chickens shows a U-shaped response to annual rainfall.

Moreover, household size correlates positively and significantly with the choice of cattle for milk, goats, sheep, oxen and chickens. This result is consistent with the assumption that larger families ordinarily have a higher labor endowment, which would allow a household to accomplish various agricultural tasks, particularly during peak seasons (Croppenstedt et al., 2003).

Conclusions and recommendations

To analyze climate change adaptation and livestock activity choice, we used a cross-sectional farm household survey dataset covering a sample of 1,000 households from five regions in the Nile Basin of Ethiopia collected during the production year of 2004/2005 by the Ethiopian Food Policy Research Institute in collaboration with the Ethiopian Development Research Institute (and enriched with secondary climate data which manifest long-term climate change).We investigate the impact of climate change on farmers' decisions about whether to engage in livestock activities and the choice of different livestock species. The likelihood of engaging in livestock activities is analyzed by employing a probit model on the entire sample. The choice of different livestock types/species is analyzed using multinomial logit and a multivariate probit model.

Results indicate that livestock holding is more sensitive to temperature than to rainfall variations. The likelihood of the farmer engaging in livestock activity is significantly and positively affected by household size and education of the household head, implying that households with larger family size are more likely to engage in livestock production. Although livestock production in pastoral regions may not be very labor-intensive, it is rather labor-intensive in smallholder farming due to the scarcity of pasture (forage) or grazing land. Even though more educated households are less likely to engage in livestock production because they have alternative income earning opportunities, education positively affects the probability of engaging in livestock production. This is because livestock production demands more than just basic skills.

In the multivariate probit model, annual temperature significantly affects the likelihood of farmers choosing cattle for milk, as well as sheep and oxen, and it exhibits a hill-shaped relationship with the probability of selecting these livestock species. In general, the probit and the multinomial logit model results suggest that there is a nonlinear relationship between the decision to engage in livestock production, choices of livestock species, and global warming, showing that farmers adapt their livestock-holding decisions and livestock species to climate change.

The results of the multinomial logit analysis of choice of species strongly suggest that with increasing warming, farmers will reduce milk cattle in preference for draught oxen. In addition, goats and sheep are less responsive to climate change than are dairy cattle and oxen, suggesting that they can hold out in harsher climate conditions. Based on the results of the multivariate probit model, the probability of choosing goats increases, while the probability of sheep-rearing decreases. The corresponding impacts of climate-related decrease in rainfall also lead to substitution of dairy cattle in place of chickens, and, likewise, goats instead of sheep.

These results suggest the need for policies that minimize the impact of climate change on production and ownership systems in the Nile Basin of Ethiopia. Such policies should include security of rights to use land for pasture, because access to pasture plays a role in livestock decisions. In general, adaptation policy should be based on comprehensive knowledge of the structure and dynamics of livestock production systems, including information about indigenous breeding strategies related to animal adaptation and management in climate-sensitive dry-lands. Research and development programs must prepare appropriate technologies, and information must be disseminated to help farmers adapt to climate change. In general, the results of this study suggest that adaptation to climate change in the Nile Basin of Ethiopia must be significant if livestock-keeping households are to counter the expected effects of climate change on their livelihoods. There is also a need to educate farmers about the vulnerability of specific livestock species to climate change. Further research is necessary concerning adaptation alternatives based on specific agroecological zones, as this may have important significance for livestock keeping and choice of livestock species, given large geographical differences and variability of both temperature and rainfall in Ethiopia.

References

Ainslie, A. 2002. Cattle Ownership and Production in the Communal Areas of the Eastern Cape, South Africa. Programme for Land and Agrarian Studies Research Report No. 10. Cape Town, South Africa: University of the Western Cape.

Apata, T.G., K.D. Samuel and A.O. Adeola. 2009. Analysis of Climate Change Perception and Adaptation among Arable Food Crop Farmers in South West Nigeria. Contributed Paper Prepared for Presentation at International Association of Agricultural Economists Conference, Beijing, China.

Baumgard, L.H., R.P. Rhoads, M.L. Rhoads, N.K. Gabler, J.W. Ross, A.F. Keating, R.L. Boddicker, S. Lenka and V. Sejian. 2012. Impact of Climate Change on Livestock Production. In *Environmental Stress and Amelioration in Livestock Production*, edited by V. Sejian, T. Ezeji, J. Lakritz and R. Lal. Heidelberg: Springer-Verlag, pp. 413–468.

Bayard, B., C.M. Jolly and D.A. Shannon. 2006. The Economics of Adoption and Management of Alley Cropping in Haiti. *Journal of Environmental Management* 85: 62–70.

Bekele, W., and L. Drake. 2003. Soil and Water Conservation Decision Behavior of Subsistence Farmers in the Eastern Highlands of Ethiopia: A Case Study of the Hunde-Lafto Area. *Ecological Economics* 46: 437–451.

Blummel, M., I.A. Wright and N.G. Hedge. 2010. Climate Change Impacts on Livestock Production and Adaptation Strategies: A Global Scenario. Paper Presented at a National Symposium on Climate Change and Rainfed Agriculture. February 18–20, CRIDA, Hyderabad.

Cameron, A.C., and P.K. Trivedi. 2005. *Microeconometrics: Methods and Applications.* New York: Cambridge University Press.

Chib, S., and E. Greenberg. 1998. An Analysis of Multivariate Probit Models. *Biometrika* 85: 347–361.

Croppenstedt, A., M. Demeke and M.M. Meschi. 2003. Technology Adoption in the Presence of Constraints: The Case of Fertilizer Demand in Ethiopia. *Review of Development Economics* 7: 58–70.

Deressa, T. 2007. Analysis of Perception and Adaptation to Climate Change in the Nile Basin of Ethiopia. *Review of Development Economics* 7(1): 58–70.

Deressa, T. 2010. Assessment of the Vulnerability of Ethiopian Agriculture to Climate Change and Farmer's Adaptation Strategies. PhD Dissertation, University of Pretoria, South Africa.

Deressa, T.T., and R.M. Hassan. 2009. Economic Impact of Climate Change on Crop Production in Ethiopia: Evidence from Cross-Section Measures. *Journal of African Economies* 18(4): 529–554.

Deressa, T.T., R.M. Hassan, C. Ringler, T. Alemu and M. Yesuf. 2009. Determinants of Farmer's Choice of Adaptation Methods to Climate Change in the Nile Basin of Ethiopia. *Global Environmental Change* 19: 248–255.

Dolisca, F., R.D. Carter, J.M. Mcdaniel, D.A. Shannon and C.M. Jolly. 2006. Factors Affecting Farmers' Participation in Forestry Management Programs: A Case Study from Haiti. *Forest Ecology and Management* 236: 324–331.

Emboga, S.A., and P. Rowlinson. 2008. Inter-Relationships between Human Reproductive Health and Livestock Keeping Is Contributing to Livelihoods Security in the Climatically Vulnerable Region of Nyanza in West Kenya. Proceedings of the International Conference on Livestock and Global Climate Change. 17–20 May, Hammamet, Tunisia, pp. 210–212.

FAO. 2009. *Enabling Agriculture to Contribute to Climate Change Mitigation.* Rome: FAO.

Fischer, G., M. Shah, N. Francesco and H. van Velhuizen. 2005. Socio-Economic and Climate Impacts on Agriculture: An Integrated Assessment, 1990–2080. *Philosophical Transactions of the Royal Society* 360: 2067–2083.

Füssel, H.M., and R.J.T. Klein. 2004. Conceptual Frameworks of Adaptation to Climate Change and Their Applicability to Human Health: PIK Report, No. 91. Potsdam, Germany.

Galvin, K.A., R.B. Boone, N.M. Smith and S.J. Lynn. 2001. Impacts of Climate Variability on East African Pastoralists: Linking Social Science and Remote Sensing. *Climate Research* 19(1): 161–172.

Garforth, C.J. 2008. Impacts on Livelihoods. Proceedings of the International Conference on Livestock and Global Climate Change. 17–20 May, Hammamet, Tunisia, pp. 25–26.

Gbetibouo, G., and R. Hassan. 2005. Economic Impact of Climate Change on Major South African Field Crops: A Ricardian Approach. *Global and Planetary Change* 47: 143–152.

Gebremedhin, B., D. Hoekstra and A. Tegegne. 2006. Improving the Competitiveness of Agricultural Input Markets in Ethiopia: Experiences since 1991. Paper Presented at the Symposium on Seed-Fertilizer Technology, Cereal Productivity and Pro-Poor Growth in

Africa: Time for New Thinking, 26th Triennial Conference of the International Association of Agricultural Economics (IAAE). August 12–18, Gold Coast, Australia.

Getahun, T. 2008. Climate Change is Posing Danger on Pastoralist: How to Manage Excess of Its Adversity. Pastoralist Forum Ethiopia. www.pfeethiopia.org/pub_files/Climate%20changes%20posing%20danger%20on%20pastoralist.pdf

Gould, B.W., W.E. Saupe and R.M. Klemme. 1989. Conservation Tillage: The Role of Farm and Operator Characteristics and the Perception of Soil Erosion. *Land Economics* 65: 167–182.

Hassan, R., and C. Nhemachena. 2008. Determinants of Climate Adaptation Strategies of African Farmers: Multinomial Choice Analysis. *African Journal of Agricultural and Resource Economics* 2(1): 83–104.

Hoffmann, I. 2008. Livestock Genetic Diversity and Climate Change Adaptation. Livestock and Global Change Conference Proceedings. 17–20 May, Hammamet, Tunisia.

Imai, K. 2003. Is Livestock Important for Risk Behavior and Activity Choice in Rural Households? Evidence from Kenya. *Journal of African Economies* 12: 271–295.

IPCC (Intergovernmental Panel on Climate Change). 2007. *Impacts, Adaptation and Vulnerability.* Cambridge, UK: Cambridge University Press.

Kabubo-Mariara, J. 2008. Climate Change Adaptation and Livestock Activity Choices in Kenya: An Economic Analysis. *Natural Resources Forum* 32: 131–141.

Kandlinkar, M., and J. Risbey. 2000. Agricultural Impacts of Climate Change: If Adaptation Is the Answer, What Is the Question? *Climatic Change* 45: 529–539.

Kassahun, M.M. 2009. Climate Change and Crop Agriculture in the Nile Basin of Ethiopia: Measuring Impacts and Adaptation Options. MSc Thesis, Addis Ababa University, Addis Ababa, Ethiopia.

Kurukulasuriya, P., and R. Mendelsohn. 2006. Crop Selection: Adaptation to Climate Change in Africa. CEEPA Discussion Paper No. 26. Pretoria, South Africa: University of Pretoria, Centre for Environmental Economics and Policy in Africa.

Luseno, W.K., J.G. Mcpeak, C.B. Barrett, D. Little and G. Gebru. 2003. Assessing the Value of Climate Forecast Information for Pastoralists: Evidence from Southern Ethiopia and Northern Kenya. *World Development* 31(9): 1477–1494.

Maddala, G.S. 1995. *Limited Dependent and Qualitative Variables in Econometrics.* New York: Cambridge University Press.

Mandleni, B., and F.D.K. Anim. 2011. Climate Change and Adaptation of Small-Scale Cattle and Sheep Farmers. Paper Presented at 85rd Annual Conference of the Agricultural Economics Society. April 18–20, Warwick University, Coventry, UK.

Mano, R., and C. Nhemachena. 2006. Assessment of the Economic Impacts of Climate Change on Agriculture in Zimbabwe: A Ricardian Approach. CEEPA Discussion Paper No. 11. Pretoria, South Africa: University of Pretoria, Centre for Environmental Economics and Policy in Africa.

McFadden, D.L. 1999. Chapter 1. Discrete Response Models. University of California at Berkeley, Lecture Notes.

MERET. 2009. Managing Environmental Resource to Enable More Transition to Sustainable Livelihood 'MERET News'. Copenhagen Climate Change Summit, Special Edition, Addis Ababa.

Nhemachena, C., and R. Hassan. 2007. Micro-Level Analysis of Farmers' Adaptation to Climate Change in Southern Africa. IFPRI Discussion Paper No. 00714. Washington, DC: International Food Policy Research Institute.

Notenbaert, A., A. Mude, J. Van de Steeg and J. Kinyangi. 2010. Options for Adapting to Climate Change in Livestock-Dominated Farming Systems in the Greater Horn of Africa. *Journal of Geography and Regional Planning* 3(9): 234–239.

Oseni, S., and O. Bebe. 2010. Climate Change, Genetics of Adaptation and Livestock Production in Low Input Systems. 2nd International Conference: Climate, Sustainability and Development in Semi-Arid Regions. August 16–29, Fortaleza-Ceará, Brazil.

Sejian, V., I. Hyder, T. Ezeji, J. Lakritz, R. Bhatta, J.P. Ravindra, C.S. Prasad and R. Lal. 2015. Global Warming: Role of Livestock. In *Climate Change Impact on Livestock: Adaptation and Mitigation*. New Delhi: Springer-Verlag, GMbH, pp. 141–170.

Seo, S.N., and R. Mendelsohn. 2007. Climate Change Adaptation in Africa: A Microeconomic Analysis of Livestock Choice. World Bank Policy Research Working Paper No. 4277. Washington, DC.

Seo, S.N., and R. Mendelsohn. 2008a. Measuring Impacts and Adaptations to Climate Change: A Structural Ricardian Model of African Livestock Management. *Agricultural Economics* 38: 1–15.

Seo, S.N., and R. Mendelsohn. 2008b. Climate Change Impacts and Adaptations on Animal Husbandry in Africa. *African Journal of Agriculture and Resource Economics* 2: 65–82.

Sherlund, S.M., C.B. Barrett and A.A. Adesina. 2002. Smallholder Technical Efficiency Controlling for Environmental Production Conditions. *Journal of Development Economics* 69(1): 85–101.

Smith, K., C.B. Barrett and P.W. Box. 2001. Not Necessarily in the Same Boat: Heterogeneous Risk Assessment among East African Pastoralists. *Journal of Development Studies* 37(5): 1–30.

Sunil, K.B.V., A. Kumarand and M. Kataria. 2011. Effect of Heat Stress in Tropical Livestock and Different Strategies for Its Amelioration. *Journal of Stress Physiology and Biochemistry* 7(1): 45–54.

Thomas, D.S.G., and C. Twyman. 2005. Equity and Justice in Climate Change Adaptation amongst Natural Resource–Dependent Societies. *Global Environmental Change* 15: 115–124.

Todaro, M.P., and S.C. Smith. 2009. *Economic Development*, 10th Edition. London: Addison-Wesley, Pearson Education.

Train, K. 2003. *Discrete Choice Methods with Simulation*. Cambridge, UK: Cambridge University Press.

Tsedeke, K. 2007. Production and Marketing of Sheep and Goats in Alaba. SNNPR. Msc Thesis. Hawassa University, Hawassa, Ethiopia.

Verbeek, M. 2008. *A Guide to Modern Econometrics*, 3rd Edition. Chichester, UK: John Wiley & Sons.

World Bank. 2016. Ethiopia Overview. www.worldbank.org/en/country/ethiopia/overview (accessed April 13, 2017).

11 The distributive effect and food security implications of biofuels investment in Ethiopia

A CGE analysis*

*Zenebe Gebreegziabher, Alemu Mekonnen,
Tadele Ferede, Fantu Guta, Jörgen Levin,
Gunnar Köhlin, Tekie Alemu, and Lars Bohlin*

Introduction

Rising prices of fossil fuels, together with apprehension about the environmental harm created by them, have resulted in increasing efforts to search for alternative energy sources. Biofuels are considered renewable and relatively cleaner substitutes for conventional energy sources. However, there is ongoing debate on the opportunities created and challenges posed by biofuel production (Azar, 2011; Worldwatch Institute, 2007). Some skeptics see the trend as land grabbing by cross-border transnational corporations and foreign governments and as the new scramble for Africa (ABN, 2007). The extreme view is that the effect of biofuel on food security amounts to a crime against humanity.[1] Others have argued that greater production of biofuels might not necessarily be harmful for the poor and that they can become more food-secure with the adoption of proper production technology (van Rheenen and Olofinbiyi, 2007).

The central question in this chapter is whether there will be positive or negative impacts on smallholder farmers and people living in rural areas as more agricultural land is used for biofuels production. The study uses firm survey data in a CGE (computable general equilibrium) analysis to assess the distributive effect and food security implications of biofuels investment. We focus on Ethiopia, using data from 15 biofuels firms and two NGOs involved in biofuel production. We find that the spillover effects of certain biofuels can increase the production of food cereals (with the effect being variable across regions) without increasing cereal prices. When spillover effects are considered, biofuel investment tends to improve the welfare of most rural poor households. Urban households benefit from returns to labour under some scenarios. These findings assume that continued

* The authors gratefully acknowledge financial support from the Swedish International Development Cooperation Agency (Sida) through the Environment for Development Initiative at the University of Gothenburg, Sweden, Department of Economics, Environmental Economics Unit.

government investment in roads allows biofuels production to expand on land that is currently unutilized, so that smallholders do not lose land.

Ethiopia is said to have tremendous potential for ethanol and biodiesel production. Some estimates put the potential area of land suitable for production of biodiesel feedstock at about 25 million hectares (Gebremeskel and Tesfaye, 2008). However, it is not obvious how much production could come from first-generation biofuels, such as cultivation of biofuel trees and cereal crops, and how much from second-generation biofuels, both from agricultural residues and industrial by-products, such as molasses.

The Ethiopian government has adopted a strategy that encourages the expanded development and utilization of biofuels as part of its energy policy. One of the justifications for this biofuels policy is the possibility of saving scarce foreign currency that is used to import fossil fuels and shifting from high-cost fossil fuels to cost-effective biofuels (MoME, 2007). In addition, Ethiopia adopted a standard of a 5% blend of biofuels in transport fuel, which was later increased to 10%. Ethiopia's First Growth and Transformation Plan (GTP I) stipulated increasing production of bioethanol to 194.9 million liters and biodiesel to 1.6 million liters over five years (i.e., 2010/11 to 2014/15), through coordination of governmental and private sugar industries (in the case of bioethanol) and through involvement of private investors and farmers (in the case of biodiesel) (MoFED, 2010). In line with this, the GTP I also envisaged oil companies increasing the number of blending facilities to 8 for ethanol and 72 for biodiesel. As of August 2017, neither of these goals under GTP I has been fully achieved; a number of planned new sugar plants are under construction but not yet operational due to various delays.

It is imperative for low-income, food-deficit countries such as Ethiopia to investigate the distributional and food security questions raised by such investments. Specifically, unintended effects on poverty and growth need to be examined. This study was the first such research conducted for Ethiopia. The questions include: Will such biofuels investment be pro-poor or will it lower the income of vulnerable people or groups? Which group in Ethiopia, if any, will be affected negatively due to increasing biofuel investment in Ethiopia? Will such investments undermine the country's food production or food security?

A review of the literature identifies the following outstanding issues. First, the empirical evidence is mixed and there is no consensus so far, particularly regarding the distributional consequences and food security implications of biofuels investment. Specifically, little is known about the actual impact of these new investments on smallholder farmers. Therefore, country-specific case studies such as this one contribute to the debate and enhance our understanding of the real effects of biofuels expansion, including its differential impacts across different regions and income groups. Second, the literature has been largely focused on developed economies; little is known and very few studies have been carried out in the context of Africa. Third, there are very few CGE models developed so far for Ethiopia. Moreover, these few CGE models have focused on other issues, including economic infrastructure, food aid, poverty reduction and livestock; this study was the first to focus on biofuels issues. Therefore, it is of interest to analyze the impact of investment in biofuels using a CGE model. Finally, this study will inform policy besides contributing to the literature.

The main objective of this study is to investigate the distributive effect and food security implications of biofuel investment in Ethiopia, using data obtained from biofuels firms on inputs and outputs in an economy-wide model. Specifically, this study assesses the effects of biofuels investment on food security in the country and examines the distributive effect in terms of the consequences for household welfare. We modify the Ethiopian SAM to include the biofuels sector, based on data collected from biofuels firms and NGOs involved in biofuels.

The rest of the chapter is organized as follows. The next section presents a literature review. The section that follows presents the model, including the dynamic structure of the CGE model, and outlines data and context. The fourth section presents simulations. Results are discussed in the fifth section. The last section concludes and draws some policy implications.

Literature review

Globally, only a few countries dominate the domestic use and export of biofuels. Ethanol, first produced in the 1970s, is still produced in much larger volumes than biodiesel, for which production started in the 1990s (Slater, 2007). The United States and Brazil are the largest producers of ethanol, accounting for over 80% of the world's total production. The EU, on the other hand, produces almost 80% of the world's biodiesel (FAPRI, 2010). Global production of biofuels has increased gradually over time. The largest future increases in production volumes are expected in Brazil, the United States, the EU, China, India, Indonesia, and Malaysia. By 2030, the total annual global production of biofuels is projected to increase to 92 Mtoe[2] under a reference scenario and to 147 Mtoe under an alternative policy scenario (IEA, 2006). Global biofuels consumption is projected to grow fourfold over the next two to three decades, accounting for 8% of transport fuel demand by 2035, up from 3% as of 2010 (IEA, 2010).

There have been continuing debates on the socioeconomic impacts of biofuel production. Proponents of biofuels production argue for the stabilizing effect that it might impose on the price of fossil fuels, its contribution to reduction of greenhouse gas emissions, and its income and growth effects in land-rich poor countries. The opponents argue that biofuel production competes with food production for land and labour and hence leads to rising food prices. Furthermore, there is concern that the increasing water demand by biofuel production might create additional pressure on already inadequate supplies of water. However, there are few studies conducted to assess these issues and quantify the impacts on growth and income distribution.

It is quite challenging to quantify the benefits of biofuels compared to oil fuels by assigning prices to effects such as climate benefits, air quality, human health, and sustainability of energy sources (Worldwatch Institute, 2007). The limited literature can be classified into three groups.

First, there are studies dealing with the effect of biofuel production on food prices (McNew and Griffith, 2005; Abbott et al., 2008; Chakravorty et al., 2009; Rosegrant, 2008; Mitchell, 2008; Worldwatch Institute, 2007; Sourie et al., 2006). Competition imposed by biofuel production could result in significant increases in prices of agricultural products, including food crops. For instance, the OECD (2006, cited in World Watch Institute, 2007) estimated significant increases in prices of sugar, vegetable oils, and cereals resulting from increased use of biofuel. Hausman et al. (2012) use a structural vector auto-regression framework and find that increased corn ethanol production during the 2006/2007 production year explains about 27% of the experienced corn price rise. Roberts and Schlenker (2010) provide a new framework to assess the impact of the US ethanol mandate on world food commodity prices and global consumers' surplus. Specifically, they employ an econometric framework – that is, an instrumental variable technique – to identify demand and supply elasticities of agricultural commodities using yield shocks – deviations from a time trend of output per area – which are predominantly caused by weather fluctuations. They then use their estimated elasticities to evaluate the impact of ethanol subsidies and mandates on world food commodity prices, quantities, and food consumers' surplus. They predicted world food prices to increase by about 30% and global consumer surplus from food consumption to decrease by USD 155 billion annually as a result of the US ethanol mandate. However, they also find that the predicted price increase scales back to 20% if a third of the biofuel calories are recycled as feedstock for livestock.

Continuing this line of research on food prices, Banse et al. (2008) analyze the trade impacts of an EU Biofuels Directive using a global CGE model – that is, a modified version of the GTAP model. They find that cereal prices actually decline in the long run, though less than they would without the directive. Using IFPRI's partial equilibrium model, Rosegrant (2008) finds that biofuels demand accounted for 39% of the increase in corn price from 2000 to 2007. Rajagopal et al. (2007) analyze the effect of the ethanol production tax credit on corn price, using a stylized partial equilibrium model. They find a 21% increase in corn price attributable to a USD 0.51 ethanol production tax credit in the United States in 2006. The US Congressional Budget Office (CBO) estimates also suggest that about 10% to 15% of the increase in food prices between April 2007 and April 2008 was attributable to the increased use of ethanol (CBO, 2009). However, Chakravorty et al. (2011) show that about two-thirds of the increase in food prices can be attributed to changes in consumption patterns and only one-third to biofuels mandates.

Second, some work has been conducted on the effect of biofuel production on farm jobs and income (Tréguer and Sourie, 2006; World Watch Institute, 2007). For instance, Tréguer and Sourie (2006) estimate the effect of the massive biofuel production decision of France in order to meet the European energy directives of 2003. Using a partial equilibrium model (the OSCAR model), their results indicate that production of biofuel crops increases farm jobs and farm income. However, some argue that whether this effect leads to net welfare gain or loss for poor farm households depends on two opposing forces (World Watch Institute, 2007). One, farm households in poor countries will receive higher prices for their agricultural products due to increased global prices resulting from the competition for resources between agriculture-based energy production and other agricultural products. Two, as discussed earlier, poor households dependent on imported food will face higher food prices and hence become poorer.

The third type of research attempts to analyze the impact of biofuel production on global trade, growth, income distribution, and poverty (Arndt et al., 2009, 2010; Peskett et al., 2007; Ugarte et al., 2007; Dufey, 2006; Birur et al., 2008). Arndt et al. (2009) quantitatively estimated the effect of biofuel investments on food security, poverty, and growth in the context of Africa, using a dynamic computable general equilibrium for Mozambique. This research indicated that biofuel investment can increase growth and reduce poverty, depending on the type of technology used in production, amid some displacement of food crops by biofuels. They find that, depending on the production technology, biofuel investment increases Mozambique's annual economic growth by about 0.6 percentage points and reduces the incidence of poverty by about 6 percentage points, over a 12-year phase-in period. They see the out-growers[3] approach to producing biofuels to be more pro-poor, due to the greater use of unskilled labour and accrual of land rents to smallholders, as compared with the more capital-intensive plantation approach. The out-grower production technology has more effect on growth and poverty reduction because it increases the income of small land holders and increases the rental value of their land. The results did not favour unrestrained biofuels development, but suggested that a carefully designed and managed biofuels policy holds the potential for substantial gains. Similarly, Arndt et al. (2010), for the case of Tanzania, showed that biofuels investment contributes positively to poverty reduction. They also argue that any trade-offs that do exist between biofuels and food production are likely to be smaller when feedstocks are produced by larger-scale farmers, because of higher yield per hectare.

The CGE models developed for Ethiopia include World Bank (2004), Gelan (2007), Diao et al. (2005), and Dorosh and Thurlow (2012). World Bank (2004) investigates how public investment can contribute to stimulate growth on a balanced path, using the economic potential of rural areas and their linkages with urban development. That paper argues that more balanced and sustained growth could be achieved by increasing economic infrastructure in high-potential zones and sectors through public investment and by reducing market distortions (including food aid effects) and factor market limitations due to property rights regimes. The results of that paper imply a reorientation of the growth strategy that is on course in Ethiopia toward high-potential zones and sectors, commercial farming, and fewer food aid–based projects.

Gelan (2007) examined impacts of food aid on domestic food production, employing a CGE model and using data from Ethiopia. His simulation experiments show that food aid has unambiguous disincentive effects on domestic food production. That is, the removal of food aid caused a modest increase in food prices but this stimulated food production. The employment and income generation effects of the removal of food aid also outweighed the adverse effect of the status quo. Overall, the removal of food aid led to improvements in aggregate household welfare. Moreover, his simulation experiments also suggest that, in reality, poor rural households and urban wage earners are the ones who benefit most in the absence of food aid, unlike the concerns observed in the food aid literature that any reduction in food aid would hurt the poor. Entrepreneurs, on the other hand, are more likely to encounter a marginal welfare decline.

Diao and Pratt (2007), using a spatially disaggregated economy-wide model, assess which agricultural subsectors have the strongest capacity to drive economic growth and poverty reduction in Ethiopia, and what kind of agricultural and nonagricultural growth is needed to achieve the millennium development goal of halving the 1990 poverty rate. Their study reveals that agriculture, primarily through growth in staple crops and livestock, has the potential to play a central role in alleviating poverty and increasing growth in Ethiopia. However, they also find that similar rates of agricultural growth have different effects on poverty, necessitating regionally based strategies for growth and poverty reduction at the subnational level. Their findings also imply that agricultural growth requires concurrent investments in roads and other market conditions.

Using a CGE model for Ethiopia based on data from the EDRI 2005/06 Ethiopia SAM (Tebekew et al., 2009), Dorosh and Thurlow (2012) analyze agricultural growth options that can support high levels of agricultural development in the context of CAADP (Comprehensive Africa Agriculture Development Programme), which Ethiopia has worked to implement, together with other African governments. As part of CAADP, Ethiopia has committed itself to meeting targets of devoting at least 10% of public expenditures to agriculture and to achieving a 6% growth rate in agricultural GDP. The CGE model results indicated that it is possible to reach and sustain the 6% agricultural growth target if Ethiopia can meet its targets for crop yields and livestock productivity. Results also suggest that achieving 6% per year of agricultural growth would reduce national poverty to 18.4%, thereby lifting an additional 3.7 million people out of poverty compared to a base simulation using medium-term growth rates.

Gelan et al. (2012) extend the Dorosh and Thurlow (2012) dynamic CGE model to integrate livestock in the CAADP framework and assess the role of livestock in rural poverty reduction. Their results reveal the important role of the livestock sector in enhancing various measures of GDP and combatting food insecurity. They find that, unlike previously held views, agricultural GDP and overall GDP growth levels resulting from livestock total factor productivity (TFP) shocks are very similar to those resulting from cereal TFP shocks. They also find that livestock sector productivity growth leads to greater factor income, particularly labour income. Their results also imply that strategies focusing on development of the cereal sector alone might be inefficient.

This study builds on these GGE models to help resolve questions about biofuels' effect on food security and poverty.

The model

This study uses a dynamic CGE model to simulate the impact of investment in biofuels. CGE models are applied to income distribution, trade strategy, and structural change in developing countries, where such models have features that make them suitable for such analysis. First of all, these models can simulate the functioning of a market economy, including markets for labour, capital, and commodities, as well as the transmission of changes in economic conditions through prices and markets. Second, the structural nature of these models allows consideration of new economic activities such as biofuels. Third, these models are specified so that all economy-wide constraints are respected. For instance, investments in biofuels may generate substantial foreign exchange earnings needed by the country in question. These investments may use a large quantity of land and demand a large amount of labour. Therefore, the models consider the balance of payments and the supply of usable land and labour. Fourth, these models provide detailed sectoral breakdowns and permit a 'simulation laboratory' to quantitatively investigate how various impact channels and feedback effects influence the structure and performances of the economy through time. Last, these models provide a theoretically consistent framework for welfare and distributional analysis.

In CGE models, economic decisions are the outcome of decentralized optimization by both producers and consumers within a coherent economy-wide framework. They include a number of substitution possibilities in response to variations in relative prices. Examples include substitution between various labour types; between capital and labour; between exports and domestic sales; and between imports and domestic goods. Market imperfections and institutional rigidities are captured by exogenously imposing immobile sector capital stocks, labour market segmentation, and home consumption. These features allow a more realistic application of these models to developing countries.

Structure and assumptions of the dynamic CGE model

Structure of the dynamic CGE model

The impacts of biofuels investment are generally economy-wide in nature and lead to strong general equilibrium feedback effects. Hence, they are amenable to CGE analysis. Moreover, shocks related to biofuels investment and production might lead to differential effects being experienced by different groups of households. Therefore, we employ a modified version of the standard CGE model for Ethiopia based on the 2005/2006 Social Accounting Matrix (SAM) of Ethiopia developed by the Ethiopian Development Research Institute (EDRI). The basic structure of the CGE framework mimics the generic IFPRI model.[4] The IFPRI standard CGE model explains all of the payments recorded in the SAM. The model essentially follows the SAM disaggregation of factors, activities, commodities, and institutions. The model is specified as a set of simultaneous equations, most of which are nonlinear, and there is no objective function. The equations define the behaviour of the different actors. For production and consumption decisions, behaviour is captured by nonlinear, first-order optimality conditions. That is, production and consumption decisions are driven by the maximization of profits and utility, respectively. The equations also include a set of constraints, such as markets (for factors and commodities) and macroeconomic aggregates (balances for savings-investment, the government, and the current account of the rest of the world). These have to be satisfied by the system as a whole but are not necessarily considered by any individual actor.

Each producer (represented by an activity) is assumed to maximize profits, defined as the difference between revenue earned and the cost of factors and intermediate inputs. Profits are maximized subject to a production technology. The technology is specified by a constant elasticity of substitution (CES) function of the quantities of value added and a Leontief function of aggregate intermediate inputs. Each activity produces one or more commodities according to fixed-yield coefficients. In addition, a commodity may be produced by more than one activity. The revenue of an activity is defined by the level of the activity, yields, and commodity prices at the producer level. As part of its profit-maximizing decision, each activity uses a set of factors up to the point where the marginal revenue product of each factor is equal to its wage (also called factor price or rent). Factor market closures are according to equilibrating supplies and demands in each factor market, with the quantity supplied of each factor fixed at the observed level. An economy-wide wage variable is free to vary to ensure that the sum of demands from all activities equals the quantity supplied. Each activity pays an activity-specific wage that is the product of the economy-wide wage and an activity-specific wage (distortion) term; these also are fixed.

In the model, institutions are represented by households, enterprises, the government, and the rest of the world. Households (disaggregated as in the SAM) receive income from the factors of production (directly or indirectly via the enterprises) and transfers from other institutions. Transfers from the rest of the world to households are fixed in foreign currency, as is the case for all transfers between the rest of the world and domestic institutions and factors. Households use their income to pay direct taxes, save, consume, and make transfers to other institutions. The direct taxes and transfers to other domestic institutions are defined as fixed shares of household income, whereas the savings share is flexible for selected households. Household consumption covers marketed commodities, purchased at market prices that include commodity taxes and transaction costs, and home commodities, which are valued at activity-specific producer prices. Household consumption is allocated across different commodities (both market and home commodities) according to linear expenditure system (LES) demand functions, derived from maximization of a Stone-Geary utility function. Enterprises may also receive transfers from other institutions. Enterprise incomes are allocated to direct taxes, savings, and transfers to other institutions. Enterprises do not consume. Government collects taxes at fixed *ad valorem* rates and receives transfers from other institutions. Government uses this income to purchase commodities for its consumption and for transfers to other institutions. Government consumption is fixed in real (quantity) terms, whereas government transfers to domestic institutions (households and enterprises) are CPI-indexed. Government savings – that is, the difference between government income and spending – is a flexible residual. The final institution is the rest of the world. Transfer payments between the rest of the world and domestic institutions and factors are all fixed in foreign currency. Foreign savings or the current account deficit is the difference between foreign currency spending and receipts.

Key assumptions of the model

In the current account, a flexible exchange rate is assumed, so that it adjusts in order to maintain a fixed level of foreign savings (i.e., the external balance is held fixed in foreign currency terms).

In this model, labour is assumed to be mobile across sectors and fully employed, which is a strong assumption. For instance, if biofuels production results in higher employment, then the trade-offs between biofuels and food production are less pronounced, as the GDP gains from the biofuels production would be larger. Full employment closure implies that expanding biofuels production reduces use of labour elsewhere in the economy. This is consistent with widespread evidence that, while relatively few people have formal jobs, a large proportion of the working-age population engages in productive activities that contribute to GDP. Therefore, employing this working-age population in biofuels production has an opportunity cost. According to the 2005 National Labour Force Survey (CSA, 2006), the country's total unemployment rate was 5.0%, with male and female unemployment rates of 2.5% and 7.8%, respectively. The urban unemployment rate of the country was 20.6% while the unemployment rate in rural areas was 2.6%. The rural unemployment rate is low because family members work on family farms when they don't have formal jobs.

The consumer price index is taken as the model's numeraire. Trade elasticities are taken from GTAP (Global Trade Analysis Project) (Dimaranan, 2006). The model is calibrated in such a way that the initial equilibrium reproduces the base-year value from the SAM.

The features of the model described so far apply to a single-period static CGE model. However, as investments in biofuels unfold over a dozen years or more, the model is made capable of producing forward-looking growth trajectories. The model is dynamized by building a set of capital accumulation and updating rules for capital stock, labour force growth by skill category, and productivity growth. In addition, a simple adaptive expectation formation is specified whereby investment is allocated according to current relative prices. This implies that investors expect current price ratios to prevail indefinitely. Crowding-in of private investment in non-biofuel sectors is not explicitly modeled, though suggested by Hausmann (2007). We opted instead to focus on the direct impact of biofuels, though we considered the potential technology spillovers.

A set of dynamic equations update various parameter values and variables from one year to another. Growth in total supply of each labour category and land is specified exogenously. In addition, growth in land supply by agro-ecological zones to biofuels sectors is specified exogenously. Sector capital stocks are adjusted each year based on investment, net of depreciation. Factor returns adjust in such a way that factor supply equals demand. This model adopts a 'putty-clay' formulation such that new investments can be directed to any sector in response to differential rates of return (Arndt et al., 2010). However, installed equipment remains immobile. Sector-specific factor productivity growth is specified exogenously. Based on these simple relationships to update key variables, we generated a series of growth trajectories for different biofuels investment scenarios.

In the modeling, we focus on the differential impacts of various biofuels production scenarios on a baseline scenario that excludes biofuels investments. Comparison with the baseline helps us evaluate whether the biofuels scenarios are reasonable. Examining the differences between the biofuels scenarios and the baseline scenario allows us to isolate the impacts of biofuels investments, thus providing clear and analytically tractable comparisons.

Data

The SAM

The SAM used in the Ethiopian CGE model contains 60 activities, including 41 agricultural and 5 non-agricultural sectors. The 2005/06 Macro-SAM of Ethiopia is presented in Table 11A1. A total of 14 agricultural and non-agricultural commodities are distinguished. The main feature of the model is that activities are classified according to agro-ecological zones (AEZs); see EDRI (2009) for details on AEZs.[5] Of the factors of production, there are three categories of labour (skilled, semiskilled, and unskilled labour types). The model also identifies agricultural capital and land, categorized in five agro-ecological zones, and non-agricultural capital. The model also distinguishes 14 household types. Rural households are classified according to their poverty status (poor and non-poor) and location (AEZs), resulting in ten rural household types. Urban households are based on urban size and poverty status, resulting in four urban household types (Table 11A2). The details in the 2005/06 SAM capture Ethiopia's economic structure and influence model results. Biofuels are expected to be either exported or used to substitute for fuel imports. As a result, a substantial increase in the quantity of biofuels will have implications for trade and foreign exchange availability. Availability of foreign exchange enables the country to import more and reduce exports of other products. As a result, one would expect that sectors with high trade share, where trade share might be taken to represent sectors with a large proportion of production exported or a high degree of import competition, will be more affected compared to non-traded sectors.

Table 11.1 provides the basic features of the Ethiopian economy in 2005/06, which is the base year for the dynamic CGE model. Agriculture generates a little less than half of the national GDP and three-fourths of total employment. The manufacturing sector, by contrast, accounts for only 13% and 7% of total GDP

Table 11.1 Structure of Ethiopia's Economy in 2005/06

	Share of total (%)				Export intensity[a] (%)	Import penetration[a] (%)
	GDP	Employment	Exports	Imports		
Total	100.00	100.00	100.00	100.00	8.96	21.88
Agriculture	48.14	73.54	42.23	4.69	11.18	3.39
Cereal crops	13.83	23.02	0.00	3.49	0.00	7.23
Cash crops	7.05	8.83	25.01	0.16	53.77	1.24
Sugarcane	0.32	0.46	0.00	0.00	0.00	0.00
Jatropha	0.00	0.00	0.00	0.00	0.00	0.00
Castor bean	0.00	0.00	0.00	0.00	0.00	0.00
Palm oil	0.00	0.00	0.00	0.00	0.00	0.00
Livestock	14.39	20.08	4.51	0.00	4.99	0.00
Other agriculture	7.80	12.49	12.71	1.05	16.82	3.73
Forestry and fisheries	4.75	8.66	0.00	0.00	0.00	0.00
Non-agriculture	51.86	26.46	57.77	95.31	7.82	28.51
Industry	11.72	7.62	17.56	70.99	6.60	41.25
Electricity	0.92	0.14	0.00	0.00	0.00	0.00
Food processing	2.30	1.66	3.01	2.42	10.53	14.92
Biofuel processing	0.00	0.00	0.00	0.00	0.00	0.00
Ethanol processing	0.00	0.00	0.00	0.00	0.00	0.00
Other industrial processing	8.49	5.81	14.56	56.40	6.38	41.58
Services	40.15	18.84	40.2	24.3	6.4	13.6

Source: Ethiopia's 2005/06 social accounting matrix (SAM).

Note:
[a] 'Export intensity' is the share of exports in domestic output. 'Import penetration' is the share of imports in total domestic demand. Sums of shares in this table and subsequent tables may not equal 100 due to rounding.

and employment, respectively. The country depends heavily on imported industrial products, accounting for 71% of total imports, while industrial exports accounted for a fifth of total exports. Note that fuel imports are quite considerable, accounting for about 12% of total imports and 18% of total industrial imports in 2005/06. The country imports more fuel than agricultural products. For example, cereals accounted for 3.5% of total imports over the same period.

With various incentives provided by the Ethiopian government, currently there are biofuel investment activities in different parts the country aimed at the production of ethanol and biodiesel. This policy addresses the problem that Ethiopia's fossil fuel demand is entirely satisfied through imports, so that the country is significantly exposed to the shocks associated with fluctuations in fossil fuel prices. The biofuel sector not only seems promising in addressing the energy security issue but also is expected to create more jobs and income, promoting the country's goal of poverty reduction. Reports indicate that over 500,000 hectares of land have already been offered by regions to national and transnational companies for

Table 11.2 Biofuel Investment by Agro-ecology

Agro-ecological zone	Type of feed stock used for the production of biofuel			
	Jatropha fruit	Castor bean seed	Palm oil	Sugarcane
AEZ 1			Yes	
AEZ 2				Yes
AEZ 3	Yes			
AEZ 4	Yes	Yes		
AEZ 5		Yes		

Source: Biofuel investment survey, 2010.

biofuels investment (ABN, 2007; Lashitew, 2008). Jatropha, castor bean, and palm (for oil) are the main biofuel crops that are being developed, particularly for biodiesel. Both foreign and domestic companies are involved (Lakew and Shiferaw, 2008). In the case of ethanol production, the feedstock comes primarily from large-scale sugarcane plantations. Currently, the sugar by-product molasses is the most favorable feedstock for large-scale ethanol production.

The country has five sugar factories in operation as of August 2017, including Wonji-Shoa, Metahara, Fincha, Tendaho, and Kessem; three of these are publicly owned. Metahara and Fincha, which are publicly owned, have ethanol production plants. Kessem, the fifth and newest sugar factory, with a production capacity of 106,000 tons of sugar per annum, went into operation in 2015. The government has commissioned the expansion of the existing factories to include ethanol production plants. This is projected to eventually raise the country's ethanol production to 80,000 cubic meters from the earlier level of 8,000 cubic meters (Gebremeskel and Tesfaye, 2008).

Biofuel activities vary across AEZs. For instance, while sugarcane plantation is undertaken in moisture-sufficient highlands (AEZ2), palm oil activity is undertaken mainly in moisture-reliable humid lowlands areas (AEZ1) (Table 11.2). Note that small-scale sugarcane production is also undertaken by smallholders in the other AEZs.[6] Jatropha and castor bean activities are produced mainly in moisture-sufficient highlands (*enset*[7]-based systems) and in drought-prone and pastoralist zones: AEZ 3 & 4 for jatropha and AEZ 4 & 5 for castor bean. This disaggregation captures some of the diversity in economic structure and potential across regions.

Survey (data collection)

A survey was conducted on biofuels investment in Ethiopia. The purpose of the survey was to generate sector and crop-specific primary data to calculate the input-output coefficients in relation to the biofuels sector for the CGE analysis. A structured instrument/questionnaire was developed to collect the relevant

data. The instrument covered questions related to time elapsed in the investment process from application and registration through land acquisition; feedstock production and utilization, including purchase price of feedstock offered
to out-growers, labour and capital inputs to feedstock production, and related
expenses; investment in plant and equipment and plant capacity; biofuels (ethanol and biodiesel) extraction (processing) and sales; and assessment of environmental and social issues. A list of over 45 companies with investment permits for
biofuels was obtained from the Ethiopian Investment Agency. Then, 15 biofuels
companies and two NGOs involved in biofuels were approached to fill out the
structured questionnaire. There were six non-responses. In fact, the data collection wasn't easy and it involved a lot of diplomacy and a number of revisits.
Besides its use in calculating the input-output coefficients, the survey also helped
to characterize the biofuels sector in Ethiopia. Table 11.3 provides an overview
of this sector from the survey results. The survey revealed that there is one com-
pany that had started exporting biodiesel and two companies that were at the
product testing stage. The survey also determined information on complementary local innovations going on in the biofuels sector, including the invention of
biodiesel stoves, processors/distilleries, and biogas-driven vehicles. All these suggest that the sector requires policy attention and could possibly be an avenue to
reducing poverty and enhancing growth. However, we also found that the sector
suffers from lack of appropriate institutional setup in terms of better regulatory
framework and follow-up, particularly at the regional level.

Table 11.3 Overview of the Biofuels Sector in Ethiopia

Indicator	Number/description
No. of firms/companies	>15 (incl. NGOs)
No. of firms already at production stage	2
No. of firms that started export	1
No. of firms at production test stage	2
Total investment (capital)	Multimillion >1.3 b ETB (>0.1 billion USD)
Investment (type)	Largely foreign but also domestic
Land (000' ha)	>308 (currently operated); >101 (additional)
Year in operation	Since 2005
Installed plant capacity	492 to 28,800 liters/day
Employment opportunities.	>17,714 (temp.), >236 (perm.)
Crop types	Sugarcane, jatropha, castor bean, palm oil
Technology	Plantation, out-growers, and community development
Regions	All regions, Oromiya, SNNPR, Amhara, etc.

Source: Results of biofuels investment survey, 2010.

Table 11.4 Biofuel Production Characteristics/Technical Coefficients

	Sugarcane and ethanol	Jatropha/castor bean diesel
Land employed (ha)	11,248.00	3,284.00
Biofuel crop production (tons)	569,168.00	200.00
Farm workers employed (in number)	5,365.00	4,384.00
Land yield	50.60	0.06
Farm labour yield	106.09	0.05
Land per capita	2.10	0.75
Capital per hectare	16.46	0.00
Labour-capital ratio	0.029	0.00
Biofuel produced (liters)	5,323,866.05	2,880.69
Processing workers employed	27	0.00
Feedstock yield (L/ton)	9.35	14.40
Processing labour yield	197,180.22	

Source: Biofuel investment survey, 2010.

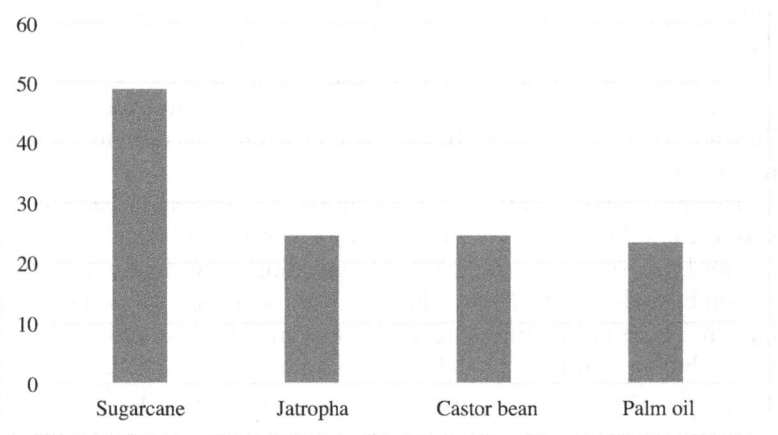

Figure 11.1 Share in Total Biofuel Crop Land by Biofuel Crop Type (%)
Source: Biofuel investment survey, 2010.

As for production characteristics, while large-scale sugarcane is mainly planta-tion-based, jatropha and castor bean production activities are undertaken by a combination of plantation-based and smallholder production through the out-growers scheme. In addition, jatropha and castor bean production activities are labour-intensive, as they require more labour per land compared with sugarcane (Table 11.4). According to this recent biofuels investment survey, sugarcane accounted for a larger share of the total land allocated to biofuel crops (Fig-ure 11.1). However, it is important to note that a small proportion of the total

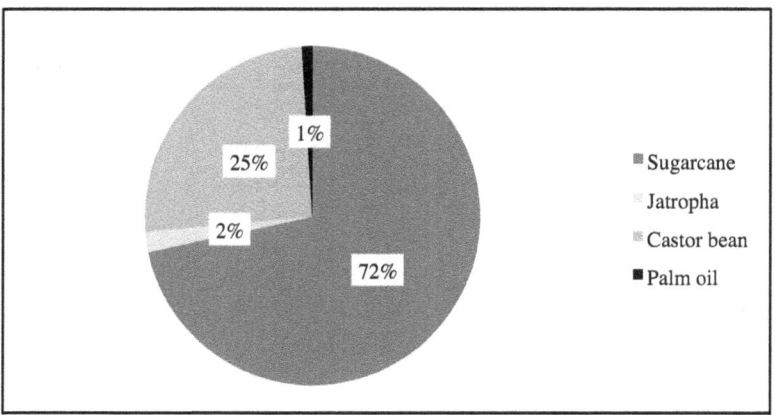

Figure 11.2 Ratio of Utilized Land to Total Land Allocated to Each Biofuel Crop (%)

land allotted to biofuels production was utilized in 2007. For instance, while a fifth of the total land allocated for biofuels was utilized in castor bean, the figures for jatropha and palm oil were very small – that is, 1.5% and 0.8%, respectively – in 2009 (Figure 11.2). A little more than half of the total land allotted to sugarcane was utilized over the same period. This suggests that there is room for further expansion of production by bringing more land into cultivation; it is unlikely that smallholders will be displaced in the short and medium term, until full-scale operation is reached.

Biofuels development in Ethiopia is unique in two important respects. First, the biofuels sector in Ethiopia is characterized by a diversity of biofuels crops (jatropha, castor bean, sugarcane, and palm oil, including indigenous trees). Second-generation biofuels – that is, by-products such as molasses – are used for ethanol production. There are also intercropping options in the case of castor beans. Secondly, the biofuels business model includes a mix of plantations, out-growers schemes, and community development models. For example, REST in Tigrai and ORDA in the Amhara region are involved in biofuels under a community development model.[8] The modeling has captured these features as much as possible.

Simulations

Scenarios

First, we produce a baseline scenario or growth path, which assumes that the economy continues to grow during 2003–2020 in line with its recent growth trajectory. For each year, we update the model to reflect changes in population, supply of labour and land, and factor productivity (see Table 11.5). Ethiopia could be considered a land-abundant country, as the proportion of land under cultivation is relatively small compared to the potential cultivable land. As a

Table 11.5 Core Macroeconomic Assumptions of the Dynamic CGE Model

	Initial (2005)	Baseline scenario (growth rates)
Population (in thousands)	70,167	2.5
GDP	100.0	2.54
Labour supply	40,479.4	7.0
Skilled labour	7.9	7.9
Professional labour-rural	5.0	5.0
Administrative labour-urban	5.0	5.0
Unskilled labour-urban	4.4	4.4
Agricultural labour-rural	4.4	4.4
Capital stock	56,455.9	10
Land supply	3.2	3.2

result, we assumed that the land supply grows at 3.2% on average in all agro-ecological zones, which is the same as the rate of cropped area expansion over the past decade. Population is assumed to grow at 2.5%, which is the same as the average rate of population growth from 1994 to 2007 (CSA, 2008). Rising skill intensity in the labour force is captured by assuming the supply and productivity of skilled and semiskilled labour forces to grow faster than that of the unskilled labour force.[9] It is also assumed that there is an unbiased technological change, which shifts the parameter on the production function (total factor productivity parameter) to grow at the rate of 2.5% in livestock and sectors that produce cereals and cash crops.

Similarly, total factor productivity in all other non-agricultural activities is assumed to grow at the rate of 2.9%. These estimates of TFP are obtained from previous studies on growth accounting in the country (e.g., Pratt and Yu, 2008; World Bank, 2004). The rate of total factor productivity growth in sugarcane activity is assumed to be 5%, which is consistent with the expansion in the sector. The total factor productivity in jatropha-producing activity grows at the rate of 3%, while that of castor bean–producing activity is assumed to grow at the rate of 3.5%. The results of these scenarios are compared with the biofuels scenarios in order to isolate the effects of biofuels investment from the effect of other factors. Given that there exists a diversity of biofuels options for Ethiopia, we considered seven biofuels scenarios (see Table 11.6), the details of which are elaborated ahead.

The sugarcane scenario, scenario 1 (S1), assumes expansion of sugarcane production through extensive cultivation – that is, by allocating more land to sugarcane production. Specifically, we increase land allocated to sugarcane by 5,116.44 hectares per year over the 2020 period.[10]

In the jatropha scenario, scenario 2 (S2), we keep increasing jatropha production by bringing more land into cultivation. Land allotted to this crop increases

Table 11.6 Scenarios for Biofuel Simulation

Scenarios	Technology	
	Plantation (capital -intensive)	*Out-grower (labour-intensive)*
(i) S1: Sugarcane	S1	
(ii) S2: Jatropha		S2
(iii) S3: Castor bean		S3
(iv) S4: Palm oil	S4	
(v) S5: Spillover effect		S2+ improvements in smallholder productivity
(vi) S6: Spillover effect		S3+ improvements in smallholder productivity
(vii) S7: Combined (i–vi)	S1+S2+S3+S4+S5+S6	

by 2,153.62 hectares per year. Given that a large proportion of land allocated to this crop is not utilized, we assume that expansion of jatropha will not affect smallholders in terms of land displacement.

In scenario 3 (S3), the castor bean scenario, castor bean production is expanded by increasing the quantity of land, which is assumed to grow by about 2,033.33 hectares per year. Notice that further expansion of land beyond this magnitude can come at the expense of smallholders – that is, smallholder land will be reduced by the amount of land allotted to castor bean production.

Scenario 4 (S4), the palm oil scenario, assumes expansion in palm oil production by increasing the quantity of land, which is assumed to grow by 2,000.00 hectares per year.

Scenario 5 (S5) includes S2 (jatropha) with improved productivity of the smallholder crop sector. This scenario intends to capture the spillover effect of biofuel technology on smallholder agriculture. Such an effect can arise, for instance, through improved farming practices or access to other agricultural inputs (e.g., chemical fertilizer, improved seeds, insecticides). Productivity growth in the cereals sector increased from 2.5% to 5% per year.

Scenario 6 (S6) is S3 (castor bean) with spillover effects of biofuel technology on smallholder crop agriculture. This induces improved productivity of the smallholder crop sector.

The last scenario, scenario 7 (S7), captures the combined effect of all biofuel interventions on the structure of the economy.

Discussion of results

Biofuels and food security

Tables 11.7 and 11.8 present the sectoral effects of biofuel expansion in Ethiopia, with particular attention to food security.[11] In general, in terms of sectoral effects, cereals (food) benefited most, especially in zones where biofuel investment is

Table 11.7 Sectoral Effects (Average Yearly Growth of GDP at Factor Cost per Sector by Region) When Land Is Assumed to Be Fully Employed and Activity-Specific

	Base	Change from (relative to) the baseline growth rate (2005–2020) (%)						
		Sugarcane	Jatropha	Castor bean	Palm oil	Jatropha + spillover	Castor bean + spillover	Combined
GDP	9.1	0	0	0	0	0.1	0.1	0
acr-z1	7.4	0	0	0	0	-1.6	-1.2	0
acr-z2	7.6	0	0	0	0	-1.6	-1.2	0
acr-z3	7.6	0	0	0	0	4.3	-1.2	0
acr-z4	7.5	0.1	0	0	0	4.3	4.6	0.1
acr-z5	7.4	0.1	0	0	0	-1.5	4.7	0.1
acc-z1	8	0.1	0	0	0	0.1	0.1	0.1
acc-z2	8.1	0	0	0	0	0	0	0
acc-z3	7.9	0.1	0	0	0	0	0	0.1
acc-z4	8.1	0	0	0	0	0	0	0
acc-z5	7.9	0.1	0	0	0	0.1	0	0.1
asc_z1	13.2	29.3	0	0	0	0	0	29.3
asc-z2	15.4	-5.1	0	0	0	0.2	0.2	-5.1
asc-z3	15.2	-3.9	0	0	0	0.1	0.1	-3.9
asc-z4	14.2	8.5	0	0	0	0.1	0.1	8.5
asc-z5	12.3	35	0	0	0	0	0	35
aj-z3	12.1	0	1.5	-0.8	-0.2	1.5	-0.8	0.4
aj-z4	12.1	0	1.5	-0.8	-0.2	1.5	-0.8	0.4
acb-z4	13.6	0	-0.6	1.6	-0.3	-0.5	1.6	0.8
acb-z5	13.6	0	-0.5	1.6	-0.2	-0.5	1.6	0.8
apo-z1	11.5	0	-0.6	-0.8	1.7	-0.6	-0.8	0.2
alv-z1	7.4	-0.1	0	0	-0.1	0	0	-0.1
alv-z2	7.3	0	0	0	0	0	0	0
alv-z3	7.4	-0.1	0	0	0	0	0	-0.1
alv-z4	7.3	0	0	0	0	0	0	0
alv-z5	7.3	0	0	0	0	0	0	0
aoa-z1	8.6	0.1	0	0	0	0.1	0.1	0.1
aoa-z2	8.5	0	0	0	0	0	0	0
aoa-z3	8.7	0.1	0	0	0	0.1	0.1	0.1
aoa-z4	8.6	0.1	0	0	0	0.1	0	0.1
aoa-z5	8.3	0	0	0	0	0	0	0
afrfs	12.7	0	0	0	0	0	0	0
afr_energ	10.7	0	0	0	0	0.1	0.1	0
aelect	7.9	0	0	0	0	0	0	0
afood	8.5	0.9	0	0	0	0.2	0.2	0.9
abdcj	11.1	0	1.5	-0.8	-0.2	1.5	-0.8	0.4
abdccb	13	0	-0.5	1.6	-0.2	-0.5	1.6	0.8
abdcpo	10.5	-0.1	-0.6	-0.9	1.6	-0.6	-0.9	0.2
Aeth	6.1	6.5	0	0	0	-0.1	-0.1	6.5
Aoip	12	-0.1	0	0	0	0	0	-0.1
Aser	9.3	0	0	0	0	0	0	0
Total	9.1	0	0	0	0	0.1	0.1	0

Table 11.8 Sectoral Effects (Average Yearly Growth of GDP at Factor Cost per Sector by Region) When Land Is Assumed to Be Fully Employed and Mobile

| | Base | Change from (relative to) the baseline growth rate (2005–2020) (%) | | | | | | |
		Sugarcane	Jatropha	Castor bean	Palm oil	Jatso	Castbso	Combined
acr-z1	7.3	0	0	0	0	−3.6	−2.8	0
acr-z2	7.6	0	0	0	0	−3.5	−2.8	0
acr-z3	7.5	0	0	0	0	6.4	−3.3	0
acr-z4	7.5	0	0	0	0	6.1	6.7	0
acr-z5	7.3	0	0	0	0	−4.1	7.2	0
acc-z1	7.9	0	0	0	0	0.8	0.7	0
acc-z2	8.3	0	0	0	0	1.6	1.3	0
acc-z3	7.6	0	0	0	0	−0.6	0.4	0
acc-z4	8	0	0	0	0	−1.6	−2	0
acc-z5	8.1	0	0	0	0	1.2	−2.7	0
asc-z1	12.8	0	0	0	0	1.5	1.4	0
asc-z2	12.6	0	0	0	0	1.5	1.3	0
asc-z3	12.1	0	0	0	0	0.2	0.7	0
asc-z4	12.6	0	0	0	0	−1.1	−1.5	0
asc-z5	11.9	0	0	0	0	2.5	−4.2	0
aj-z3	12	0	0	0	0	−0.1	0.2	0
aj-z4	12	0	0	0	0	−0.1	−0.3	0
acb-z4	13.8	0	0	0	0	0	0.2	0
acb-z5	13.8	0	0	0	0	−0.1	−0.3	0
apo-z1	11.3	0	0	0	0	0.4	0.2	0
alv-z1	7.3	0	0	0	0	0	0.1	0
alv-z2	7.3	0	0	0	0	0	0	0
alv-z3	7.3	0	0	0	0	0	0	0
alv-z4	7.3	0	0	0	0	0	0	0
alv-z5	7.3	0	0	0	0	0	0	0
aoa-z1	8.7	0	0	0	0	0.7	0.6	0
aoa-z2	8.4	0	0	0	0	0.9	0.8	0
aoa-z3	8.9	0	0	0	0	−0.5	0.2	0
aoa-z4	8.7	0	0	0	0	−1	−1.3	0
aoa-z5	7.9	0	0	0	0	0.3	−0.8	0
afrfs	12.7	0	0	0	0	0	0	0
afr_energ	10.7	0	0	0	0	0.1	0.1	0
aelect	7.9	0	0	0	0	0	0	0
afood	9	0	0	0	0	0.2	0.2	0
abdcj	11.2	0	0	0	0	−0.1	−0.1	0
abdccb	13.5	0	0	0	0	0	0	0
abdcpo	10.4	0	0	0	0	0.4	0.3	0
aeth	8.8	0	0	0	0	−0.1	−0.1	0
aoip	12	0	0	0	0	0	0	0
aser	9.3	0	0	0	0	0	0	0
Total	9.1	0	0	0	0	0.2	0.2	0

located. In particular, the jatropha and castor bean scenarios that involved spillover effects affected cereal production, with the effect being variable across regions. For example, under the jatropha with spillover scenario, cereal production in $z3$ and $z4$ (AEZ 3 & 4) increased by 4.3%; it decreased by 1.6% in $z1$ and $z2$ and by 1.5% in $z5$. Similarly, under the castor bean with spillover scenario, cereal production in $z4$ and $z5$ respectively increased by 4.6% and 4.7% but decreased by 1.2% elsewhere.

Moreover, the effect on cereal production (food security) of the jatropha and castor bean scenarios that involved spillover effects is larger when land is fully employed but mobile (Table 11.8). This might be because the forward and backward linkages between cereals production and biofuels investment are stronger than the linkages between biofuels and the other-agriculture sector. Smallholders in regions where biofuels crops are located can benefit from biofuel expansion in different ways, including wage employment, improved farm practices (e.g., technology transfer), infrastructure (e.g., roads, markets), and so forth. The spillover effects of these biofuels crops on cereal production are significant in magnitude, even though these biofuels crops are located in AEZs where there is very minimal cereal production (e.g., AEZs 3 and 4 for jatropha and AEZs 4 and 5 for castor bean). However, there are negative effects on cereal (food) production elsewhere. These negative effects might be due to the price-reducing effect of increased cereal (food) production in the biofuels regions.

When the spillover effects are not included, jatropha, castor bean, and palm oil scenarios are found to have no effect on cereals. The effect of the sugarcane scenario is positive, even without including spillover effects, however, at only .1 percentage point, especially in AEZ4 and AEZ5. Unlike the jatropha and castor bean scenarios that involved spillover effects, the sugarcane scenario has no impact elsewhere.

Similarly, the combined scenario, which includes spillover effects for some biofuel crops, has a positive effect on cereals, especially in AEZ4 and AEZ5, but at a small magnitude of 1 percentage point.

The effect of biofuels investment on cash crops also turns out to be either neutral or mild and positive. However, the effect of biofuels investment on livestock is found to be neutral or negative, perhaps because biofuel crops compete for land with livestock to some extent. For example, whereas the sugarcane, palm oil, and combined scenarios negatively affect livestock, the jatropha and castor bean scenarios involving spillover effects, as well as the other scenarios (jatropha, castor bean, and palm oil scenarios without spillover effects), are found to have no effect on livestock. Moreover, the effect of biofuels expansion on other agriculture is found to be either neutral or positive.

Only the jatropha and castor bean scenarios that involved spillover effects have any effect in the simulations where land is assumed to be fully employed but mobile. The effect on cereal (food) production of these two scenarios is variable across regions (Table 11.8). Moreover, the effect is relatively larger in magnitude as compared to the simulations where land is fully employed but activity-specific. Furthermore, except for livestock, the effects on cash crop production and other agriculture are also mixed or variable across regions. Table 11.9 also shows the trade-off or the

Table 11.9 Trade-Off (Difference in Average Yearly Growth of GDP at Factor Cost per Sector When Land Is Assumed to Be Fully Employed and Mobile, and Fully Employed and Activity-Specific)

	Base	Sugarcane	Jatropha	Castor bean	Palm oil	Jatso	Castbso	Combined
acr-z1	−0.09	−0.04	0.00	0.00	0.00	−2.00	−1.60	−0.04
acr-z2	0.08	−0.04	0.00	0.00	0.00	−2.04	−1.64	−0.04
acr-z3	−0.08	−0.04	0.00	0.00	0.00	2.12	−2.04	−0.04
acr-z4	−0.03	−0.04	0.00	0.00	0.00	1.75	2.16	−0.04
acr-z5	−0.09	−0.04	0.00	0.00	0.00	−2.54	2.54	−0.04
acc-z1	−0.09	−0.06	0.00	0.00	0.00	0.70	0.60	−0.06
acc-z2	0.20	−0.06	0.00	0.00	0.00	1.58	1.33	−0.06
acc-z3	−0.28	−0.06	0.00	0.00	0.00	−0.70	0.35	−0.06
acc-z4	−0.03	−0.03	0.00	0.00	0.00	−1.61	−2.02	−0.03
acc-z5	0.19	−0.06	0.00	0.00	0.00	1.15	−2.76	−0.06
asc-z1	−0.40	−29.31	0.00	0.00	0.00	1.50	1.36	−29.31
asc-z2	−2.87	5.15	0.00	0.00	0.00	1.35	1.21	5.15
asc-z3	−3.04	3.90	0.00	0.00	0.00	0.00	0.57	3.90
asc-z4	−1.66	−8.48	0.00	0.00	0.00	−1.12	−1.50	−8.48
asc-z5	−0.35	−34.98	0.00	0.00	0.00	2.45	−4.29	−34.98
aj-z3	−0.16	0.02	−1.44	0.84	0.27	−1.49	1.09	−0.34
aj-z4	−0.12	0.02	−1.45	0.84	0.27	−1.60	0.47	−0.35
acb-z4	0.18	0.02	0.57	−1.56	0.26	0.53	−1.31	−0.78
acb-z5	0.22	0.02	0.57	−1.57	0.26	0.43	−1.94	−0.78
apo-z1	−0.21	0.02	0.58	0.82	−1.66	0.98	1.08	−0.21
alv-z1	−0.01	0.01	0.00	−0.01	0.01	−0.02	−0.02	0.01
alv-z2	−0.02	0.02	0.00	0.00	0.00	−0.01	−0.01	0.02
alv-z3	−0.02	0.02	0.00	0.00	0.00	−0.01	−0.01	0.02
alv-z4	−0.02	0.01	0.00	0.00	0.00	−0.01	−0.01	0.01
alv-z5	−0.02	0.02	0.00	0.00	0.00	−0.01	−0.01	0.02
aoa-z1	0.10	−0.04	0.00	0.00	0.00	0.65	0.51	−0.04
aoa-z2	−0.06	−0.04	0.00	0.00	0.00	0.89	0.71	−0.04
aoa-z3	0.18	−0.04	0.00	0.00	0.00	−0.53	0.22	−0.04
aoa-z4	0.09	−0.04	0.00	0.00	0.00	−1.07	−1.37	−0.04
aoa-z5	−0.34	−0.05	0.00	0.00	0.00	0.28	−0.86	−0.05
Afrfs	−0.03	0.04	0.00	0.00	0.00	−0.01	−0.01	0.04
afr_energ	0.00	0.00	0.00	0.00	0.00	−0.01	−0.01	0.00
aelect	0.06	−0.06	0.00	0.00	0.00	0.01	0.01	−0.06
Afood	0.44	−0.83	0.00	0.00	0.00	0.06	0.05	−0.83
Abdcj	0.03	0.02	−1.43	0.83	0.27	−1.53	0.78	−0.34
abdccb	0.48	0.02	0.57	−1.56	0.26	0.48	−1.62	−0.78
abdcpo	−0.06	0.02	0.58	0.83	−1.67	0.99	1.09	−0.21
Aeth	2.66	−6.52	0.00	0.00	0.00	0.06	0.03	−6.52
Aoip	−0.02	0.09	0.00	0.00	0.00	−0.01	−0.01	0.09
Aser	0.03	0.01	0.00	0.00	0.00	0.00	0.00	0.01
Total	0.01	−0.05	0.00	0.00	0.00	0.05	0.05	−0.05

difference between the two simulations – that is, when land is assumed to be fully employed but mobile and fully employed but activity-specific.

Overall, contrary to the notion that increased biofuels production might undermine the food security objectives of developing countries, our simulations suggest they can have some positive effects – or the opposite – depending on the region. In fact, production in both cereals (food) and cash crops increases in four of the seven scenarios, with some negative effects in two of the four scenarios. Biofuels expansion also enhances other agriculture, although the effects tend to be mild. However, the effects of biofuels on cereals (food) varied by regions. Moreover, results also varied depending on whether land is activity-specific or is allowed to be mobile. This suggests that biofuel investment initiatives can have a 'win-win' outcome that can improve smallholder productivity and increase cereals production depending on the region.

In general, biofuel expansion complements the growth of the agriculture sector, with important linkages to other sectors. That is, a big rise in agricultural income implies that demand for local manufactures goes up (Adelman and Robinson, 1978). Note that this result is based on the assumption that there is no displacement of smallholders – that is, that biofuels investments take place on unutilized land that is not occupied by smallholders. Given the government's huge investment in road infrastructure in the country (see MoFED, 2010), access to unused land will no longer be constrained by inadequate road infrastructure. Therefore, further biofuel investment is likely to be undertaken on unoccupied lands, at least in the short to medium term.

Another important observation from the results in Tables 11.7 and 11.8 is that the different biofuels crops tend to compete with each other for the available land. The degree of competition tends to be strong, especially in the case of the sugarcane and combined scenarios. Note also that, because a significant proportion of land is utilized in sugarcane production, there is less scope for sugarcane production to increase through land expansion, in comparison to the opportunity for land expansion of other biofuels crops. That explains the higher degree of competition in the sugarcane and combined scenarios. In addition, only the combined scenario and the sugarcane scenario significantly increase the production of biodiesel and ethanol, respectively.

The study also assesses the impact of biofuels investment on the price of cereals for the two different simulations. The spillover scenarios, with higher productivity growth in cereals, result in much less effect on cereal prices when land is fully employed but activity-specific. The other biofuel scenarios do not have any impact on cereals prices. On the other hand, when land is fully employed but mobile, the sugarcane scenario results in less effect on cereal prices. The spillover scenarios, with higher productivity growth in cereals, result in a significant fall in cereal prices. The other biofuel scenarios do not have any impact on cereals prices.

Welfare (distributive) effects of biofuel investment

We also examine welfare and distributive effects of biofuels investment. The results are presented in Table 11.10. Note that the effects on household welfare

Table 11.10 Effects of Biofuel Investment on Household Welfare (Income) When Land Is Fully Employed but Activity-Specific (in %)

	Initial spending (2005/06)	Base	Sugar-cane	Jatropha	Castor bean	Palm oil	Jatso	Castbso	Combined
rural poor-z1	480.81	1.42	0.10	0.01	0.01	0.01	1.78	1.41	0.11
rural poor-z2	9,584.28	27.33	−0.01	0.00	0.00	0.00	2.28	1.79	−0.01
rural poor-z3	4,436.87	12.81	0.26	0.00	0.00	0.00	2.88	1.21	0.27
rural poor-z4	8,349.40	22.87	0.24	0.00	0.00	0.00	4.31	3.93	0.24
rural poor-z5	1,304.67	3.26	1.25	0.00	0.00	0.00	2.91	3.27	1.25
rural nonpoor-z1	583.15	2.07	0.47	0.00	0.00	−0.01	1.41	1.12	0.47
rural nonpoor-z2	31,003.13	95.42	0.12	0.00	0.00	0.00	1.08	0.81	0.12
rural nonpoor-z3	12,536.06	39.82	−0.09	0.00	0.00	0.00	2.59	0.91	−0.09
rural nonpoor-z4	24,150.76	70.55	−0.01	0.00	0.00	0.00	4.53	4.38	−0.01
rural nonpoor-z5	3,079.40	8.88	0.87	0.00	0.00	0.00	2.49	2.17	0.87
urban poor-s	2,549.88	12.60	0.96	0.00	0.00	0.00	1.83	1.45	0.97
urban poor-b	1,694.07	5.46	1.36	0.00	0.00	0.00	1.49	1.17	1.36
urban nonpoor-s	12,611.32	50.52	0.98	0.00	0.00	0.00	1.04	0.82	0.98
urban nonpoor-b	11,057.75	29.91	1.25	0.00	0.00	0.00	0.58	0.45	1.25

Source: CGE simulation results.

Note: z stands for AEZs; s stands for small; and b stands for big.

are measured in terms of changes in equivalent variation (EV). The EV captures what income change at the current price would be equivalent to the proposed change in terms of its impact on utility (welfare), using the current prices as the base. As can be seen, in the case of the simulation results where land is assumed to be fully employed but activity-specific (see Table 11.10), the sugarcane and combined scenarios, as well as the jatropha and castor bean scenarios that involved spillover effects, affect both poor and non-poor households in both rural and urban areas.

In the case of the sugarcane and combined scenarios, most of the rural poor gain some benefits, but the urban households benefit most. Except for those residing in AEZs 1, 2, and 5, the welfare of rural non-poor households elsewhere is negatively affected under both scenarios. Factor returns is one of the transmission mechanisms through which the benefits of biofuel investment can affect household welfare. Returns to factors depend, among other things, on the rate of return to labour, which in turn depends on factors such as human capital, physical assets, and natural capital, such as land (Osmani, 2005). This means that urban households are better able than rural households to exploit opportunities created by biofuels investments, due to either a greater quantity of employment or greater earnings per unit of employment (or both) compared to rural households. In addition, the welfare effect is greater for urban households compared with poor rural households under both the sugarcane and combined scenarios. This may be because of differences in both quality and quantity of factor endowments between the two groups of households. The impact on the welfare of rural households also varies by agro-ecology in both cases.

The combined scenario – that is, when all biofuels investments are undertaken simultaneously – has more or less the same effect on household welfare as the sugarcane scenario. However, the welfare effects, as measured by changes in the equivalent variation, tend to be stronger in the biofuels scenarios involving spill-over effects, because of the lower cereal prices that these scenarios generate.

The jatropha, castor bean, and palm oil scenarios without spillover effects turn out to have no significant effects on the welfare of households, unlike the sugar-cane and combined scenarios. On the other hand, the jatropha and castor bean scenarios involving spillover effects have a significant positive effect on the wel-fare of all households involved.

In the case of simulations where land is fully employed but mobile, only the jatropha and castor bean scenarios involving spillover effects affect household wel-fare. Both scenarios affect household welfare positively everywhere (Table 11.11).

We also examine the trade-off or difference between the two simulations – that is, when land is fully employed but activity-specific and when land is fully employed but mobile. The results indicate that the nature and magnitude of the welfare impact of the scenarios considered can vary depending on whether land is assumed to be activity-specific or mobile (Table 11.12).

Table 11.11 Effects of Biofuel Investment on Household Welfare (Income) When Land Is Fully Employed but Mobile (in %)

	Base	Sugarcane	Jatropha	Castor bean	Palm oil	Jatso	Castbso	Combined
hhd-pz1-r	1.42	0.00	0.00	0.00	0.00	1.99	1.59	0.00
hhd-pz2-r	27.33	0.00	0.00	0.00	0.00	2.46	1.94	0.00
hhd-pz3-r	12.81	0.00	0.00	0.00	0.00	3.90	1.26	0.00
hhd-pz4-r	22.87	0.00	0.00	0.00	0.00	5.50	5.34	0.00
hhd-pz5-r	3.27	0.00	0.00	0.00	0.00	3.34	3.47	0.00
hhd-npz1-r	2.08	0.00	0.00	0.00	0.00	2.02	1.64	0.00
hhd-npz2-r	95.65	0.00	0.00	0.00	0.00	1.90	1.48	0.00
hhd-npz3-r	39.61	0.00	0.00	0.00	0.00	2.24	1.35	0.00
hhd-npz4-r	70.59	0.00	0.00	0.00	0.00	4.07	3.80	0.00
hhd-npz5-r	8.90	0.00	0.00	0.00	0.00	2.80	2.36	0.00
hhd-sp-u	12.65	0.00	0.00	0.00	0.00	2.02	1.63	0.00
hhd-bp-u	5.48	0.00	0.00	0.00	0.00	1.66	1.33	0.00
hhd-snp-u	50.72	0.00	0.00	0.00	0.00	1.15	0.93	0.00
hhd-bnp-u	30.01	0.00	0.00	0.00	0.00	0.65	0.52	0.00

Table 11.12 Trade-Off (Difference between the Simulations)

	Base	Sugarcane	Jatropha	Castor bean	Palm oil	Jatso	Castbso	Combined
hhd-pz1-r	0.00	−0.10	−0.01	−0.01	−0.01	0.21	0.18	−0.11
hhd-pz2-r	0.00	0.01	0.00	0.00	0.00	0.17	0.15	0.01
hhd-pz3-r	−0.01	−0.26	0.00	0.00	0.00	1.02	0.05	−0.27
hhd-pz4-r	0.00	−0.24	0.00	0.00	0.00	1.19	1.41	−0.24
hhd-pz5-r	0.01	−1.25	0.00	0.00	0.00	0.43	0.20	−1.25
hhd-npz1-r	0.00	−0.47	0.00	0.00	0.01	0.60	0.52	−0.47
hhd-npz2-r	0.22	−0.12	0.00	0.00	0.00	0.82	0.67	−0.12
hhd-npz3-r	−0.21	0.09	0.00	0.00	0.00	−0.35	0.43	0.09
hhd-npz4-r	0.03	0.01	0.00	0.00	0.00	−0.46	−0.58	0.01
hhd-npz5-r	0.03	−0.87	0.00	0.00	0.00	0.31	0.19	−0.87
hhd-sp-u	0.05	−0.96	0.00	0.00	0.00	0.19	0.18	−0.97
hhd-bp-u	0.02	−1.36	0.00	0.00	0.00	0.17	0.16	−1.36
hhd-snp-u	0.20	−0.98	0.00	0.00	0.00	0.11	0.11	−0.98
hhd-bnp-u	0.10	−1.25	0.00	0.00	0.00	0.07	0.07	−1.25

Conclusion and policy implications

This study provides empirical evidence regarding the economy-wide effects of biofuel investment in Ethiopia. Specifically, this study investigates the distributive effect and food security implications of biofuel investment in Ethiopia using an economy-wide CGE model. One of the key features of the study is that it captures the impact of biofuels investment by agro-ecology zones. Another feature is its use of data from firms and NGOs involved in biofuels production.

The model results indicate that biofuels can provide the country with an opportunity to accelerate economic growth and improve household welfare. Two important conclusions can be drawn. One, biofuels can provide an opportunity to accelerate agricultural growth and enhance food security. In particular, in the sugarcane scenario and the jatropha and castor bean scenarios with spillover effects, as well as the combined scenario, production activities are projected to increase overall agriculture production and food security. Two, biofuels expansion can improve household welfare. Moreover, the benefits of biofuels investment are magnified if such investment is accompanied by technology spillovers to other agricultural crops, such as cereals. This suggests that biofuel investment complements the growth of the agriculture sector. Rural households tend to be the main beneficiaries of such investment, particularly in the case of the scenarios that involved spillover effects. Urban households benefited most in the case of the sugarcane and the combined scenarios.

Contrary to the notion that increased biofuel production runs counter to the food security objectives of developing countries, our simulations suggest the opposite. In fact, both cereals and cash crops productions increase in all scenarios.

As long as such schemes are undertaken on unutilized land, biofuel investment initiatives can have a 'win-win' outcome that can improve smallholder productivity and increase food crop production. Assuming that government investments in road infrastructure in Ethiopia will continue in the years to come, access to unused land will no longer be constrained by inadequate road infrastructure, and further biofuel investment will be undertaken on unoccupied lands, at least in the short to medium term.

An important implication is that, to maximize the benefits of biofuel investment, it is important to expand infrastructure. This will help expand and attract biofuel investment in areas not occupied by smallholders. This approach will not only improve food security and enhance household welfare but also help to stabilize the macroeconomy.

Appendices

Table 11A1 The 2005/06 Macro-SAM for Ethiopia (billion ETB)

	ACT	COM	LAB	CAP	ENT	HHD	GOV	DTAX	MTAX	STAX	S-I	ROW	TOTAL
ACT		187.3											187.3
COM	65.1	46.2				114.8	15.9				31.8	16.8	290.5
LAB	68.8												68.8
CAP	53.5												53.5
ENT				48.2			-5.4						42.9
HHD			68.8	5.5	41.5		1.5					15.7	133.0
GOV								4.1	7.0	3.1		3.3	17.5
DTAX					1.3	2.7							4.1
MTAX		7.0											7.0
STAX		3.1											3.1
S-I						15.5	5.4				3.6	10.9	35.5
ROW		46.9		-0.2									46.7
TOTAL	187.3	290.5	68.8	53.5	42.9	133.0	17.5	4.1	7.0	3.1	35.5	46.7	

Source: Tebekew et al. (2009).

Table 11A2 Household Income Distribution

	Share (%)						Total (%)
	Labour	Land	Livestock	Capital	Govt	ROW	
All groups	44.4	5.0	3.3	30.4	1.5	15.4	100.0
Rural	46.8	8.6	5.6	32.6	0.6	5.7	100.0
Poor	66.4	3.6	7.3	15.1	1.2	6.4	100.0
Non-poor	40.2	10.3	5.0	38.5	0.4	5.5	100.0
Urban	41.0			27.3	2.7	29.0	100.0
Poor	62.8			8.0	3.9	25.3	100.0
Non-poor	37.4			30.5	2.5	29.6	100.0
Poor	65.8	3.0	6.1	14.0	1.6	9.5	100.0
Non-poor	39.4	7.3	3.5	36.2	1.0	12.5	100.0

Source: Ethiopia SAM 2005/06.

Notes

1 The UN's special rapporteur on the right to food, Jean Ziegler, stated at a press conference in New York that it is a crime against humanity to divert arable land to the production of biofuels crops as a replacement for petrol (Mathews, 2008).
2 Mtoe stands for metric ton of oil equivalent.
3 An out-growers scheme is a contract farming arrangement in which a firm enters into a binding agreement with individuals or groups of farmers to grow a certain crop and through which the firm ensures its supply of agricultural products (Felgenhauer and Wolter, 2008).
4 See Lofgren et al. (2002) for a detailed description.
5 The five AEZs include: moisture-reliable humid lowlands (AEZ1); moisture-sufficient highlands (cereal-based systems) (AEZ2); moisture-sufficient highlands (enset-based systems) (AEZ3); drought-prone (AEZ4); and pastoralist (AEZ5).
6 See, for instance, agriculture sample survey of CSA (Central Statistical Agency) (various issues).
7 Enset (i.e., *Enset ventricosum*) is an edible crop (plant). It is the edible species of a separate genus of the banana family, thus named 'false banana', but the *enset* fruit is not edible. The plant is cut before flowering, the pseudostem (false stem) and leaf midribs are scraped, the pulp is fermented for 10–15 days and finally steam-baked flatbread is prepared out of it and used as food (Shank and Ertiro, 1996).
8 REST is an abbreviation for 'Relief Society of Tigrai' and ORDA is an abbreviation for 'Organization for Rehabilitation and Development of Amhara'.
9 Skilled labour of youth and adults is assumed to grow at a rate of 7.9% per year, which is consistent with expansion of higher education in the country. While semi-skilled labour is assumed to grow at the rate of 5% per year, unskilled labour is assumed to grow at 4.4% per year, a little more slowly than the rate at which semi-skilled is assumed to grow. According to data from national labour force surveys, the labour force grows faster than the rate of population growth. The most recent Population Census (CSA, 2007) indicates that the age composition of the population is skewed towards the young and adult population, suggesting that the labour force grows faster than the population growth rate.
10 The recent biofuel investment survey indicates that, of the total land allocated to sugarcane, jatropha, caster bean, and palm oil, about 34,058.42, 31,804.33, 24,500,

and 29,775 hectares of land are not utilized, in that order. In the biofuel scenarios, we evenly distribute this unutilized land over the 15 periods, which implies no displacement of smallholders.

11 Notice that we make use of Table 11.2 in introducing biofuel shocks. In other words, in the sugarcane scenario, we expand land area in all zones, as this crop is grown by smallholders, as indicated in the EDRI SAM. In the palm scenario, land expansion occurs in AEZ 1 only. We expand land for biofuel crops in the jatropha and castor bean scenarios in AEZ 3 & 4 and AEZ 4 & 5, respectively.

References

Abbott, P.C., C. Hurt and W.E. Tyner. 2008. What's Driving Food Prices? Farm Foundation, July.

ABN (African Biodiversity Network). 2007. AGRO Fuels in Africa: The Impacts on Land, Food and Forests. www.africanbiodiversity.org/sites/default/files/PDFs/Agrofuels%20in%20Africa-Impacts%20on%20land%2C%20foods%20%26%20forests.pdf (accessed May 30, 2011).

Adelman, I., and S. Robinson. 1978. *Income Distribution Policy in Developing Countries: A Case Study of Korea.* Stanford and Oxford: Stanford University Press and Oxford University Press.

Arndt, C., R. Benfica, F. Tarp, J. Thurlow and R. Uaiene. 2009. Biofuels, Poverty, and Growth: A Computable General Equilibrium Analysis of Mozambique. *Mozambique Environment and Development Economics* 15: 81–105.

Arndt, C., K. Pauw and J. Thurlow. 2010. Biofuels and Economic Development in Tanzania. IFPRI Discussion Paper No. 00966. Washington, DC: IFPRI.

Azar, C. 2011. Biomass for Energy: A Dream Come True . . . or a Nightmare? *Wiley Interdisciplinary Reviews: Climate Change* 2(3): 309–323.

Banse, M., H. van Meijl, A. Tabeau and G. Woltjer. 2008. Will EU Biofuel Policies Affect Global Agricultural Markets? *European Review of Agricultural Economics* 35(2): 117–141.

Birur, D.K., T.W. Hertel and W.E. Tyner. 2008. Impact of Biofuel Production on World Agricultural Markets: A Computable General Equilibrium Analysis. GTAP Working Paper No. 53.

CBO (US Congressional Budget Office). 2009. The Impact of Ethanol Use on Food Prices and Greenhouse-Gas Emissions. Congress of the United States Congressional Budget Office Pub. No. 3155.

Chakravorty, U., M. Hubert, M. Moreaux and L. Nostbakken. 2011. Will Biofuel Mandate Raise Food Prices? University of Alberta, Department of Economics Working Paper No. 2011–01.

Chakravorty, U., M. Hubert and L. Nostbakken. 2009. Fuel versus Food. *Annual Review of Resource Economics* 1: 645–663.

CSA (Central Statistical Agency). 2006. Report on the 2005 National Labour Force Survey. Statistical Bulletin 365, Addis Ababa, Ethiopia: CSA.

CSA (Central Statistical Authority). 2008. Summary Statistical Report of the 2007 Population and Housing Census. Addis Ababa, Ethiopia: CSA.

Diao, X., P. Hazell, D. Resnick and J. Thurlow. 2005. *The Role of Agriculture in Pro-Poor Growth in Sub-Saharan Africa.* Washington, DC: IFPRI.

Diao, X., and A.N. Pratt. 2007. Growth Options and Poverty Reduction in Ethiopia: An Economy-Wide Model Analysis. *Food Policy* 32(2): 205–228.

Dimaranan, B. (Ed.). 2006. *Global Trade, Assistance, and Production: The GTAP 6 Database*. Center for Global Trade Analysis. West Lafayette, IN: Purdue University.

Dorosh, P., and J. Thurlow. 2012. Implications of Accelerated Agricultural Growth on Household Incomes and Poverty in Ethiopia: A General Equilibrium Analysis. In *Food and Agriculture in Ethiopia: Progress and Policy Challenges*, edited by P. Dorosh and S. Rashid. Philadelphia, PA: PENN (University of Pennsylvania Press), pp. 219–255.

Dufey, A. 2006. Biofuels Production, Trade and Sustainable Development: Emerging Issues. International Institute for Environment and Development (IIED). Sustainable Markets Discussion Paper No. 2.

EDRI (Ethiopian Development Research Institute). 2009. Ethiopia Input-Output Table and Social Accounting Matrix (SAM) 2005/06. Addis Ababa: EDRI.

FAPRI (Food and Agricultural Policy Research Institute). 2010. US and World Agricultural Outlook, FAPRI Staff Report 10-FSR 1. Ames: Iowa State University and Columbia: University of Missouri.

Felgenhauer, K., and D. Wolter. 2008. Outgrower Schemes: Why Big Multinationals Link up with African Smallholders. Paris: OECD Development Center.

Gebremeskel, L., and M. Tesfaye. 2008. A Preliminary Assessment of Socioeconomic and Environmental Issues Pertaining to Liquid Biofuel Development in Ethiopia. In *Agrofuel Development in Ethiopia: Rhetoric, Reality and Recommendations*, edited by T. Heckett and N. Aklilu. Addis Ababa: Forum for Environment.

Gelan, A. 2007. Does Food Aid Have Disincentive Effects on Local Production? A General Equilibrium Perspective on Food Aid in Ethiopia. *Food Policy* 32: 436–458.

Gelan, A., E. Engida, A.S. Caria and J. Karugia. 2012. Integrating Livestock in the CAADP Framework: Policy Analysis Using a Dynamic Computable General Equilibrium Model for Ethiopia. IFPRI/ESSP (Ethiopia Strategy Support Program) II Working Paper No. 34.

Hausman, C., M. Auffhammer and P. Berck. 2012. Farm Acreage Shocks and Food Prices: An SVAR Approach to Understanding the Impacts of Biofuels. http://ssrn.com/abstract=1605507 (accessed October 31, 2012).

Hausmann, R. 2007. Biofuels Can Match Oil Production. *Financial Times* November 6.

IEA (International Energy Agency). 2006. World Energy Outlook 2006: The Outlook for Biofuels. Paris: OECE/IAE.

IEA (International Energy Agency). 2010. World Energy Outlook 2010, Executive Summary. Paris: OECE/IAE.

Lakew, H., and Y. Shiferaw. 2008. Rapid Assessment of Biofuels Development Status in Ethiopia. In *Proceedings of National Workshop on Environmental Impact Assessment and Biofuels*, edited by T. Anderson and M. Belay. Addis Ababa: MELCA Mahiber.

Lashitew, A.A. 2008. Competition between Biofuel and Food Production in Ethiopia: A Partial Equilibrium Analysis. MSc Thesis. Wageningen University, Wageningen, Netherlands.

Lofgren, H., R. Lee and S. Robinson. 2002. *Standard Computable General Equilibrium Model in GAMS*. Microcomputers in Policy Research 5. Washington, DC: International Food Policy Research Institute.

Mathews, J.A. 2008. Opinion: Is Growing Biofuel Crops a Crime against Humanity? *Biofuels, Bioproducts & Biorefining* 2: 97–99.

McNew, K., and D. Griffith. 2005. Measuring the Impact of Ethanol Plants on Local Grain Prices. *Review of Agricultural Economics* 27(2): 164–180.

Mitchell, D. 2008. A Note on Rising Food Prices. Policy Research Working Paper No. 4682. Washington, DC: The World Bank.

MoFED (Ministry of Finance and Economic Development). 2010. The Five Year (2010/11–2014/15) Growth and Transformation Plan. Addis Ababa: MoFED.

MoME (Ministry of Mines and Energy). 2007. The Biofuel Development and Utilization Strategy of Ethiopia. Addis Ababa: MoME.

Osmani, S. 2005. *The Employment Nexus between Growth and Poverty: An Asian Perspective.* Swedish International Development Agency Studies 15. Stockholm: SIDA.

Peskett, L., R. Slater, C. Stevens and A. Dufey. 2007. Biofuels, Agriculture and Poverty Reduction. Natural Resource Perspectives No. 107. London: Overseas Development Institute.

Pratt, N.A., and B.B. Yu. 2008. An Updated Look at the Recovery of Agricultural Productivity in Sub-Saharan Africa. Discussion Paper No. 00787. Washington, DC: IFPRI.

Rajagopal, D., S.E. Sexton, D. Roland-Holst and D. Zilberman. 2007. Challenge of Biofuel: Filling the Tank without Emptying the Stomach? *Environmental Research Letters* 2: 1–9.

Roberts, M.J., and W. Schlenker. 2010. Identifying Supply and Demand Elasticities of Agricultural Commodities: Implications for the US Ethanol Mandate. NBER Working Paper No. 15921.

Rosegrant, M.W. 2008. *Biofuels and Grain Prices: Impacts and Policy Responses.* Washington, DC: IFPRI.

Shank, R., and C. Ertiro. 1996. *A Linear Model for Predicting Enset Plant Yield and Assessment of Kocho Production in Ethiopia.* Addis Ababa: UNDP and WFP.

Slater, R. 2007. Biofuels: Starving People to Feed Cars? Or Growing Our Way out of Poverty? ODI. www.odi.org.uk/events/2007/06/27/203-presentation-rachel-slater.pdf (accessed September 19, 2009).

Sourie, J.-C., D. Tréguer and S. Rozakis. 2006. Economic Impact of Biofuel Chains in France. Mixed Unit of Research INRA and INA P-G. Paris: Economie Publique.

Tebekew, T., A. Amoge, B. Teferra, Z. Seyoum, M. Amha, H. Beyene, E. Fisseha, E. Tsehaye, H.A. Ahmed, S. Robinson, D. Willenbockel, P. Dorosh and S. McDonald. 2009. Ethiopia Input-Output Table and Social Accounting Matrix 2005/06. Addis Ababa and University of Sussex: EDRI (Ethiopian Development Research Institute) and IDS (Institute of Development Studies).

Tréguer, D., and J.-C. Sourie. 2006. The Impact of Biofuel Production on Farm Jobs and Income: The French Case. Article Presented for the 96th EAAE Seminar in Tanikon, Switzerland.

Ugarte, D., G. De La Torre, B.C. English, C.M. Hellwinckel, R.J. Menard and M.E. Walsh. 2007. Economic Implications to the Agricultural Sector of Increasing the Production of Biomass Feedstocks to Meet Biopower, Biofuels, and Bioproduct Demands. The University of Tennessee, Department of Agricultural Economics. Research Series 08–01.

Van Rheenen, T., and T. Olofinbiyi. 2007. Policy Making and Land Use Changes: Facing New and Complex Realities. In *Development Economics between Markets and Institutions: Incentives for Growth, Food Security and Sustainable Use of the Environment,* edited by E. Bulte and R. Ruben. Wageningen, Netherlands: Wageningen Academic.

World Bank. 2004. Ethiopia Country Economic Memorandum. Background Paper. Quantitative Framework for Public Investment Policies in Ethiopia. Draft. http://site resources.worldbank.org/INTETHIOPIA/Resources/PREM/QuantitativeFrameworkfor PublicInvestmentPoliciesInEthiopia.pdf (accessed September 20, 2009).

Worldwatch Institute. 2007. *Biofuels for Transport: Global Potential and Implications for Sustainable Energy and Agriculture.* London: EarthScan.

12 Climate change and post-harvest agriculture

Martin J. Chegere

Introduction

Agriculture-based livelihood systems such as those in sub-Saharan Africa (SSA) are especially vulnerable to the adverse effects of climate change. Climate change will directly and indirectly affect all the stages of the food chain and impose a threat to food security. Its impacts will be felt in the short term, resulting from shifts in rainfall patterns and more frequent and intense extreme weather events, and in the long term, caused by changing temperatures and precipitation patterns. Until recently, most assessments of the impact of climate change on food systems and the agriculture sector have focused on the implications for production and global supply of food, with less consideration of the post-harvest value chain issues, such as harvesting and drying, primary processing,[1] storage, secondary processing[2] and marketing (FAO, 2008a). This chapter aims at exploring the linkages between climate and post-harvest agriculture.

The role of agriculture in producing food and generating income is vital, but the entire food chain is important in improving incomes and ensuring food security (FAO, 2013). Over the years, significant effort and resources have been directed toward increasing food output and productivity. However, increasing agricultural productivity may not be sufficient. Currently, food production expansion is faced with challenges, such as limited land and water resources and increased weather variability due to climate change (World Bank, 2011; Aulakh and Regmi, 2013). However, there is an additional factor that exacerbates food insecurity which has received less attention in the literature: post-harvest losses (PHL) (World Bank, 2011).

It is estimated that 10%–20% of the total grain produced in SSA suffers post-harvest physical losses. The value of this loss could potentially reach USD 4 billion a year (World Bank, 2011). According to the same Word Bank report, this amount is equivalent to the annual calorific requirement of 48 million people (at 2,500 kcal per person per day); equates to the annual value of cereal imports of SSA; and exceeds the value of total food aid SSA received over the first decade of the twenty-first century. Recent estimates the Food and Agriculture Organization of the United Nations (FAO) suggest that up to 47% of the USD 940 billion needed for investment to eradicate hunger in SSA by 2050 should be geared toward reduction of PHL, by investing in cold and dry storage, rural roads, rural and wholesale market facilities, and first-stage processing (FAO and World Bank, 2010).

PHL cause not only the loss of the economic value of the food produced but also the waste of scarce resources, such as labour, land and water, as well as non-renewable resources, such as fertilizer and energy, all of which are used to produce, process, handle and transport food (FAO, 2011). Production of food that will not be consumed results in unnecessary greenhouse gas emissions, which may accelerate climate change (FAO, 2011; World Bank, 2011). According to Vermeulen et al. (2012), food systems contribute 19%–29% of global anthropogenic greenhouse gas emissions. Increasingly, it is recognized that PHL reduction can provide an environmentally sustainable and cost-effective contribution to food security and income improvement, compared to sole reliance on increasing production in a world with limited natural resources, and in an era of high and volatile food prices (FAO and World Bank, 2010; Aulakh and Regmi, 2013).

Post-harvest losses: definitions, overview and reduction efforts

Post-harvest losses are defined as measurable quantitative and qualitative food loss from the time of harvest to the time the food reaches the end consumers (Hodges et al., 2011; World Bank, 2011). The chain comprises interconnected activities, such as harvesting, threshing, shelling, drying, processing, storage, packaging, transportation, milling, marketing and consumption. Food *loss* is the subset of PHL which accounts for the loss of an edible share of food that was available for consumption but ended up not being consumed (Hodges et al., 2011; Aulakh and Regmi, 2013). Food *waste* is a subset of food loss due to human action or inaction and is potentially recoverable for human consumption (Hodges et al., 2011; Aulakh and Regmi, 2013).

Post-harvest losses can be quantitative or qualitative. Quantitative PHL is defined as reduction in physical weight of food available for human consumption and other utilization (FAO, 1980). Quantitative loss can be caused by spillage, broken grain, rodent and pest damage, and spoilage due to temperature changes, chemical changes and humidity content (Aulakh and Regmi, 2013; FAO, 1980). The reduction in weight due to shrinkage of food grain after drying to allow for its storage for a longer period is not counted as a loss because it does not involve any food loss (FAO, 1980). Similarly, losses due to theft are not recorded as quantitative PHL. Qualitative PHL refers to loss in the nutritional value, edibility, caloric value, acceptability or other intrinsic feature of the food (FAO, 1980; World Bank, 2011). Qualitative losses can be due to factors such as contamination by microorganisms, pest and rodents attacks, humidity content, chemical changes, broken grain, contamination by mycotoxin, and pesticides residues. If qualitative deterioration of food makes it unfit for human consumption, leading to eventual rejection, this will be counted as a quantitative loss (FAO, 1980).

Attention to the concept of post-harvest food loss reduction as a significant means to increase food availability was drawn, for the first time, by the World Food Conference held in Rome in 1974 after the food crisis[3] of that time (FAO, 1981). The 7th Special Session of the U.N. General Assembly in 1975 resolved to reduce PHL by 50% by 1985 and declared that it should be undertaken as a matter of priority in developing countries (FAO, 1981). This was followed by ample development investment on PHL reduction for staple foods (World Bank, 2011). Initially, the focus targeted reducing losses of grains; in later years, the coverage extended to roots, tubers, fruits and vegetables (Affognon et al., 2015). After food prices stabilized and due to low adoption of PHL technologies promoted in various SSA countries, the importance of PHL in the African grain sector seemed to be forgotten. International programs which were involved, such as FAO's Prevention of Food Losses Program and the Global Postharvest Forum (PhAction), became dormant (World Bank, 2011). Recently, the discussion on PHL reduction has been revitalized following the food price surge in 2008 and continuing challenges facing expansion of food production.

Causes and estimates of post-harvest losses

Causes of post-harvest losses

Causes of PHL can be classified into two groups: primary and secondary causes.

Primary causes of losses are those which affect the food directly. They can be categorized into the following groups:

1. *Biological:* Damage can be due to consumption by either large or microscopic organisms, such as monkeys, rodents, birds, insects, fungi and bacteria, causing direct food loss. The damage also could come from contamination of food by the excreta or toxic substances produced by the organisms, causing rotting or other defects, thus rendering the food unfit for human consumption.
2. *Chemical:* Food may become unfit for consumption by contact with chemical substances, such as pesticides, preservatives, oil or chemical constituents naturally present in stored foods.
3. *Physical and mechanical:* This is due to bruising, extreme heat conditions or humidity, which spoils food.
4. *Psychological:* This is due to food aversion, not consuming available food before its expiry date, or serving meal sizes beyond one's ability to consume.

Secondary causes are those that lead to conditions that encourage a primary cause of loss.

1. *Post-harvest handling practices:* Inadequate knowledge on handling, processing, storage and transportation skills.
2. *Facilities and necessary infrastructure:* Lack of adequate storage and transportation facilities, drying equipment and smooth marketing systems.

Estimates of post-harvest losses

Various studies have used different and interchangeable definitions for the same terms, thus making it difficult to compare the loss figures from different studies. In addition, it is difficult to ensure perfect uniformity across countries or regions because what might be regarded as leftovers or damaged food to discard in one region could be counted as food fit for human consumption somewhere else. In other instances, leftovers from processing or a particular kind of damage to food may make it unfit for human consumption but it may be used for other purposes, such as feeding livestock. This kind of loss may be recorded up to a certain limit.

In the 1980s, Lipton (1982) queried the conventional wisdom that PHL of cereals were as high as 30%–50%. His reanalysis found that on-farm grain losses, between harvest and consumption, were typically 5%–8% and very rarely above 10%–12%. He also explained that these losses caused suffering only when grain is scarcest, downplaying the need for PHL reduction interventions. This may explain why, after receiving considerable attention following the food crisis in the mid-1970s, PHL reduction seemed to be forgotten by the late 1980s (Kaminski and Christiaensen, 2014).

The growing food demands from increasing and more affluent populations, as well as food price volatility in recent years, have increased concerns about food insecurity (FAO, 2011). This situation called for a critical review of all the food supply and demand components, including physical and economic PHL (Kaminski and Christiaensen, 2014). The actual levels of PHL were probed again. A recent study by Kaminski and Christiaensen (2014) explains that the divergence in PHL estimates in different studies is largely definitional. The estimates may vary because of different metrics used (e.g., calories versus weight), type of crop and the stage in the food chain.

FAO estimates that about one-third of food produced is lost or wasted globally (FAO, 2011). Food losses can occur during production or post-harvest stages. Figure 12.1 shows proportions of food lost at different stages for cereals in different regions. For all regions, PHL are more than two-thirds of the total food loss. Though food losses in developed countries are as high as in developing countries, in the latter, the largest proportion of food losses is at post-harvest, storage and processing levels, and in the former the food losses occur mostly at retail and consumer levels (FAO, 2011).

In developing countries, the supply chain is less mechanized and thus inefficient post-harvest agricultural systems, such as premature harvesting, poor storage facilities, lack of infrastructure, lack of processing facilities, and inadequate market facilities, are the main reasons for losses. For these reasons, losses occur mainly in the early and middle stages of the food supply chains, during drying, storage, processing and transportation. In contrast, developed countries have more efficient farming systems, and better transport, management, storage and processing facilities, which ensure that a larger proportion of harvested output is delivered to the markets. Therefore, losses are generally low in the middle stages of the supply chain (Hodges et al., 2011).

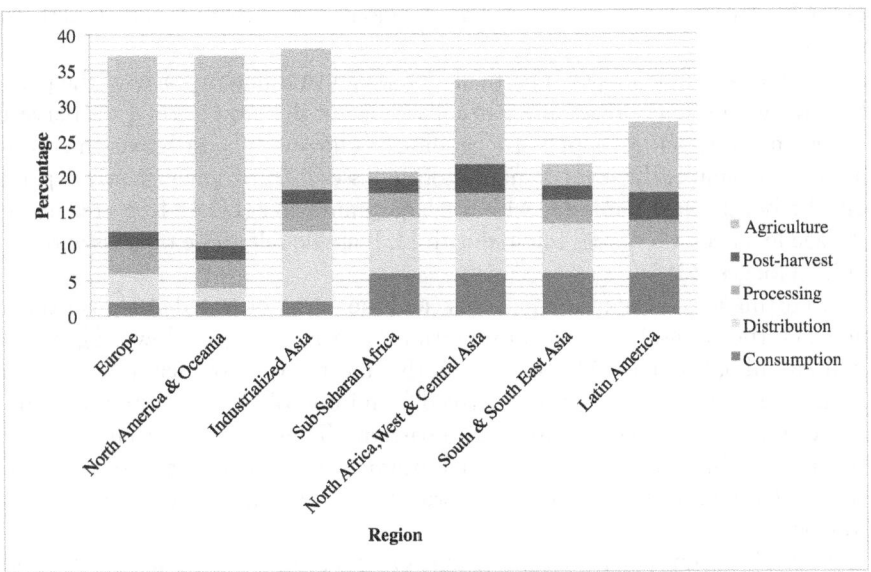

Figure 12.1 Proportion of Food Lost at Different Stages for Cereals in Different Regions
Source: Food and Agriculture Organization of the UN (2011).

Table 12.1 Sub-Saharan Africa PHL by Cereal (% of Total Annual Production)

Cereal	2006	2007	2008	2009	2010	2011	2012	2013	2014	2015
Wheat	9.9	12.8	12.6	15.1	14	13.1	12.9	15.2	13.3	13.3
Maize	17.3	18.4	19.9	17.8	18.6	17.9	17.9	18	17.8	17.6
Rice	11.8	11.8	12.1	12.1	12.6	12.1	14.2	11.9	12.5	12.4
Sorghum	12.3	12.3	13	12.5	12.5	12.4	12.4	12.4	12.5	12.5
Barley	9.4	9.4	9.5	10.9	10.1	9.7	9.7	10.5	–	–
Millet	9.9	10.1	10.3	9.6	9.4	9.4	9.5	9.7	10	10.3

Source: The African Post-Harvest Losses Information System (APHLIS).

In developed countries, more losses are experienced at the consumer level. This is driven by a growing consumer intolerance of substandard foods or non-essential defects, such as bruised products, as well as consumer purchases of more food than they consume, leaving it on the plate after a meal, or letting it pass its expiry date (Hodges et al., 2011). In less developed countries, the low-quality food remaining in markets at the end of the day finds its way to the very poor (Hodges et al., 2011).

The African Post-Harvest Losses Information System[4] provides estimates of PHL of major cereal crops in Africa, as shown in Table 12.1. For the past ten years, the PHL of these major cereal crops have been consistently high, ranging between

10% and 20%. Because the majority of the farmers in Africa engage in production of these major food crops, these are large economic losses that threaten their food security.

The levels of food loss for a particular crop vary at different stages of the post-harvest system. Even losses at a particular stage of the crop in the post-harvest system may vary across countries, depending on climate and post-harvest methods used. For example, the level of maize storage losses in Africa may greatly depend on whether the area is infested with the large grain borer (LGB). Reported maize storage loss figures for areas infested with LGB are double those of areas without LGB (Hodges, 2012).

Adjusting for consumption patterns (i.e., the outlet of stored food over time, due to either household self-consumption or sales) may give lower figures of PHL. Rembold et al. (2011) pointed out that most of the loss measures reported in the literature, especially those derived from interpolations or extrapolations, do not take into account consumption patterns. They estimated that, on average, the unadjusted PHL for maize in Tanzania was about 11% and the average adjusted for consumption patterns was 5.4% for a nine-month period of storage.

It has also been shown that maize losses in Tanzania differ between large and small farmers, with losses experienced by large farmers recorded at 6%, whereas those by small farmers are at 11% (AGRA, 2013).

Linkages between climate change and post-harvest agriculture

During and after harvesting, agricultural products are subjected to physical and biological deterioration. The rate at which this deterioration occurs is highly determined by environmental conditions, including climate. The projected increases in mean temperatures and precipitation will not occur through constant gradual changes; instead, they will be experienced as increased frequency, duration and intensity of hot spells and precipitation events (FAO, 2008a). Whereas the mean temperatures are expected to increase in all parts of the globe, the mean global increase in precipitation is not expected to be uniformly distributed around the world. It is projected that wet regions will become wetter and dry regions dryer (FAO, 2008a). It is also expected that there will be greater weather variability and less predictability. Ahead is a discussion on how different climate change features will affect post-harvest systems.

Effect of climate change on post-harvest losses

Global warming will expand pest and disease territories due to pest outbreaks in new areas which previously were not favorable for pests and diseases and will increase the frequency of outbreaks of field and storage pests and disease invasions (Epstein, 2001). As a result, more losses of food will occur while the crop is in the field and during storage. Studies by D'Souza et al. (2004) and Kovats et al. (2005) document that crop diseases reports are more frequently preceded by weeks of higher local temperature.

Higher temperatures cause shorter life cycles of insect pests and diseases, which may foster reproduction and buildup of field and storage pests and diseases. Rising mean temperature will also increase the rate of crop drying in the field, which will reduce the opportunity for pest attacks but also increase the risk that the grain will become too dry, which makes it shriveled and hard to shell (Stathers et al., 2013). Overly dry grain becomes brittle and can crack after threshing or during milling and thus has low viability.

Higher temperatures may also lead to heat stress affecting farmers during laborious activities, such as harvesting, threshing and drying, thus decreasing their effectiveness in performing these activities.

Increased temperatures and moist weather conditions could result in grain being harvested with more than the 12%–14% moisture required for stable storage (FAO, 2008a). Generally moist, humid conditions favour growth of mould. Furthermore, some fungi perform better at higher temperatures. Therefore, increasing average temperatures and moist conditions following periods of heavy precipitation or floods would be expected to favour mould growth and could lead to changes in the range of latitudes at which certain fungi are able to compete (FAO, 2008b). This increases the risk of fungal growth and mycotoxin contamination of stored crops.

Higher humidity and temperature may reduce the effectiveness of the active ingredients of some commercial grain protectants and may increase chemical and bio-deterioration of the crop (Stathers et al., 2013). These climate change events may also increase chemical and bio-deterioration of stored products, leading to shorter shelf life.

Unseasonal rains can dampen the matured crop before harvesting and result in mould growth, which may later reduce the grain quality, cause some of the grain to be discarded, and increase the associated risk of aflatoxin or other mycotoxin contamination (Hodges et al., 2011; USAID Rwanda, 2012). Chegere (2017) has shown that harvesting when weather conditions are damp (compared to when the weather is sunny) correlates with higher pre-storage and storage losses. In case unfavorable damp weather conditions occur during and after crop harvesting, drying becomes difficult, which may cause rotting or germination and increased risk of mycotoxin contamination.

Various climate change events may increase the cost of grain management in different forms: (1) damage to storage structures, which cannot withstand the new and erratic weather conditions, and increased cost for improving storage structures to suit the new weather conditions; (2) the need for re-drying when the harvest gets wet during threshing due to unseasonal rains; (3) the need for re-winnowing, re-sorting and repeating treatment with protectants in the midpoint of storage due to increased pest reproduction and mobility; and (4) increased difficulty in predicting the likely storage duration and planning for post-harvest management investment (Stathers et al., 2013).

Post-harvest losses as an accelerator of climate change

The relationship between climate change and post-harvest losses is not one-way only. Food loss has a lot to do with climate change. Roughly one-third of food produced for human consumption is lost or wasted globally (FAO, 2011). Recent figures released by FAO suggest that food loss and waste account for about 4.4 gigatons of greenhouse gas emissions per year, an amount surpassed only by the emissions of China and the United States (FAO, 2015). Food loss and waste generate about 8.2% of total anthropogenic greenhouse gas emissions, a share which is second after road transport (which contributes about 10%), and about five times as much annually as aviation (Hanson et al., 2015). Cereals contribute one-third of the total greenhouse gas emissions due to food loss and waste, followed by vegetables and meat, which each contribute about 21% (FAO, 2015). In Africa, the contribution of cereals will be even higher because of the bigger proportion of cereals in food production and diet composition.

Food loss and waste require energy, land, water, time, fuel, natural as well as human resources, money and a certain amount of polluting inputs to be produced, transported, processed, packaged, stored, sold, bought, transported again and stored at home (Hudson and Messa, 2015). The production of this food generates greenhouse emissions that contribute to changing the climate. Hanson et al. (2015) documents a variety of sources that contribute to the greenhouse gas emissions associated with producing food that is ultimately lost or wasted. These include (1) on-farm agriculture emissions (e.g., emissions from the digestive systems of ruminant animals), livestock manure, on-farm energy use and fertilizer emissions; (2) the production of electricity and heat used to manufacture and process food; (3) the energy used to transport, store and cook food; (4) the landfill emissions from decaying food; and (5) the emissions from land use change and deforestation.

Post-harvest agriculture adaptation to climate change

Different stakeholders, including governments, NGOs and individuals, are increasingly responding to the need to reduce the human contribution to the changing climate (Stathers et al., 2013). However, the effects of climate change are already being experienced and the mitigation actions will not make a significant difference to the effects which will be experienced within the next few decades. Therefore, adaptation measures to climate changes are paramount (Lemos et al., 2007).

Adaptation measures to climate change for post-harvest agriculture involve mainly employing 'good' post-harvest management practices: (1) prompt harvesting and sorting to reduce the risk of carrying over pests from the field to stores; (2) better drying practices and re-drying in case the produce gets damp or insect-damaged, to reduce mycotoxin contamination; (3) protecting drying produce from rain by using covered or roofed drying structures; (4) improved storage management practices, such as hygiene, maintenance and monitoring, and improved skills in performing these practices; and (5) treating grains and pulses to be stored for more than three months with an appropriate grain protectant or storing them in airtight containers (FAO, 2008a, 2008b; Stathers et al., 2013).

A number of these adaptation strategies are already known and practised to some extent (Morton, 2010). Many of these post-harvest adaptation practices are appropriate responses to several climate change trends (e.g., hotter weather and more frequent extreme weather); they also would provide benefits even without climate change (Stathers et al., 2013). Chegere et al. (2017) identify 'good' post-harvest handling practices, such as timely harvesting, proper immediate handling, sorting maize from dirt, proper drying, and use of storage protectants; these practices significantly correlate with lower PHL. A cost-benefit analysis of each of those practices shows that the adoption of most of the PHL mitigating practices is on average economically beneficial. However, the net gains of adoption per ton of maize were small for most of the practices. This implies a low economic motivation to adopt for some of the farmers.

The addition of training can increase the benefits. Chegere et al. (2017) showed that the use of airtight bags, together with training, can be economically effectively adopted by small-scale farmers to reduce storage losses in maize. The airtight bags technology is an example of cheap and easy-to-use technology that can be promoted to reduce post-harvest losses among small-scale farmers.

There are other measures that can be employed to create or improve the enabling environment for adaptation to climate change for PHL reduction. Government and other organizations should (1) support research activities to monitor changes in storage pest population dynamics, diseases, damage, geographical spread, host ranges and farmers' management and adaptation responses; (2) support investment in processing and value addition opportunities and equipment; (3) expand weather and seasonal forecasting and early warning and emergence response systems; and (4) promote climate change risk management and insurance schemes (FAO, 2008a, 2008b; Stathers et al., 2013).

Conclusion

Climate change will seriously affect all the stages of food chains. The impacts of climate change on food systems are expected to be widespread, complex, spatially variable and highly influenced by socioeconomic conditions. Scientific projections and research studies show that agriculture-based economic systems, such as sub-Saharan Africa, will more adversely affected. Consequently, a lot of resources and efforts have already been used to develop adaptation mechanisms to increase food production in sub-Saharan Africa. However, food production expansion is limited by scarcity of land and water resources, rapid population growth, food price volatility, and inadequate adoption of climate-smart agricultural practices. In addition to that, high levels of PHL also offset the efforts to increase food production.

The reduction of PHL can potentially increase the food available for consumption at household and national levels and thus improve food security. It can also increase the real income of both consumers (especially the poor, who devote a high percentage of their disposable income to staple foods) and producers (especially in the regions where crop production contributes a significant proportion of incomes, such as rural SSA) (World Bank, 2011). Moreover, reduction in PHL will imply a more judicious use of scarce resources and reduced externalities from greenhouse gas emissions that are associated with lost and wasted food.

The effects the changing climate will have on post-harvest agriculture are vivid. This calls for investment in PHL mitigation. Application of 'good' post-harvest management practices, such as timely harvesting, proper immediate handling, sorting maize from dirt, proper drying, and use of storage protectants, provides resilience against climate change. Already the FAO and World Bank suggest that about USD 442 billion, which is just less than half of the amount needed for investment to eradicate hunger in SSA by 2050, should be directed toward PHL (FAO and World Bank, 2010). To facilitate PHL mitigation decisions at national and lower levels, there is a need to increase awareness among farmers, extension officers, NGOs working with farmers, technology developers and other stakeholders about the extent of PHL, and what these losses imply in terms of lost revenue and food security. Training and extension services on harvesting techniques and good post-harvest handling practices that can potentially reduce PHL should be supported. Investment in development of affordable modern storage technologies and promotion of usage of these technologies should be emphasized.

Notes

1 Primary processing involves several activities designed to clean, sort and remove the inedible fractions from the grains. It involves activities such as cleaning, grading, hulling, shelling and threshing.
2 Secondary processing takes place at an advanced stage of the post-harvest system, to add variety to the diet or add value to the crop for marketing; it involves activities such as milling, fermentation, baking, frying and extrusion.
3 This food crisis exploded in 1973 and 1974 and was characterised by rapid food price increases in the West and by famines in Africa and Asia. The main causes were bad weather, rising agricultural input prices, grain export bans, and hoarding of food purchases.
4 The African Postharvest Losses Information System was created within the framework of the project 'Postharvest Losses Database for Food Balance Sheet Operations', initiated and financed by the European Commission's Joint Research Centre, led by the national natural resources experts.

References

Affognon, H., C. Mutungi, P. Sanginga and C. Borgemeister. 2015. Unpacking Postharvest Losses in Sub-Saharan Africa: A Meta-Analysis. *World Development* 66: 49–68.
Alliance for a Green Revolution in Africa (AGRA). 2013. Establishing the Status of Post-Harvest Losses and Storage for Major Staple Crops in Eleven African Countries (Phase I). Nairobi, Kenya: AGRA.
Aulakh, J., and A. Regmi. 2013. Post-Harvest Food Losses Estimation: Development of Consistent Methodology. Paper Presented at the Agricultural and Applied Economics Association. AAEA & CAES Joint Annual Conference. Washington, DC.

Chegere, M.J. 2017. Post-Harvest Losses Reduction by Small-Scale Maize Farmers: The Role of Handling Practices. In Chegere, M.J. *Post-Harvest Losses, Intimate Partner Violence and Food Security in Tanzania*. PhD Thesis, Economic Studies No. 230, University of Gothenburg, Sweden.

Chegere, M.J., H. Eggert and M. Söderbom. 2017. How Economically Effective Are Hermetic Bags in Maize Storage? A RCT with Small-Scale Farmers. In Chegere, M.J. *Post-Harvest Losses, Intimate Partner Violence and Food Security in Tanzania*. PhD Thesis, Economic Studies No. 230, University of Gothenburg, Sweden.

D'Souza, R.M., N.G. Becker, G. Hall and K.B.A. Moodie. 2004. Does Ambient Temperature Affect Foodborne Disease? *Epidemiology* 15: 86–92.

Epstein, P.R. 2001. Climate Change and Emerging Infectious Diseases. *Microbes and Infection* 3: 747–754.

FAO. 1980. *Assessment and Collection of Data on Post-Harvest Food Grain Losses*. Rome: Food and Agriculture Organization of the UN.

FAO. 1981. Food Loss Prevention in Perishable Crops. FAO Agricultural Services Bulletin. www.fao.org/docrep/s8620e/S8620E00.htm#Contents (accessed May 30, 2017).

FAO. 2008a. Climate Change and Food Security: A Framework Document. Rome: Food and Agriculture Organization. www.fao.org/docrep/010/k2595e/k2595e00.htm

FAO. 2008b. Climate Change: Implications for Food Safety. Rome: Food and Agriculture Organization. www.fao.org/docrep/010/i0195e/i0195e00.htm

FAO. 2011. The State of Food and Agriculture; Women in Agriculture: Closing the Gender Gap for Development. Rome: Food and Agriculture Organization.

FAO. 2013. The State of Food and Agriculture; Food Systems for Better Nutrition. Rome: Food and Agriculture Organization.

FAO. 2015. Food Wastage Footprint and Climate Change. Food and Agriculture Organization. www.fao.org/3/a-bb144e.pdf (accessed July 2017).

FAO and World Bank. 2010. FAO/World Bank Workshop on Reducing Post-Harvest Losses in Grain Supply Chains in Africa; Lessons Learned and Practical Guidelines. Rome, Italy.

Hanson, C., B. Lipinski, J. Friedrich, C. O'Connor and K. James. 2015. What's Food Loss and Waste Got to Do with Climate Change? A Lot, Actually. World Resource Institute. www.wri.org/blog/2015/12/whats-food-loss-and-waste-got-do-climate-change-lot-actually (accessed July 2017).

Hodges, R.J. 2012. Postharvest Weight Losses of Cereal Grains in Sub-Saharan Africa. Mimeo.

Hodges, R.J., J.C. Buzby and B. Bennett. 2011. Postharvest Losses and Waste in Developed and Less Developed Countries: Opportunities to Improve Resource Use. *Journal of Agricultural Science* 149: 37–45.

Hudson, U., and M. Messa. 2015. Position Paper on Food Losses and Waste. Slow Food. www.slowfood.com/sloweurope/wp-content/uploads/ING-position-paper-foodwaste.pdf

Kaminski, J., and L. Christiaensen. 2014. Post-Harvest Loss in Sub-Saharan Africa: What Do Farmers Say? *Global Food Security* 3(3): 149–158.

Kovats, R.S., S.J. Edwards, D. Charron, J. Cowden, R.M. D'Souza, K.I. Ebi, C. Gauci, P. Gerner-Smidt, S. Hajit, S. Hales, G. Hernandez Pezzi, B. Kriz, K. Kutsar, P. McKeown, K. Mellou, B. Meene, S. O'Brien, W. van Pelt and H. Schmid. 2005. Climate Variability and Campylobacter Infection: An International Study. *International Journal of Biometeorology* 49: 207–214.

Lemos, M.C., E. Boyd, E.L. Tompkins, H. Osbahr and D. Liverman. 2007. Developing Adaptation and Adapting Development. *Ecology and Society* 12(2): 26.

Lipton, M. 1982. Post-Harvest Technology and the Reduction of Hunger. *The IDS Bulletin* 13(3): 4–11.

Morton, J. (2010). Strengthening Local Agricultural Innovation in the Face of Climate Change. Presentation at the Annual Meeting of the American Association for the Advancement of Science. 18–22 February, San Diego, CA.

Rembold, F., R. Hodges, M. Bernard, H. Knipschild and O. Léo. 2011. The African Postharvest Losses Information System (APHLIS): An Innovative Framework to Analyse and Compute Quantitative Postharvest Losses for Cereals under Different Farming and Environmental Conditions in East and Southern Africa. JRC Scientific and Technicak Reports, EUR 24712 EN. Luxembourg: Publications Office of the European Union.

Stathers, T., R. Lamboll and B.M. Mvumi. 2013. Postharvest Agriculture in Changing Climates: Its Importance to African Smallholder Farmers. *Food Security* 5(3): 361–392.

USAID Rwanda. 2012. Post-Harvest Handling and Storage (phhs) Project Final Report. Annex ii: Maize Post-Harvest Practices, Revised Loss Estimates and Training Needs Seasons a and b, 2012. Technical Report. United States Agency for International Development of the United States Government.

Vermeulen, S.J., B.M. Campbell and J.S.I. Ingram. 2012. Climate Change and Food Systems. *Annual Review of Environment and Resources* 37: 195–222.

World Bank. 2011. Missing Food: The Case of Post-Harvest Grain Losses in Sub-Saharan Africa. Report No. 60371-AFR.

Part IV

Gender issues

13 Contribution of smallholder agriculture to daily calories, macronutrients, minerals and vitamins in male- and female-headed farm households in sub-Saharan Africa

Byela Tibesigwa, Martine Visser,
Razack Lokina and Richard Zadocky Jacob

Introduction

Food security exists when all people, at all times, have physical, social and economic access to sufficient, safe and nutritious food that meets dietary needs and food preferences for an active and healthy life (FAO, 2001). Despite significant economic and agricultural improvement over the last ten years in sub-Saharan Africa (SSA), the rates of household poverty and poor nutrition have not substantially decreased. For example, the prevalence of undernourishment in 2012–2014 stood at approximately 23% of the total population (FAO, IFAD and WFP, 2015). While this represents a decrease in the proportion of undernourished people from 33% in 1990–1992 (FAO, IFAD and WFP, 2015), it is the highest prevalence worldwide. In addition, while the hunger rate has decreased, the number of under-nourished people has increased by 44 million since 1990, reflecting the region's high population growth rate, with the population reaching approximately 936 million in 2013 (FAO, IFAD and WFP, 2014). Regionally, this translates to hunger for 20% of the population in Southern Africa, 23% in Western Africa, 28% in Eastern Africa and 31% in the middle region of Africa (OECD/FAO, 2016). Added to this, the population in the African continent is increasing exponentially, and eight of the ten countries with the highest average growth rate in world are located in Africa (FAO, IFAD and WFP, 2014). This rapidly growing population makes it even harder to achieve food security in the region.

In addition to the Sustainable Development Goals (SDGs) and World Food Summit (WFS), sub-Saharan African countries have strategically initiated different policy responses geared to reaching food security and nutrition targets. For instance, in the East Africa Community (EAC), there exist both the EAC Food Security Action Plan (2011–2015) and the EAC Food and Nutrition Security Policy (FNSP), with efforts concentrated on boosting food productivity and increasing food access. So far, sub-Saharan Africa has managed to reduce the prevalence of hunger by 31% between 1990 and 2015 (FAO, IFAD and WFP,

2015). A total of 18 out of 40 countries in the region have reached their hunger targets, and 7 have achieved both MDGs and World Food Summit targets. Progress has been remarkable in West Africa, where food insecurity has been successfully reduced by 63% – that is, from 24.2% in 1990/92 to 9% in 2014/16. This is equal to 13 million West Africans; however, such progress is insufficient to reach World Food Security targets (OECD/FAO, 2016). East and Southern Africa countries have also made some modest progress toward the goals, but countries in Central Africa are lagging behind.

Smallholder agriculture is a great food security booster in most households in the developing regions of the world. In sub-Saharan Africa, women form the backbone of smallholder agriculture, and are responsible for 60%–80% of food production in most of the countries. For example, the time contribution of women to agricultural production ranges from about 30% in Gambia to 60%–80% in Cameroon. Women also play multiple roles, serving as producers, laborers, processors and traders within largely domestic markets. In addition, they dominate in household-level food production and preparation. These women in sub-Saharan Africa achieve all this despite unequal access to land and inputs, such as fertilizers. With equal access to resources and human capital, female farmers could achieve productivity equal to or even higher than that of men. For instance, the FAO estimates that, if gender barriers were removed, women could raise productivity by 20%–30%, increase crop production by 2.5%–4% and reduce food insecurity by 12%–17%.

Therefore, removing gender barriers and supporting smallholder agriculture could potentially be a starting point toward alleviating food insecurity in the region, as well as narrowing the gender gap. However, most of the sub-Saharan African countries have failed to address major weaknesses in their food security policies. One of the weaknesses is lack of emphasis on the role that gender plays in smallholder agricultural activities – and by extension food security – especially in, but not limited to, rural areas. This chapter provides information to support policy formulation in the region. This is achieved by reviewing and summarizing the relevant literature and using data from the World Bank Living Standard Measurement Survey (LSMS) from Ethiopia, Malawi, Nigeria, South Africa, Tanzania and Uganda. We do this by showing the role that smallholder crop farming plays in providing energy, macronutrients, minerals and vitamins in sub-Saharan African households. We show the consumption gaps that exist between male- and female-headed farm households and further show that male-female gender gaps often depend on whether a household is de jure female-headed (i.e., the female head is divorced, widowed or never married) or de facto female-headed (i.e., the male spouse is not living at home).

First, we find that smallholder agriculture is beneficial to households in sub-Saharan Africa, where crop farmers produce an average of 1,748 kilograms (kg) per year; of this, approximately half is consumed by the households. The consumed crops generate approximately 1,034 calories per day for each household member. Going by the Food and Agriculture Organisation of the United Nations (FAO) estimates, this is close to half of the daily recommended nutrient intake (RNI). In addition to caloric benefits, each household member receives macronutrients (i.e., 28 g of protein and 23 g of fat), minerals (i.e., 9 g of iron, 29 mg zinc, 0.4 mg of riboflavin, 3 mg of thiamin and 9 mg of niacin) and vitamins (i.e., 0.083 mg of vitamin B12, 5 mg of vitamin B6, 7 mg of vitamin B9 and 42 mg of vitamin C) from smallholder crop farming. At the country level, although the production levels are somewhat similar in all countries, we observe wide variation in the amount of crops that are kept by the households for consumption. This variation is then reflected in the caloric energy, macronutrients, minerals and vitamins available for household members. Second, we do find gender differences, such that male-headed households enjoy a consumption advantage over female-headed households, and the gaps are higher between male- and de facto female-headed households than they are between male- and de jure female-headed households. These differences are 42% versus 52% for calories, 40% versus 48% for macronutrients, and 43.6% versus 52.5% for minerals. Interestingly, the gaps are much lower for vitamins: 12.3% versus 22.1%. This indicates that members in male- and female-headed households consume somewhat similar amounts of vitamins, and suggest that they benefit almost equally from fruit tree farming, while the farming of crops that are more likely to produce macronutrients and minerals – for example, wheat – is more likely to be male-dominated. This is a reflection of the fact that male and female farmers are likely to grow different crops, which are correlated with whether the crops are food crops or cash crops (see, e.g., Carr, 2008; World Bank, 2009; FAO, 2010). Third, upon comparing urban and rural areas, we find that, in rural areas, the gaps between male-headed and de facto female-headed households pertaining to calories, macronutrients, minerals and vitamins are higher than the gaps between male-headed households and de jure female-headed households. In urban areas, by contrast, the male-female gaps are somewhat similar regardless of whether the female head is de facto or de jure. An exception is vitamins, for which the gender gap for de jure households is much higher in urban (32%) than rural (1.4%) areas. The gaps continue to be smallest among vitamins in both rural and urban areas, especially in rural areas among the de jure female-headed households.

The results from our study can be used to inform agricultural policies. We suggest that smallholder agriculture needs re-emphasis and increased support. Agricultural policies are likely to be more effective if disaggregated by food and nutrition issues and by urban and rural areas. In addition, these aggregates need to be accompanied by the gender differences embedded in them.

The rest of the chapter is structured as follows. The next section summarizes the literature on food security in sub-Saharan Africa. The following section describes data and methodology. The fourth section reveals the empirical results. The concluding section focuses on the policy message.

Smallholder agriculture and food security in sub-Saharan Africa

Studies on smallholder agriculture and food security in sub-Saharan Africa continue to increase. Most recently, Flatø et al. (2017) found that de jure female-headed households were more vulnerable to rainfall variation, which in turn affected their household income. Ngigi et al. (2017) look at intra-household climate change coping strategies between husbands and wives, finding that wives used crop-related strategies while husbands adopted livestock and agroforestry strategies. Asfaw and Maggio (2017) found that weather shocks affect female-managed plots more than male-managed plots in Malawi, thus affecting female-headed households' food consumption. Tibesigwa and Visser (2016) reveal that agriculture contributes to food security of female-headed households more than male-headed households, especially in rural areas of South Africa. They further show that male-headed households are more food-secure than female-headed households and that the food security gap is wider in rural than in urban areas. Making a similar observation in the city of Tshwane, South Africa, Adeyefa (2016) notes that food insecurity is most prevalent in female-headed households, with severe food insecurity of 23%. The study also shows that the larger the family size, the more food-insecure it becomes. In Malawi, Kassie et al. (2015) examines the gendered food security gap in rural areas. They observe that, if women were given the same resources as men, the food security of female-headed household would greatly improve. Similarly, Habyarimana (2015) observes that 23% of households living in rural Rwanda are food-insecure, and that rural household food insecurity is more serious in households headed by women than in households headed by men. Moving to East Africa, Kabubo-Mariara et al. (2016) assess the effects of climate change on kilocalories produced by smallholder farming in Kenya. The results confirm that climate variability will increase food insecurity; the authors recommend strong policies on climate change mitigation and adaptation. Silvestri et al. (2015) analyse household food security in rural areas of the East African region; the results show the existence of high levels of food insecurity – that is, 62% in Rakai (Uganda), 80% in Lushoto (Tanzania) and 85% in Wote (Kenya). They further find that, of these, many are female-headed households.

In South Africa, Tibesigwa, Visser, Hunter et al. (2015) conducted a study on rural households and examined the gender differences in climate risk, food security and adaptation. The results show the existence of a gender gap, with male-headed households appearing to be more food-secure than female-headed households. Providing further evidence, Kassie et al. (2014) investigate the relationship between gender inequality and household food security in rural Kenya. They find that there are differences in food security between female- and male-headed households due to observable and unobservable characteristics. Zakari et al. (2014) reveal that female-headed households are more vulnerable to food insecurity compared to their male counterparts in southern Niger. They also find that 92.6% fear running out of food in the coming days. The study recommends more assistance to female-headed households. Following a similar trend, Tefera (2014) finds that women are the major participants in agricultural activities that increase food security in the Amhara region of Ethiopia.

In another study, Tibesigwa et al. (2015b) examine the impact of agriculture-related shocks on the consumption pattern of rural households in South Africa.

The findings put emphasis on the use of informal social capital and natural capital, which are cheaper and easier coping strategies than more capital-intensive strategies, which are likely to be relatively more costly. Moving again to West Africa, Adepoju and Adejare (2013) show that male-headed households have a slightly higher incidence (0.46) of food insecurity than female-headed households (0.33) in rural Nigeria. They further discovered that households with married heads of households were more vulnerable to food insecurity compared to single heads of households; this is influenced by the fact that the latter households are likely to be larger. Ndobo (2013) examines food security status in Kwakwatsi, South Africa. The study found that 29.3% of households are food-insecure, with female-headed households being more food-insecure than male-headed households. Severe food insecurity is mostly associated with marital status – that is, household heads who are not married compared to those who are married. In yet another study, Ndobo and Sekhampu (2013) examine household food security. The findings show that 49% of households were prone to food insecurity, and that female-headed households were more vulnerable compared to male-headed households. They also observe that food insecurity increases with age and decreases with household income and employment status of men, while, in female-headed households, it increases with marital status, age of household head, and household size and decreases with household income. Another study in South Africa by Ndobo (2013) points out that natural disasters – that is, floods and drought – in conjunction with structural problems, such as gender, race and ethnicity, income inequality, poor policies and overdependence on agricultural production for survival in rural areas, were responsible for food insecurity.

In Ghana, Hjelm and Dasori (2012) report three main causes of food insecurity: persistent poverty, fluctuation in food prices and limited agricultural outputs. However, these problems appear to be more structural and lack the gender dimensions that have been observed in other studies in sub-Saharan Africa. In Ethiopia, it is observed that increasing crop diversity (Di Falco et al., 2010) and adapting to climate change (Di Falco et al., 2011) increase food security of smallholder farmers in Ethiopia. Modirwa and Oladele (2012) analyses the differences in food security between female- and male-headed households in South Africa. The findings indicate significant differences in their food security level. This may be due to inequality in ownership and allocation among women and men. Gürkan and Sanogo (2011) report on data collected in Tanzania, Laos, Cameroon, Madagascar and Mauritania. They show that female-headed households are food-insecure and that Madagascar and Cameroon have higher levels of food insecurity among female-headed households compared to other countries. Similarly, Babatunde et al. (2008) observe that female-headed households were more vulnerable to food insecurity than male-headed households. In contrast, but similar to Adepoju and Adejare (2013), Nyariki et al. (2002) observe that female-headed households were more food- and nutrition-secure compared to male-headed households in Dryland, Kenya. This finding is in line with Haddad et al. (1994) and Carter (1997).

Data and methodology

In providing evidence from sub-Saharan Africa, we use data from the World Bank Living Standards Measurement Study (LSMS). These are nationally representative household surveys that elicit information on households' economic activities, education, agriculture, health and well-being. In our study, we use the most recent sub-Saharan Africa LSMS that covers Ethiopia, Malawi, Nigeria, Tanzania and Uganda. The LSMS used for Ethiopia and Tanzania is for the year 2013, while the LSMS survey used for Nigeria is from 2012. For Uganda, we use 2010. We do not include Niger, Burkina Faso, Côte d'Ivoire and Mali because the surveys are not in English. We look at the gender differences by outlining the consumption patterns in male-headed farm households, and compare these with de jure and de facto female-headed households. The de facto households have female heads who are married, but their spouses are away most of the time. De jure heads, on the other hand, are women who are not married – that is, divorced, widowed or never married.

Our empirical model is $y_{it} = \mathbf{X}_{it}\beta + \varepsilon_{it}$, where y_{it} is household consumption, $\mathbf{X}_{it}$ is a vector of household characteristics and ε_{it} is the random error term. Following Oaxaca and Ransom (1994; 1999) and Neuman and Oaxaca (2004), we use decomposition to explain the gender differences in smallholder agriculture. This is achieved by estimating y_{it} for male- and female-headed households separately, and thereafter we obtain the differential in consumption by decomposition: $\bar{y}_m - \bar{y}_f = (\bar{\mathbf{X}}_m - \bar{\mathbf{X}}_f)\hat{\beta}_m + \bar{\mathbf{X}}_f(\hat{\beta}_m - \hat{\beta}_f)$, where $\bar{y}_m$ and $\bar{y}_f$ are the average consumption in male- and female-headed households, respectively. $\bar{\mathbf{X}}_m$ ($\bar{\mathbf{X}}_f$) are household characteristics for male- and female-headed households. $\hat{\beta}_m$ and $\hat{\beta}_f$ are the estimated coefficients. $(\bar{\mathbf{X}}_m - \bar{\mathbf{X}}_f)\hat{\beta}_m$ is the explained component from household characteristics, and $\bar{\mathbf{X}}_f(\hat{\beta}_m - \hat{\beta}_f)$ is the residual. Our measure of household consumption (y_{it}) is derived by aggregating the total amount of crops produced in the year in kilograms. Because we are interested only in consumption, we deduct the kilo-quantity amount that was sold by the households. This then gives us the total kilo-quantity that was retained by the households for consumption. Thereafter, the kilo-quantity amount of each crop is converted to calories, macronutrients, minerals and vitamin values. For example, a kilogram of groundnut produces 5980 kcal, 492 g of fat, 46 g of iron, 121 mg of niacin, 258 g of protein, 1 mg of riboflavin, 6 mg of thiamine, 3 mg of vitamin B12 and 33 mg of zinc. We use FAO (1998) estimates for the conversion. We then sum the macronutrients, minerals and vitamins across all crops to obtain the total nutrients derived in a year. To obtain the individual or per capita values, we divide the total nutrients by the number of days in a year, and further by the number of members in each household. Note that it is unlikely that all retained crops are actually consumed. That is, some of the crops retained are likely to be given away as gifts or used for making non-cash payments. Also note that in this study we concentrate on the key macronutrients, minerals and vitamins, although crop farming is likely to produce other nutrients not covered in this study.

Results

Descriptive statistics

Table 13.1 reports the descriptive statistics. Column 1 shows the statistics from the aggregated data, and Columns 2–7 show the country-level statistics. From Column 1, the average head of household is about 47 years old, and most of them are men (76%). Most of these households are located in rural areas, and produce an average of 1,748 kg per year from crop farming; of this, approximately half (956 kg) is retained for consumption. As previously mentioned, it is unlikely that all retained crops are consumed – that is, some retained crops are likely to be given away as gifts or for making non-cash payments. For example, a study by Tibesigwa, Visser and Turpie (2015) found that 12% of the crops are given away as gifts in South Africa.

When we disaggregate by country (Columns 2–7), we find somewhat similar household characteristics. That is, the age of the head of household ranges between 43–55 years, with the youngest heads of household in Malawi and the oldest in South Africa. The percentage of female heads of households ranges from 11% to 29%, with the smallest percentage found in Nigeria, and the largest in Ethiopia. The number of married heads of households (both male and female) ranges between 60%–85%, with 85% in Nigeria. Only 4.7% of the women are de facto heads of households and 19.1% are de jure heads of households. Excluding South Africa, the highest percentages of both de facto and de jure heads of households are found in Ethiopia and the lowest in Nigeria.

This is consistent with the current statistics on female headship in sub-Saharan Africa. A study by Tibesigwa, Visser and Turpie (2015), using the World Bank indicators, shows that the number of female-headed households is highest in the Southern Africa region and lowest in the western parts of sub-Saharan Africa. More specifically, 22.9% of households in Benin are female-headed; this statistic is 26.8% and 25.5% in Burundi and Cameron, respectively. Moving to Ethiopia, 26.1% are female-headed households, while 30%, 28% and 35% are observed in Gabon, Malawi and Mozambique. The number of female-headed households is similar in Tanzania and Senegal, at about 24%, while in Rwanda this is slightly higher at 33%. There are fewer female-headed households in Guinea, Côte d'Ivore and Burkina Faso, with 17%, 18% and 9% respectively. When we move to Southern Africa, we observe a large spike in the statistics – for example, in Zimbabwe, 44% of the households are headed by women, and, according to Statistics South Africa, this figure is 41% in South Africa. Generally, it appears that the number of female-headed households is on the rise (Bongaarts, 2001; Horrell and Krishnan, 2007).

Going back to Table 13.1 and moving to the household size, the number of household members is between 5 and 7, with the minimum number found in Ethiopia and the maximum in Uganda. With regard to crop farming, Table 13.1 shows that Nigerian households produce the largest quantity (3,138 kg), while Malawian households produce the least (951 kg). There is large variation in the

Table 13.1 Descriptive Statistics

	1 Total		2 Ethiopia		3 Tanzania		4 Uganda		5 Malawi		6 Nigeria	
	Mean	SD	Mean	SD	Mean	SD	Mean	SD	Mean	SD	Mean	SD
Age of HH	47.351	21.603	44.258	15.637	48.38	15.51	47.795	15.478	43.913	16.207	51.503	38.187
HH is female	0.258	0.438	0.293	0.455	0.229	0.421	0.282	0.450	0.238	0.426	0.114	0.317
HH is married	0.716	0.451	0.692	0.462	0.768	0.422	0.755	0.430	0.774	0.418	0.852	0.356
De facto HH	0.072	0.259	0.067	0.249	0.05	0.218	0.084	0.278	0.043	0.202	0.004	0.065
De jure HH	0.187	0.390	0.227	0.419	0.182	0.386	0.199	0.3996	0.195	0.396	0.11	0.314
HH members	5.533	2.968	4.582	2.402	6.392	2.124	7.399	3.531	5.191	2.334	6.322	3.182
Urban	0.203	0.402	0.368	0.482	0.117	0.322	0.122	0.327	0.141	0.348	0.116	0.321
Total quantity (kg/year)	1627.9	5676.9	1008.2	5026.8	1441	14706	3536.2	11577.5	951.1	1973.3	3138.6	5760.5
Total consumption (kg/year)	891.6	2540.4	338.4	1209.3	350.5	6743	2025.3	3775.7	826.2	1810.7	2122.0	4204.9
% consumption	54.8		33.6		24.3		57.3		86.9		67.6	
n (households)	16936		5103		2565		2021		3204		2814	

amount of crops retained for household consumption, with the highest amount recorded in Malawi (86.9%), followed by Nigeria (67.6%), Uganda (57.3%), Ethiopia (33.6%) and Tanzania (24.3%).

Smallholder crops

Smallholder farmers grow a wide variety of crops, with the most common being maize, beans, nuts and fruits, as shown in Appendix A.

In Tanzania, certain crops are grown during the short and long rains. In addition, there are crops that are grown on a permanent basis, which mainly include fruit trees. The long rains are experienced during the March–May season; the short rains occur in October–December. Therefore, in Tanzania, crop production is classified according to (1) long-rains crops, (2) short-rains crops, (3) fruit crops and (4) other permanent crops. In Appendix A, Tables 13A1 through 13A4 show the different types of crops grown in Tanzania. Table 13A1 shows the 41 different crops grown during the long rains in 2013, with maize (38.8%), beans (10.9%), rice paddy (10.2%), groundnuts (5.9%) and sweet potatoes (4.5%) being the most common crops. The crops grown Tanzania in the short-rains season are shown in Table 13A2. A total of 39 different crops are grown during the short rains; the majority of these crops are maize (43.5%), beans (21.1%), groundnuts (5.9%), sweet potatoes (5.9%) and rice paddy (3.4%). Table 13A3 provides details of the fruit crops, where we observe a total of 31 different fruit crops, with the most common being mangoes (22.6%), bananas (25.3%), pawpaws (9.0%), oranges (8.1%) and avocadoes (7.0%). Finally, the crops that are grown on a permanent basis in Tanzania are shown in Table 13A4; these are 31 in number, with the majority being cassava (31.6%), cashew nuts (14.7%), coconut (8.8%), coffee (6.4%) and sugarcane (3.9%).

Smallholder farmers in Malawi plant crops in both the rainy season and the dry season, while other crops are grown on a permanent basis. Accordingly, there are three types of crops in Malawi: (1) rainy-season crops, (2) dry-season crops and (3) crops grown on a permanent basis. Tables 13A5 through 13A7 provide details of the crops grown in Malawi in 2013. The rainy season in Malawi produces approximately 24 different crops. The most common crops include maize (42.1%, with about five different varieties), groundnuts (11.3%, also with about five different varieties), pigeon pea (11.7%) and *nkhwani*[1] (6.9%). Table 13A5 presents more details on the types of rainy-season crops in Malawi. The dry season produces fewer crops, with only about 18 varieties; see Table 13A6. The most common crops grown during the dry season include maize (22.9%), tomatoes (15.4%), *nkhwani* (15.1%) and beans (12.0%). Table 13A7 shows the 15 different crops grown on a permanent basis in Malawi. Mangoes (35.6%) are the most common crops, followed by cassava (14.3%), bananas (12.3%) and pawpaws (10.1%).

Uganda has two farming seasons; hence the crops are categorized as (1) season 1 and (2) season 2 crops. We provide detailed information about these crops in Tables 13A8 and 13A9. There are 41 different crops grown in the first season: beans (16.9%), bananas (16.9%), cassava (15.9%), maize (15.9%) and sweet potatoes (7.5%). The same crops that are common in the first season are also common in the second season: bananas (19.1%), beans (19.9%), maize (15.9%), cassava (10.7%) and sweet potatoes (8.6%). There are many other crops planted in Uganda as well, for a total of 37 crop varieties.

In Nigeria, we observe a total of 69 crops produced in 2012; see Table 13A10. However, these crops have not been disaggregated by planting season. Table 13A10 lists these crops; we find that cassava (16.6%), groundnuts (13.0%), guinea corn (11.9%), yam (10.7%) and beans (9.1%) are the most common crops grown in Nigeria.

Table 13A11 shows the crops grown in Ethiopia; here we find the greatest varia-tion in the types of crops in comparison to other countries. That is, there are a total of 84 different crops. However, as in Nigeria, these crops have not been classified by any growing season. Maize (11.1%), sorghum (9.1%), a small grain called *teff* (7.8%), coffee (7.1%) and *enset*/false banana (6.1%) are the most com-monly grown crops in Ethiopia.

Household-level daily calorie, macronutrient, mineral and vitamin consumption from smallholder crop farming

Table 13.2 reports the contribution of smallholder crops to household food (calo-ries, macronutrients, minerals and vitamins), overall and for each country. These values are then compared to the daily recommended nutrient intake (RNI). We observe that, overall, crop farming provides approximately 1,034 kcal per household member per day. According to the FAO (1998), an average individual requires about 2,650 kcal per day; this implies that crop farming provides about 39% of daily energy requirements. Regarding macronutrients, crop farming pro-vides 28 g of protein and 23 g of fat, which is about half the daily RNI. Among the minerals, each household member is likely to receive about 9 g of iron, 29 mg zinc, 0.4 mg of riboflavin, 3 mg of thiamin and 9 mg of niacin. This is more than half the daily RNI, but the availability of thiamin exceeds the daily RNI. It appears that consumption of vitamins also exceeds the daily RNI, with the exception of vitamin C. Specifically, each household member is likely to con-sume about 0.083 mg of vitamin B12, 5 mg of vitamin B6, 7 mg of vitamin B9 and 42 mg of vitamin C. Figure 13.1 shows the percentage contribution of small-holder crop farming to the daily RNI.

In Table 13.2, the daily calories, macronutrients, minerals and vitamins are also portrayed for each country. Because most of the crops in Ugandan, Malawian and Nigerian households were retained for consumption, this implies that they receive more nutrients from crops than households in either Ethiopia or Tanzania, who sold relatively more. For example, while the caloric energy from households' own production ranges between 1,240 and 2,395 in Uganda, Malawi and Nigeria, with Nigeria having the highest caloric intake, Tanzania and Ethiopia households con-sume 452 and 366 calories respectively from their own production. When we compare our caloric energy values with past studies, we observe some similarities. For example, a study in Kenya by Kabubo-Mariara et al. (2016) shows that

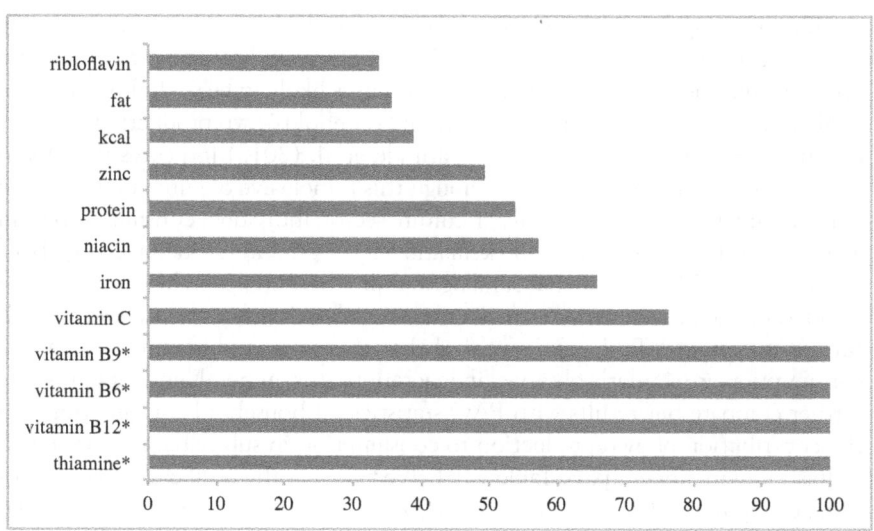

Figure 13.1 Percentage of the Daily RNI for Calories, Macronutrients, Minerals and Vita-
minsNote: Vitamin B9, B6, B12 and thiamin exceed the daily recommended nutrient
intake (RNI).

Table 13.2 Daily per Capita Calorie, Macronutrients, Minerals and Vitamins Obtained
from Crop Farming

	RNI	Total	Ethiopia	Tanzania	Uganda	Malawi	Nigeria
Energy & macronutrients							
Kcal	2650.0	1034.3	366.4	452.6	1240.1	1359.2	2395.7
Protein (g)	54.000	28.982	12.845	5.645	32.753	46.105	55.616
Fat (g)	65.000	23.170	9.569	4.553	21.468	53.624	32.843
Minerals							
Iron (g)	13.700	9.030	4.616	1.357	9.226	10.028	22.209
Zinc (mg)	59.000	29.050	8.997	5.974	22.901	34.979	79.963
Riboflavin (mg)	1.300	0.439	0.201	0.093	0.608	0.688	0.753
Thiamin (mg)	1.600	3.113	0.851	0.916	3.489	5.722	5.449
Niacin (mg)	16.000	9.153	4.785	1.577	9.451	14.830	17.301
Vitamins							
Vitamin B12 (mg)	0.002	0.083	0.009	0.014	0.058	0.349	0.000
Vitamin B6 (mg)	1.700	5.054	1.289	1.603	6.059	8.767	9.158
Vitamin B9 (mg)	0.400	7.833	3.276	2.156	27.792	8.616	4.146
Vitamin C (mg)	55.000	42.0	10.539	8.308	67.405	48.466	105.918

Note: RNI is recommended nutrient intake; mg represents micrograms.

smallholder crop farming produces about 12,654,510 kcal per year. Considering that there are about eight members per household, as indicated by the study, and on the assumption that about half of the output is likely to be sold, this translates to about 2,126 kcal per day per person from households' own production. A study of farm households in Nigeria by Babatunde et al. (2010) indicates that about 2,427 kcal are consumed per day, although this is inclusive of purchased food and gifts. In Sudan, in the traditional agriculture sector, individual consumption from all sources is low, at roughly 1,466 kcal and 54 g of protein (Abdalla, 2014). Benfica and Kilic (2015) show that rural agricultural households in Malawi consume about 2,425 kcal, although it is unclear how much of it is from household agriculture. Returning to Table 13.2, we find that the country differences in caloric consumption are similarly observed in macronutrients, minerals and vitamins. We further compare our results with FAO statistics on household consumption and the contribution of own production to consumption in sub-Saharan Africa; this is illustrated in Appendix B, Table 13B1. We find that the total caloric consumption per capita per day ranges between 1,798.67 and 2,461.29. When we consider only the caloric consumption from own production, we find that this ranges between 171 and 1,535, which is somewhat consistent with our results.

Our results show that smallholder agriculture is important in the provision of some, but not all, caloric energy, as previously highlighted in the literature and shown in this study, but it is also important in providing essential daily macronutrients, minerals and vitamins. It is likely that these smallholder farming households receive other nutrients not covered in this study; for example, smallholder farmers are likely to engage in livestock keeping in addition to crop farming. However, overall, crop farming is practiced more in comparison to smallholder livestock keeping; the livestock is mostly animals such as chickens (kept for eggs and for sale), pigs and goats, kept mainly for cash payments, while the livestock products – for example, eggs – are kept for consumption. For example, a study by Tibesigwa, Visser and Turpie (2015) found that, in South Africa, while the majority of the crops are consumed, most of the livestock are sold. Hence it is unlikely that livestock keeping can contribute as much as crop farming to household nutrition.

Table 13.3 shows the determinants of production, caloric energy, macronutrients, minerals and vitamins. In Table 13.3, Panel A uses the kilo-quantity produced from crop farming as the outcome. Column 2 is based on the kilo-quantity of crops kept for consumption. The daily consumption of caloric energy, protein and fat, per household member, is shown in Columns 3–5. Panel B is based on minerals as outcomes – that is, daily consumption per household member of iron, zinc, riboflavin, thiamine and niacin, shown in Columns 1–5 respectively. Panel C shows the determinants of vitamins, where we separately regress vitamin A, vitamin B12, vitamin B6, vitamin B9 and vitamin C on household characteristics. Overall we find that the head of household's age and gender, region, and number of household members are significant determinants. That is, the number of household members is negatively associated with energy and nutrition, as expected, and age has a non-linear relationship. We also observe a negative and significant coefficient on the urban dummy, which indicates that households in urban areas

Table 13.3 Factors Determining Calorie, Macronutrient, Mineral and Vitamin Consumption from Crop Farming

Variables	Panel A: calories and macronutrients				
	(1)	(2)	(3)	(4)	(5)
	Prod	Consu	Kcal	Protein	Fat
Age	3.170	2.254*	2.229	0.0213	0.0665
	(3.562)	(1.164)	(1.447)	(0.0443)	(0.0522)
age2	−0.00473	−0.00337***	−0.00371**	−6.18e-05	−9.42e-05*
	(0.00348)	(0.00116)	(0.00149)	(4.39e-05)	(5.03e-05)
HH members	220.0***	123.3***	−81.49***	−1.924***	−1.714***
	(37.43)	(14.56)	(8.027)	(0.235)	(0.262)
De facto	−828.8***	−300.9***	−352.1***	−9.865***	−8.703***
	(86.91)	(48.78)	(48.70)	(1.852)	(1.973)
De jure	−575.7***	−248.1***	−418.8***	−12.07***	−9.890***
	(132.7)	(40.68)	(46.13)	(1.409)	(1.747)
Rural	−563.6***	−198.6***	−448.8***	−15.31***	−12.55***
	(138.9)	(43.48)	(42.29)	(1.287)	(1.631)
Tanzania	−182.6*	−338.1***	−121.0***	−7.456***	−7.850***
	(107.7)	(29.66)	(31.29)	(1.017)	(0.996)
Uganda	2,215***	1,447***	1,324***	33.01***	22.93***
	(241.9)	(79.63)	(53.90)	(1.632)	(1.946)
Malawi	−359.8***	352.9***	918.1***	30.30***	41.84***
	(94.91)	(37.63)	(59.00)	(2.120)	(2.931)
Constant	268.5	−167.9*	936.9***	30.01***	22.15***
	(217.2)	(101.8)	(77.50)	(2.423)	(2.822)
Observations	15,707	15,707	15,707	15,707	15,707
R-squared	0.050	0.114	0.108	0.075	0.046

Variables	Panel B: minerals				
	(1)	(2)	(3)	(4)	(5)
	Iron	Zinc	Riblo	Thia	Niac
Age	0.00643	−0.0624	0.00114*	0.00594	0.0360*
	(0.0140)	(0.0617)	(0.000687)	(0.00619)	(0.0194)
age2	−1.74e-05	4.55e-06	−1.58e-06**	−7.12e-06	−4.61e-05**
	(1.50e-05)	(6.05e-05)	(6.89e-07)	(6.44e-06)	(1.90e-05)
HH members	−0.668***	−1.457***	−0.0282***	−0.181***	−0.724***
	(0.0759)	(0.367)	(0.00364)	(0.0338)	(0.0891)
De facto	−3.176***	−10.87***	−0.160***	−1.180***	−2.932***
	(0.538)	(1.608)	(0.0277)	(0.238)	(0.655)
De jure	−4.157***	−13.00***	−0.168***	−1.215***	−3.675***
	(0.408)	(1.632)	(0.0216)	(0.204)	(0.563)

(Continued)

Table 13.3 (Continued)

Variables	Panel B: minerals				
	(1)	*(2)*	*(3)*	*(4)*	*(5)*
	Iron	Zinc	Riblo	Thia	Niac
Rural	−4.852***	−12.21***	−0.218***	−1.100***	−5.202***
	(0.382)	(1.644)	(0.0185)	(0.215)	(0.492)
Tanzania	−3.459***	−0.608	−0.107***	0.531***	−3.899***
	(0.341)	(1.317)	(0.0176)	(0.167)	(0.433)
Uganda	8.927***	24.30***	0.597***	3.678***	9.286***
	(0.483)	(1.942)	(0.0293)	(0.222)	(0.587)
Malawi	4.509***	23.37***	0.446***	4.654***	9.182***
	(0.521)	(2.066)	(0.0304)	(0.317)	(0.802)
Constant	10.39***	26.68***	0.413***	2.213***	9.532***
	(0.775)	(3.725)	(0.0370)	(0.350)	(0.917)
Observations	15,707	15,707	15,707	15,707	15,707
R-squared	0.100	0.071	0.065	0.041	0.054

Variables	Panel C: vitamins				
	(1)	*(2)*	*(3)*	*(4)*	*(5)*
	Vita	Vitb12	Vitb6	Vitb9	Vitc
Age	1.054	−0.000252	0.0120	0.00594	0.659***
	(0.749)	(0.000519)	(0.0105)	(0.0270)	(0.123)
age2	−0.00113	2.31e-07	−1.35e-05	−1.06e-05	−0.000714***
	(0.000730)	(4.94e-07)	(1.10e-05)	(2.56e-05)	(0.000123)
HH members	−8.853	−0.00538***	−0.292***	−0.836***	−6.464***
	(8.829)	(0.00187)	(0.0579)	(0.135)	(0.718)
De facto	−145.1***	−0.0380***	−1.966***	−0.262	−5.418
	(41.70)	(0.0119)	(0.403)	(2.527)	(3.590)
De jure	−23.41	−0.00816	−1.869***	−2.746***	−10.44***
	(39.74)	(0.0212)	(0.343)	(0.778)	(3.958)
Rural	−103.2***	−0.0483***	−1.614***	−3.912***	−11.46***
	(21.96)	(0.0107)	(0.363)	(0.518)	(3.653)
Tanzania	−165.1***	−0.000455	1.154***	1.168**	−3.743**
	(19.70)	(0.00916)	(0.284)	(0.592)	(1.600)
Uganda	953.7***	0.0906***	6.368***	29.07***	78.93***
	(110.4)	(0.0129)	(0.381)	(1.531)	(5.186)
Malawi	−61.57***	0.331***	7.162***	4.984***	39.05***
	(21.27)	(0.0280)	(0.526)	(1.073)	(3.412)
Constant	208.6***	0.0670***	3.310***	8.852***	19.57***
	(68.51)	(0.0251)	(0.592)	(1.129)	(6.009)
Observations	15,707	15,707	15,707	15,707	15,707
R-squared	0.024	0.030	0.036	0.047	0.044

Note: Robust standard errors in parentheses *** $p<0.01$,
** $p<0.05$,
* $p<0.1$.

consume fewer calories and nutrients from crop farming. The country coefficients are also significant but differ in the level and type of influence. Additional, being a female-headed household (whether de jure or de facto) implies less consumption from crop farming in comparison to male-headed households. This is evident in the negative and statistically significant coefficient for female head; however, the level of significance disappears for vitamins, which suggests somewhat equal consumption between male- and female-headed households.

Gendered household-level daily calorie, macronutrient, mineral and vitamin consumption from smallholder crop farming

Next, we compare the consumption of macronutrients, minerals and vitamins between male- and female-headed households using the Oaxaca-Blinder decomposition. The results are reported in Tables 13.4 and 13.5. Table 13.4 compares male- and de facto female-headed households, while Table 13.5 compares male- and de jure female-headed households. In both Tables 13.4 and 13.5, we observe positive and statistically significant gaps, suggesting that male-headed households enjoy higher consumption than female-headed households. This finding is similar to past studies that have found that male-headed households are more food-secure than female-headed households; see, for example, Mallick and Rafi (2010), Kakota et al. (2011), Kassie et al. (2014), Tibesigwa and Visser (2016) and Debela and Abebe (2017). Further, we observe that the consumption gaps are larger between male- and de facto female-headed households (Table 13.4) than they are between male- and de jure female-headed households (Table 13.5). This suggests that de jure female-headed households benefit more than de facto female-headed households from crop farming. Figure 13.2 summarizes the results from Tables 13.4 and 13.5. In addition to the consumption gap differences between de facto and de jure female-headed households, it is quite evident that the gaps differ between macronutrients, minerals and vitamins – that is, the vitamin gaps are much smaller compared to the macronutrient and mineral gaps. This indicates that members of male- and female-headed households consume somewhat similar amounts of vitamins. Because vitamins are mainly from fruits, this may suggest that male- and female-headed households benefit almost equally from fruit tree farming, while, in the farming of other crops that are more likely to produce macronutrients and minerals – for example, wheat – we are more likely to see male dominance. This reflects the fact that male and female farmers are likely to grow different crops, as documented in the literature, and this is highly correlated with whether the crops are food crops or cash crops (see, e.g., Carr, 2008; World Bank, 2009; FAO, 2010). Figure 13.3 shows the gaps for each country.

Our next interest is to look at differences in consumption between male- and female-headed households in urban and rural areas. The results are reported in Tables 13.6–13.9. In Tables 13.6 and 13.7, we outline the gaps in urban and rural areas respectively for male and de facto female-headed households, while Tables 13.8 and 13.9 compare the gap between male- and de jure female-headed

Table 13.4 Daily per Capita Calorie, Macronutrient, Mineral and Vitamin Gap and Decomposition – De Facto vs Male

Daily per capita	Macronutrients					Minerals					Vitamins				
	(1)	(3)	(5)	(7)	(9)	(11)	(13)	(15)	(17)	(19)	(21)	(23)	(25)	(27)	(29)
	Prod	Consu	Kcal	Prot	Fat	Iron	Zinc	Ribl	Thia	Niac	Vita	Vitb12	Vitb6	Vitb9	Vitc
Male-headed household	2,028***	1,101***	1,158***	32.51***	25.76***	10.27***	34.00***	0.485***	3.489***	10.12***	237.1***	0.0860***	5.647***	7.944***	43.09***
	(55.80)	(24.87)	(23.08)	(0.705)	(0.866)	(0.219)	(0.991)	(0.0107)	(0.106)	(0.267)	(19.04)	(0.00673)	(0.180)	(0.344)	(1.837)
Female-headed household	856.1***	528.1***	545.6***	17.26***	14.05***	4.650***	11.43***	0.271***	1.684***	5.531***	200.4***	0.0519***	2.637***	10.57***	27.91***
	(55.76)	(39.50)	(41.86)	(1.676)	(1.706)	(0.482)	(1.286)	(0.0241)	(0.205)	(0.584)	(22.62)	(0.00909)	(0.347)	(2.462)	(2.954)
Household consumption gap	1,172***	572.6***	612.6***	15.25***	11.71***	5.623***	22.57***	0.215***	1.805***	4.586***	36.72	0.0341***	3.010***	−2.631	15.18***
	(78.88)	(46.68)	(47.81)	(1.818)	(1.913)	(0.529)	(1.624)	(0.0264)	(0.230)	(0.642)	(29.57)	(0.0113)	(0.391)	(2.485)	(3.478)
% per nutrient %	57.8	52.0	52.9	46.9	45.5	54.8	66.4	44.3	51.7	45.3	15.5	39.7	53.3	−33.1	35.2
%					46.2					52.5					22.1
Explained	371.0***	292.3***	280.4***	5.995***	3.392***	2.646***	12.98***	0.0624***	0.674***	1.713***	−113.9***	−0.00100	1.106***	−2.886***	8.714***
	(34.74)	(23.94)	(16.37)	(0.450)	(0.478)	(0.157)	(0.730)	(0.00702)	(0.0655)	(0.152)	(20.05)	(0.00246)	(0.114)	(0.284)	(1.555)
Unexplained	800.7***	280.3***	332.2***	9.255***	8.314***	2.977***	9.589***	0.152***	1.130***	2.872***	150.6***	0.0351***	1.904***	0.256	6.461*
	(85.52)	(48.59)	(48.17)	(1.845)	(1.962)	(0.535)	(1.566)	(0.0277)	(0.237)	(0.652)	(43.52)	(0.0115)	(0.402)	(2.531)	(3.569)
Observations	12,700	12,700	12,700	12,700	12,700	12,700	12,700	12,700	12,700	12,700	12,700	12,700	12,700	12,700	12,700

Note: Robust standard errors in parentheses. *** p<0.01,
** p<0.05,
* p<0.1

Table 13.5 Daily per Capita Calorie, Macronutrient, Mineral and Vitamin Gap and Decomposition – De Jure vs Male

Daily per capita	Macronutrients					Minerals					Vitamins				
	(1) Prod	(3) Cons	(5) Kcal	(7) Prot	(19) Fat	(9) Iron	(11) Zinc	(13) Ribl	(15) Thia	(17) Niac	(21) Vita	(23) Vitb12	(25) Vitb6	(27) Vitb9	(29) Vitc
Male-headed household	2,028***	1,101***	1,158***	32.51***	25.76***	10.27***	34.00***	0.485***	3.489***	10.12***	237.1***	0.0860***	5.647***	7.944***	43.09***
	(55.79)	(24.87)	(23.08)	(0.705)	(0.866)	(0.219)	(0.991)	(0.0107)	(0.106)	(0.267)	(19.04)	(0.00673)	(0.180)	(0.344)	(1.837)
Female-headed household	889.0***	489.9***	666.8***	18.70***	16.26***	5.194***	14.92***	0.308***	2.131***	6.347***	243.1***	0.0848***	3.537***	6.648***	38.66***
	(90.26)	(24.43)	(35.12)	(1.080)	(1.367)	(0.290)	(1.224)	(0.0159)	(0.160)	(0.414)	(34.84)	(0.0180)	(0.267)	(0.526)	(3.115)
Household consumption gap	1,139***	610.8***	491.4***	13.81***	9.499***	5.079***	19.08***	0.177***	1.358***	3.770***	-5.977	0.00113	2.109***	1.296**	4.429
	(106.1)	(34.87)	(42.03)	(1.290)	(1.619)	(0.364)	(1.575)	(0.0192)	(0.191)	(0.493)	(39.71)	(0.0192)	(0.322)	(0.629)	(3.616)
% per nutrient %	56.2	55.5	42.4	42.5	36.9	49.5	56.1	36.5	38.9	37.3	-2.5	1.3	37.3	16.3	10.3
%					39.7					43.6					12.5
Explained	570.0***	364.7***	69.82***	1.736***	-0.321	0.901***	6.052***	0.00884	0.136	0.0882	-30.78**	-0.00622	0.225	-1.435***	-6.414***
	(76.03)	(26.82)	(19.31)	(0.599)	(0.701)	(0.186)	(0.825)	(0.00915)	(0.0878)	(0.255)	(13.68)	(0.00509)	(0.150)	(0.384)	(1.569)
Unexplained	568.8***	246.0***	421.6***	12.08***	9.820***	4.178***	13.03***	0.169***	1.222***	3.681***	24.80	0.00734	1.884***	2.731***	10.84***
	(134.1)	(40.76)	(45.99)	(1.403)	(1.750)	(0.405)	(1.600)	(0.0216)	(0.204)	(0.565)	(39.70)	(0.0213)	(0.344)	(0.784)	(3.976)
Observations	14,962	14,962	14,962	14,962	14,962	14,962	14,962	14,962	14,962	14,962	14,962	14,962	14,962	14,962	14,962

Note: Robust standard errors in parentheses. *** p<0.01, ** p<0.05, * p<0.1.

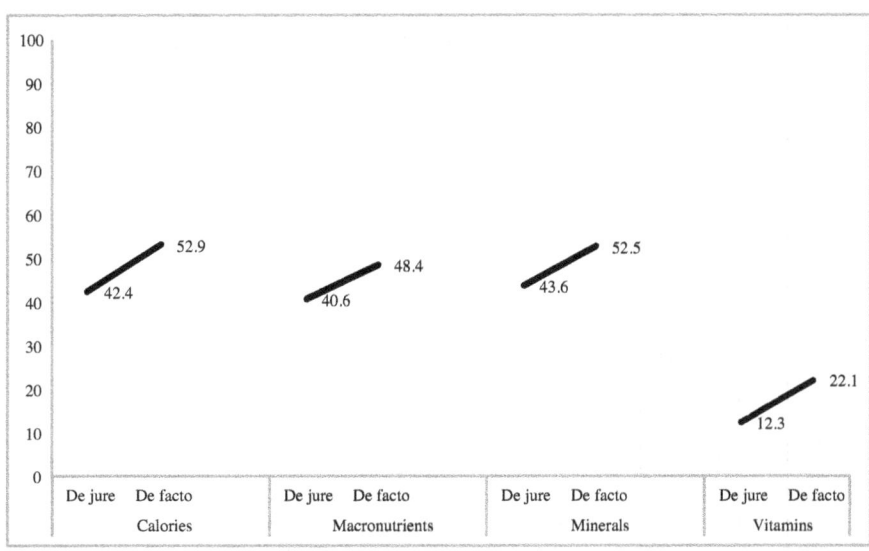

Figure 13.2 Daily per Capita Calorie, Macronutrient, Mineral and Vitamin Gap and Decomposition – Comparing between De Facto vs Male and De Jure vs Male

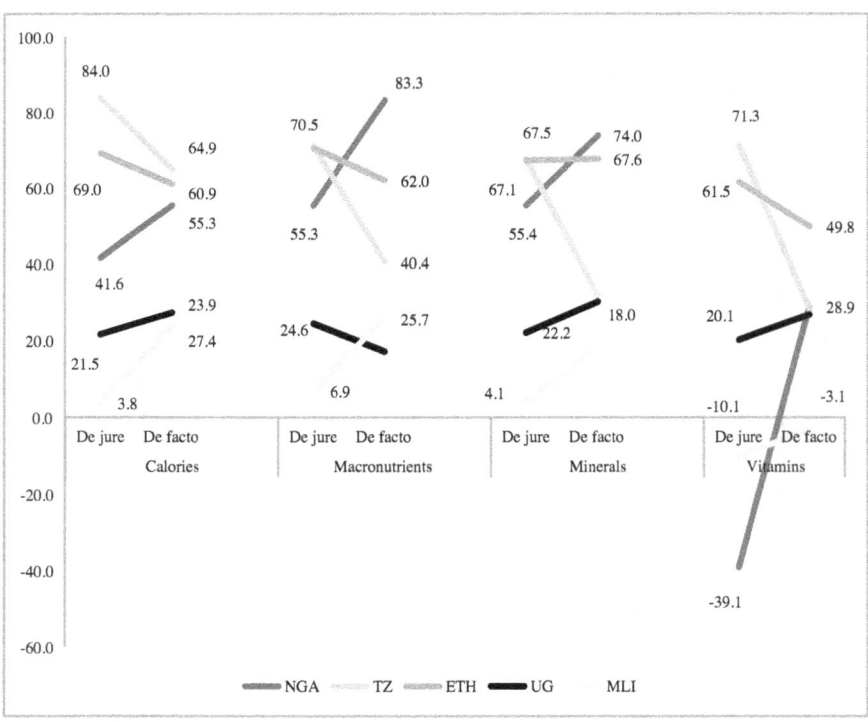

Figure 13.3 Daily per Capita Calorie, Macronutrient, Mineral and Vitamin Gap and Decomposition – Comparing between De Facto vs Male and De Jure vs Male by Country

Table 13.6 Daily per Capita Calorie, Macronutrient, Mineral and Vitamin Gap and Decomposition – De Facto vs Male in Urban

Daily per capita	(1) Kcal	(3) Prot	(5) Fat	(7) Iron	(9) Zinc	(11) Ribl	(13) Thia	(15) Niac	(17) Vita	(19) Vitb12	(21) Vitb6	(23) Vitb9	(25) Vitc
Male-headed household	803.0*** (30.06)	20.90*** (0.810)	14.89*** (0.961)	6.265*** (0.249)	17.75*** (1.013)	0.368*** (0.0164)	2.444*** (0.152)	6.236*** (0.278)	444.8*** (62.03)	0.0403*** (0.00495)	4.196*** (0.261)	12.03*** (0.666)	41.99*** (3.250)
Female-headed household	484.9*** (35.46)	14.02*** (1.322)	10.31*** (1.646)	3.909*** (0.361)	9.285*** (0.964)	0.200*** (0.0160)	1.195*** (0.112)	3.937*** (0.438)	324.7*** (43.53)	0.0313*** (0.00950)	1.884*** (0.200)	12.59*** (2.135)	25.07*** (2.404)
Household consumption gap	318.2*** (46.49)	6.874*** (1.550)	4.582** (1.906)	2.355*** (0.438)	8.463*** (1.398)	0.169*** (0.0229)	1.250*** (0.189)	2.299*** (0.519)	120.1 (75.78)	0.00894 (0.0107)	2.312*** (0.329)	-0.565 (2.236)	16.93*** (4.043)
% per nutrient %	39.6	32.9	30.8	37.6	47.7	45.9	51.1	36.9	27.0	22.2	55.1	-4.7	40.3
%			31.8					43.8					28.0
Explained	67.81*** (20.89)	1.078** (0.533)	1.984*** (0.598)	0.492*** (0.152)	3.436*** (0.697)	0.0135 (0.00922)	0.310*** (0.105)	0.606*** (0.184)	-75.25** (31.31)	0.00856*** (0.00322)	0.512*** (0.181)	-3.281*** (0.372)	2.872 (2.641)
Unexplained	250.3*** (46.27)	5.796*** (1.548)	2.598 (1.871)	1.863*** (0.442)	5.027*** (1.240)	0.155*** (0.0244)	0.940*** (0.162)	1.693*** (0.516)	195.3** (95.96)	0.000380 (0.0104)	1.800*** (0.284)	2.716 (2.269)	14.06*** (4.328)
Observations	3,703	3,703	3,703	3,703	3,703	3,703	3,703	3,703	3,703	3,703	3,703	3,703	3,703

Note: Robust standard errors in parentheses. *** p<0.01,
** p<0.05,
* p<0.1.

Table 13.7 Daily per Capita Calorie, Macronutrient, Mineral and Vitamin Gap and Decomposition – De Facto vs Male in Rural

Daily per capita	(1) Kcal	(3) Prot	(5) Fat	(7) Iron	(9) Zinc	(11) Ribl	(13) Thia	(15) Niac	(17) Vita	(19) Vitb12	(21) Vitb6	(23) Vitb9	(25) Vitc
Male-headed household	1,297*** (29.82)	37.06*** (0.927)	30.02*** (1.144)	11.85*** (0.289)	40.37*** (1.319)	0.531*** (0.0134)	3.899*** (0.134)	11.64*** (0.354)	155.7*** (10.49)	0.104*** (0.00916)	6.215*** (0.229)	6.342*** (0.400)	43.52*** (2.216)
Female-headed household	595.4*** (69.91)	19.92*** (2.831)	17.13*** (2.758)	5.259*** (0.821)	13.19*** (2.195)	0.329*** (0.0416)	2.086*** (0.360)	6.841*** (0.992)	98.26*** (19.65)	0.0688*** (0.0144)	3.255*** (0.606)	8.915*** (4.107)	30.25*** (4.950)
Household consumption gap	702.0*** (76.01)	17.14*** (2.979)	12.89*** (2.986)	6.587*** (0.870)	27.18*** (2.560)	0.202*** (0.0437)	1.812*** (0.384)	4.797*** (1.053)	57.42*** (22.27)	0.0351** (0.0171)	2.960*** (0.648)	-2.573 (4.127)	13.27** (5.423)
% per nutrient %	54.1	46.2 / 44.6	42.9	55.6	67.3	38.0	46.5	41.2 / 49.7	36.9	33.8	47.6	-40.6	30.5 / 21.6
Explained	321.4*** (20.90)	5.554*** (0.616)	0.170 (0.746)	3.015*** (0.209)	14.6*** (0.973)	0.0610*** (0.00863)	0.558*** (0.0871)	1.249*** (0.224)	-43.27*** (8.778)	-0.0343*** (0.00450)	1.021*** (0.151)	-1.024*** (0.240)	13.61*** (1.631)
Unexplained	380.7*** (76.78)	11.59*** (3.024)	12.72*** (3.135)	3.571*** (0.875)	12.61*** (2.523)	0.141*** (0.0449)	1.254*** (0.395)	3.548*** (1.078)	100.7*** (25.83)	0.0694*** (0.0191)	1.940*** (0.666)	-1.550 (4.118)	-0.346 (5.317)
Observations	8,997	8,997	8,997	8,997	8,997	8,997	8,997	8,997	8,997	8,997	8,997	8,997	8,997

Note: Robust standard errors in parentheses. *** p<0.01,
** p<0.05,
* p<0.1.

Table 13.8 Daily per Capita Calorie, Macronutrient, Mineral and Vitamin Gap and Decomposition – De Jure vs Male in Urban Areas

Daily per capita	(1) Kcal	(3) Prot	(5) Fat	(7) Iron	(9) Zinc	(11) Ribl	(13) Thia	(15) Niac	(17) Vita	(19) Vitb12	(21) Vitb6	(23) Vitb9	(25) Vitc
Male-headed household	803.0***	20.90***	14.89***	6.265***	17.75***	0.368***	2.444***	6.236***	444.8***	0.0403***	4.196***	12.03***	41.99***
	(30.06)	(0.810)	(0.961)	(0.249)	(1.013)	(0.0164)	(0.152)	(0.278)	(62.03)	(0.00495)	(0.261)	(0.666)	(3.250)
Female-headed household	491.3***	13.07***	9.203***	3.769***	11.54***	0.233***	1.499***	4.123***	344.0***	0.0290***	2.548***	8.408***	25.32***
	(51.68)	(1.609)	(1.848)	(0.413)	(2.241)	(0.0209)	(0.236)	(0.551)	(47.32)	(0.00990)	(0.378)	(0.711)	(3.509)
Household consumption gap	311.8***	7.822***	5.690***	2.496***	6.205**	0.135***	0.945***	2.113***	100.8	0.0113	1.648***	3.622***	16.67***
	(59.78)	(1.802)	(2.083)	(0.483)	(2.460)	(0.0266)	(0.281)	(0.617)	(78.02)	(0.0111)	(0.459)	(0.974)	(4.783)
%	38.8	37.4	38.2	39.8	35.0	36.7	38.7	33.9	22.7	28.0	39.3	30.1	39.7
		37.8						36.8					32.0
Explained	128.3***	2.944***	2.027***	0.916***	3.021***	0.0571***	0.415***	0.929***	97.57***	0.00671	0.719***	0.910	4.801***
	(18.52)	(0.591)	(0.770)	(0.172)	(0.703)	(0.00942)	(0.0907)	(0.211)	(31.30)	(0.00430)	(0.156)	(0.580)	(1.541)
Unexplained	183.5***	4.879***	3.663*	1.580***	3.184	0.0779***	0.530*	1.185*	3.240	0.00462	0.929**	2.713**	11.87**
	(60.15)	(1.872)	(2.204)	(0.506)	(2.261)	(0.0270)	(0.286)	(0.644)	(73.03)	(0.0116)	(0.468)	(1.287)	(4.775)
Observations	4,516	4,516	4,516	4,516	4,516	4,516	4,516	4,516	4,516	4,516	4,516	4,516	4,516

Note: Robust standard errors in parentheses. *** p<0.01,
** p<0.05,
* p<0.1.

Table 13.9 Daily per Capita Calorie, Macronutrient, Mineral and Vitamin Gap and Decomposition – De Jure vs Male in Rural Areas

Daily per capita	(1) Kcal	(3) Prot	(5) Fat	(7) Iron	(9) Zinc	(11) Ribl	(13) Thia	(15) Niac	(17) Vita	(19) Vitb12	(21) Vitb6	(23) Vitb9	(25) Vitc
Male-headed household	1,297*** (29.82)	37.06*** (0.927)	30.02*** (1.144)	11.85*** (0.289)	40.37*** (1.319)	0.531*** (0.0134)	3.899*** (0.134)	11.64*** (0.354)	155.7*** (10.49)	0.104*** (0.00916)	6.215*** (0.229)	6.342*** (0.400)	43.52*** (2.216)
Female-headed household	775.3*** (46.93)	22.17*** (1.435)	20.63*** (1.893)	6.076*** (0.394)	17.00*** (1.411)	0.354*** (0.0223)	2.522*** (0.212)	7.723*** (0.577)	180.7*** (48.16)	0.119*** (0.0285)	4.149*** (0.363)	5.559*** (0.730)	46.90*** (4.546)
Household consumption gap	522.1*** (55.60)	14.89*** (1.708)	9.394*** (2.212)	5.769*** (0.488)	23.37*** (1.932)	0.177*** (0.0260)	1.377*** (0.251)	3.915*** (0.677)	-25.00 (49.28)	-0.0155 (0.0299)	2.066*** (0.429)	0.783 (0.832)	-3.387 (5.057)
%	40.3	40.2 35.7	31.3	48.7	57.9	33.3	35.3	33.6 6.8	-16.1	-14.9	33.2	12.3	-7.8 1.4
Explained	-32.28 (28.63)	-1.222 (0.880)	-4.300*** (1.043)	0.228 (0.276)	4.400*** (1.197)	-0.0374*** (0.0134)	-0.242* (0.131)	-1.087*** (0.382)	-39.32*** (10.55)	-0.0311*** (0.00676)	-0.346 (0.224)	-1.576*** (0.465)	-12.59*** (2.328)
Unexplained	554.4*** (64.53)	16.11*** (1.968)	13.69*** (2.526)	5.541*** (0.571)	18.97*** (2.168)	0.214*** (0.0301)	1.618*** (0.280)	5.002*** (0.814)	14.32 (42.94)	0.0156 (0.0327)	2.412*** (0.476)	2.358** (0.966)	9.198 (5.598)
Observations	10,446	10,446	10,446	10,446	10,446	10,446	10,446	10,446	10,446	10,446	10,446	10,446	10,446

Note: Robust standard errors in parentheses. *** p<0.01,
** p<0.05,
* p<0.1.

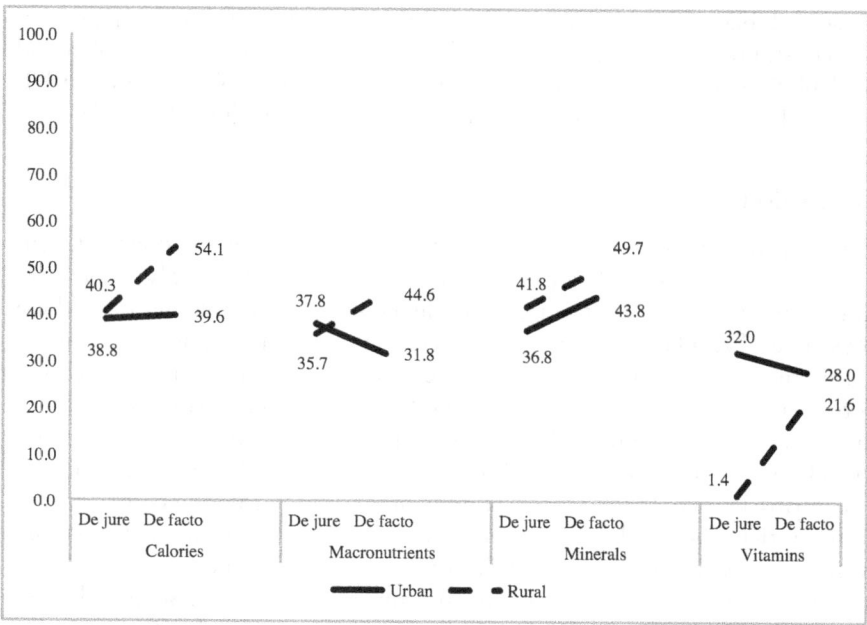

Figure 13.4 Daily per Capita Calorie, Macronutrient, Mineral and Vitamin Gap and Decomposition – Comparing between De Facto vs Male and De Jure vs Male in Urban and Rural Areas

households in urban and rural areas. In rural areas, the gender gap pertaining to calories, macronutrients, minerals and vitamins is higher for de facto female-headed households than for de jure female-headed households. In urban areas, the gender gap is similar regardless of de facto and de jure status; see Figure 13.4. In rural areas, the gender gap for calories is 40% for de jure and 54% for de facto female-headed households. The gap for macronutrients in rural areas is 37% for de jure and 44% for de facto. For minerals, the gap in rural areas is 41% for de jure and 49% for de facto. The de jure female-headed households gaps are similar in rural and urban areas, except for vitamins, for which the gender gap is much higher in urban (32%) than rural (1.4%) areas. The gaps continue to be smallest among vitamins in both rural and urban areas, especially in rural areas among the de jure female-headed households.

Speculatively, some of the reasons for the consumption advantage in male-headed households, as suggested by past studies, include lack of access to and control over resources by women (e.g., income and land) and limited human capital (i.e., education and agricultural knowledge) (Hjelm and Dasori, 2012). Babatunde et al. (2008), Tibesigwa and Visser (2016) and Owusu et al. (2011) agree that off-farm income is responsible for gendered food insecurity in many sub-Saharan African countries. They also reveal the factors determining food security in male-headed households as being essential in determining food security in female-headed households. These factors are household size, climate variables, household income, marital status and age. Kebede (2009) outlines crop production constraints, such as access to land, land size, soil fertility, access to labor and availability of inputs in Southern Ethiopia. This is similarly observed by Bezabih and Holden (2010), Asfaw and Maggio (2017) and Theriault et al. (2017).

Conclusion

This study documents the role of smallholder agriculture in increasing food security of farming households located in sub-Saharan Africa. We also show how crop farming provides energy, macronutrients, minerals and vitamins for these households. We additionally portray the consumption gap (specific to consumption from crop production) between male- and female-headed farm households in rural and urban areas. This is made possible by the World Bank Living Standard Measurement Study (LSMS) for Ethiopia, Malawi, Nigeria, Tanzania and Uganda. We find the following results. First, agriculture plays a significant role in boosting household food security, providing not only caloric energy but also macronutrients, minerals and vitamins. Approximately 1,034 calories per day are available for each household member from crop farming, which is about 39% of the daily caloric energy requirements. In addition, crop farming supplies households with about 53% of the daily recommended nutrient intake (RNI) for proteins, 65% of the RNI for iron, 49.2% for zinc, 33.8% for riboflavin, 57% for niacin, 35% for fats, 76% for vitamins, and more than the daily RNI for thiamine and for vitamins B12, B6 and B9. Second, we find that male-headed households have higher consumption from crop farming and are therefore more likely to be food-secure in comparison to female-headed households. The differences are 42% versus 52% for calories, 40% versus 48% for macronutrients, 43.6% versus 52.5% for minerals and relatively smaller gaps for vitamins (12.3% versus 22.1%). This is a reflection of the fact that male and female farmers are likely to grow different crops, correlated with whether the crops are food crops or cash crops (see, e.g., Carr, 2008; World Bank, 2009; FAO, 2010). Third, upon comparing urban and rural areas, we find that the de facto female-headed household gender gaps pertaining to calories, macronutrients, minerals and vitamins are higher than the de jure female-headed household gender gaps in rural areas, while in urban areas these gaps are similar for de facto and de jure households. Also, the de jure female-headed households gaps are similar in rural and urban areas, except for vitamins, where the gap is much higher in urban than rural areas. Lastly, the gaps continue to be smallest among vitamins in both rural and urban areas, especially in rural areas among the de jure female-headed households. The results from our study suggest the need for agricultural policies to emphasize food security. This chapter also sheds more light on nutrition as well as the gender differences that go along with it.

Appendix A
Smallholder crops in sub-Saharan Africa

Tanzania

Table 13A1 Long-Rains Crops in Tanzania

	Freq.	Percentage
Maize	3,079	38.81
Beans	871	10.98
Paddy	805	10.15
Groundnut	472	5.95
Sweet potatoes	364	4.59
Sorghum	324	4.08
Pigeon pea	276	3.48
Sunflower	246	3.1
Cowpeas	244	3.08
Cotton	204	2.57
Cocoyams	122	1.54
Sesame	114	1.44
Green gram	85	1.07
Bulrush millet	75	0.95
Bambara nuts	71	0.89
Irish potatoes	66	0.83
Tobacco	62	0.78
Pumpkins	56	0.71
Tomatoes	45	0.57
Cassava	36	0.45
Chickpeas	29	0.37
Yams	28	0.35
Finger millet	27	0.34
Wheat	24	0.3
Amaranths	22	0.28
Fiwi	22	0.28

(Continued)

Table 13A1 (Continued)

	Freq.	Percentage
Pyrethrum	19	0.24
Onions	16	0.2
Okra	14	0.18
Soyabeans	12	0.15
Spinach	10	0.13
Field peas	9	0.11
Cucumber	9	0.11
Seaweed	8	0.1
Eggplant	8	0.1
Cabbage	7	0.09
Watermelon	5	0.06
Bilimbi	5	0.06
Carrot	3	0.04
Chili	2	0.03
Coffee	1	0.01
Other (specify)	37	0.47
Total	7,934	100

Note: The frequency represents plot-level production of crops produced during the long rains in Tanzania.

Table 13A2 Short-Rains Crops in Tanzania

	Freq.	Percentage
Maize	967	43.46
Beans	470	21.12
Groundnut	131	5.89
Sweet potatoes	129	5.8
Paddy	76	3.42
Cowpeas	76	3.42
Cocoyams	63	2.83
Green gram	55	2.47
Cotton	51	2.29
Sorghum	46	2.07
Pigeon pea	22	0.99
Sunflower	18	0.81
Irish potatoes	15	0.67
Tomatoes	14	0.63
Amaranth	11	0.49
Yams	7	0.31
Tobacco	6	0.27
Pumpkins	6	0.27

	Freq.	Percentage
Okra	6	0.27
Bambara nuts	5	0.22
Sesame	5	0.22
Finger millet	4	0.18
Cassava	4	0.18
Cabbage	4	0.18
Seaweed	3	0.13
Chickpeas	3	0.13
Spinach	3	0.13
Cucumber	3	0.13
Onions	2	0.09
Eggplant	2	0.09
Bulrush millet	1	0.04
Field peas	1	0.04
Coconut	1	0.04
Soyabeans	1	0.04
Carrot	1	0.04
Chili	1	0.04
Watermelon	1	0.04
Cauliflower	1	0.04
Fiwi	1	0.04
Other (specify)	9	0.4
Total	2,225	100

Note: The frequency represents plot-level production of crops produced during the short rains in Tanzania.

Table 13A3 Fruit Crops in Tanzania

	Freq.	Percentage
Mango	1,449	25.56
Banana	1,434	25.3
Pawpaw	512	9.03
Orange	458	8.08
Avocado	399	7.04
Guava	293	5.17
Lemon	168	2.96
Pineapple	155	2.73
Jackfruit	152	2.68
Custard apple	103	1.82
Mandarin	88	1.55

(Continued)

Table 13A3 (Continued)

	Freq.	Percentage
Passion fruit	86	1.52
Lime	61	1.08
Peaches	35	0.62
Plum	24	0.42
Plums	19	0.34
Peaches	19	0.34
Pears	16	0.28
Star fruit	14	0.25
Pomegranate	12	0.21
Apples	9	0.16
Bread fruit	7	0.12
Date	6	0.11
Rambutan	4	0.07
Cassava	3	0.05
Bilimbi	2	0.04
Cocoyams	1	0.02
Malay apple	1	0.02
Pomelo	1	0.02
Durian	1	0.02
Mitobo	1	0.02
Other (specify)	136	2.4
Total	5,669	100

Note: The frequency represents plot-level production of other crops produced permanently in Tanzania.

Table 13A4 Other Permanent Crops in Tanzania

	Freq.	Percentage
Cassava	1,475	31.55
Cashew nut	686	14.67
Timber	563	12.04
Firewood/fodder	471	10.07
Coconut	411	8.79
Coffee	301	6.44
Sugarcane	185	3.96
Palm oil	158	3.38
Pigeon pea	70	1.5
Sisal	56	1.2
Clove	38	0.81

	Freq.	Percentage
Tamarind	26	0.56
Cocoa	23	0.49
Cinnamon	22	0.47
Cocoyams	20	0.43
Cardamom	19	0.41
Fence tree	19	0.41
Bamboo	17	0.36
Monkeybread	15	0.32
Black pepper	11	0.24
Medicinal plants	11	0.24
Tea	9	0.19
Yams	7	0.15
Kapok	5	0.11
Sweet potatoes	2	0.04
Grapes	2	0.04
Rubber	1	0.02
Wattle	1	0.02
Banana	1	0.02
Pineapple	1	0.02
Durian	1	0.02
Other (specify)	48	1.03
Total	4,675	100

Note: The frequency represents plot-level production of other crops produced permanently in Tanzania.

Malawi

Table 13A5 Rainy-Season Crops in Malawi

	Freq.	*Percentage*
Maize (local, composite, hybrid, hybrid recycled, others)	4,197	42.13
Tobacco (burley, flue cured, NNDF, others)	329	3.30
Groundnut (chalimbana, cg7, manipinta, mawanga, jl24, others)	1122	11.26
Rice (local, faya, pussa, tcg10, senga, wambone, kilombero, mtupatupa, others)	195	1.96
Ground bean (nzama)	30	0.30
Sweet potato	173	1.74
Irish (Malawi) potato	30	0.30
Finger millet (Mawere)	32	0.32
Sorghum	367	3.68
Pearl millet (Mchewere)	57	0.57
Beans	458	4.60
Soyabean	402	4.03
Pigeon pea (Nandolo)	1,167	11.71
Cotton	167	1.68
Sunflower	55	0.55
Sugarcane	3	0.03
Cabbage	1	0.01
Tanaposi	5	0.05
Nkhwani	694	6.97
Therere/okra	50	0.50
Tomatoes	28	0.28
Onion	9	0.09
Pea	218	2.19
Paprika	7	0.07
Other (specify)	167	1.68
Total	9,963	100.00

Note: The frequency represents plot-level production of crops produced in the rainy season in Malawi.

Table 13A6 Dry Season Crops in Malawi

	Freq.	*Percent*
Maize (local, composite, hybrid, hybrid recycled)	281	22.96
Tobacco flue cured	1	0.08
Rice (faya, tcg10, senga, kilombero, mtupatupa, others)	10	0.82
Ground bean (nzama)	1	0.08

	Freq.	Percent
Sweet potato	53	4.33
Irish (Malawi) potato	82	6.70
Sorghum	1	0.08
Beans	147	12.01
Pigeonpea (nandolo)	2	0.16
Sugarcane	13	1.06
Cabbage	17	1.39
Tanaposi	146	11.93
Nkhwani	172	14.05
Therere/okra	14	1.14
Tomatoes	188	15.36
Onion	11	0.90
Pea	16	1.31
Paprika	3	0.25
Other (specify)	66	
Total	1,224	

Note: The frequency represents plot-level production of crops produced in the dry season in Malawi.

Table 13A7 Permanent/Tree Crops in Malawi

	Freq.	Percentage
Cassava	387	14.26
Tea	3	0.11
Mango	965	35.56
Orange	90	3.32
Pawpaw/papaya	298	10.98
Banana	335	12.34
Avocado	117	4.31
Guava	135	4.97
Lemon	46	1.69
Naartje (tangerine)	25	0.92
Peach	42	1.55
(Custard apple) poza	42	1.55
Mexican apple (masuku)	102	3.76
Masau	56	2.06
Pineapple	4	0.15
Other (specify)	43	1.58
Total	2,714	100

Note: The frequency represents plot-level production of permanent crops in Malawi.

Uganda

Table 13A8 First-Season Crops in Uganda

	Freq.	Percentage
Beans	2,252	16.95
Banana (food, beer, sweet)	2,243	16.88
Maize	2,113	15.90
Cassava	2,110	15.88
Sweet potatoes	1,036	7.80
Coffee all	650	4.89
Groundnuts	591	4.45
Sorghum	482	3.63
Finger millet	281	2.11
Soyabeans	122	0.92
Yam	115	0.87
Field peas	105	0.79
Irish potatoes	93	0.70
Rice	91	0.68
Simsim	89	0.67
Pigeon peas	87	0.65
Sunflower	87	0.65
Sugarcane	73	0.55
Jackfruit	49	0.37
Avocado	47	0.35
Pumpkins	45	0.34
Green gram	45	0.34
Tomatoes	44	0.33
Cotton	43	0.32
Tobacco	36	0.27
Mango	32	0.24
Oranges	31	0.23
Cabbage	28	0.21
Pawpaw	28	0.21
Pineapples	27	0.20
Cowpeas	20	0.15
Onions	18	0.14
Eggplant	18	0.14
Dodo	17	0.13
Cocoa	15	0.11
Passion fruit	12	0.09
Vanilla	7	0.05
Tea	5	0.04
Carrots	2	0.02
Wheat	1	0.01
Other	100	0.75
Total	13,290	100.0

Note: The frequency represents plot-level production of first-season crops in Uganda.

Table 13A9 Second-Season Crops in Uganda

	Freq.	Percentage
Banana (food, beer, sweet)	1,748	19.12
Beans	1,822	19.93
Maize	1,459	15.96
Cassava	976	10.68
Sweet potatoes	785	8.59
Coffee all	516	5.64
Groundnuts	325	3.56
Finger millet	242	2.65
Sorghum	195	2.13
Simsim	165	1.81
Irish potatoes	87	0.95
Yam	87	0.95
Soyabeans	80	0.88
Cotton	76	0.83
Field peas	59	0.65
Sugarcane	49	0.54
Rice	47	0.51
Sunflower	47	0.51
Jackfruit	46	0.50
Tomatoes	41	0.45
Avocado	37	0.40
Cowpeas	30	0.33
Pigeon peas	27	0.30
Mango	25	0.27
Pumpkins	22	0.24
Pawpaw	22	0.24
Cabbage	20	0.22
Oranges	16	0.18
Onions	13	0.14
Pineapples	13	0.14
Eggplant	11	0.12
Cocoa	10	0.11
Passion fruit	8	0.09
Dodo	7	0.08
Tea	5	0.05
Vanilla	3	0.03
Tobacco	2	0.02
Other	18	0.20
Total	9,141	100.00

Note: The frequency represents plot-level production of second-season crops in Uganda.

Nigeria

Table 13A10 Crops Produced in Nigeria

	Freq.	Percentage
Cassava old	2,128	16.62
Groundnuts/peanuts (unshelled, shelled)	1664	13.00
Guinea corn/sorghum	1,533	11.97
Yam (white, yellow, water, three leave)	1368	10.68
Beans/cowpea	1,160	9.06
Millet/maiwa	910	7.11
Maize (cob, grain, popcorn)	492	3.84
Melon (unshelled, shelled)	350	2.73
Rice (rice paddy)	341	2.66
Okra	321	2.51
Cocoyam	305	2.38
Pumpkin (leave, fruit, seed)	257	2.01
Oil palm tree	241	1.88
Cocoa (pod, beans)	198	1.55
Soyabeans	191	1.49
Plantain	172	1.34
Pepper	160	1.25
Beeni-seed/sesame	136	1.06
Kolanut	123	0.96
Tomatoes	85	0.66
Orange	73	0.57
Potato	66	0.52
Sweet potato	62	0.48
Banana	48	0.37
Bambara nut	46	0.36
Green vegetable	42	0.33
Onion	32	0.25
Cotton	27	0.21
Acha	19	0.15
Garden egg	19	0.15
Pineapple	19	0.15
Cashew nut	19	0.15
Pear	17	0.13
Agbono (oro seed)	15	0.12
Pigeon pea	13	0.10

	Freq.	Percentage
Coconut	12	0.09
Pawpaw	12	0.09
Ginger	9	0.07
Small pepper	9	0.07
Palm oil	9	0.07
Carrot	8	0.06
Cabbage	8	0.06
Sugarcane	8	0.06
Avocado pear	8	0.06
Zobo	7	0.05
Mandarin/tangerine	6	0.05
Mango	6	0.05
Seed cotton	5	0.04
Cucumber	4	0.03
Lettuce	3	0.02
Bitter kola	3	0.02
Lime	3	0.02
Oil bean	3	0.02
Wheat	2	0.02
Grapefruit	2	0.02
Guava	2	0.02
Rubber	2	0.02
Other spices/vanilla	1	0.01
Gum arabic	1	0.01
Dry leaves (kuka)	1	0.01
Rizga	1	0.01
Tobacco	1	0.01
Walnut	1	0.01
Apple	1	0.01
Coffee	1	0.01
Jute	1	0.01
Locust bean	1	0.01
Cherry (agbalumo)	1	0.01
Eru	1	0.01
Other (specify)	8	0.06
Total	12,803	100.00

Ethiopia

Table 13A11 Crops Produced in Ethiopia

	Freq.	Percentage
Maize	3,270	11.06
Sorghum	2,690	9.1
Teff	2,306	7.8
Coffee	2,106	7.12
Enset	1,793	6.06
Wheat	1,383	4.68
Barley	1,313	4.44
Chat	1,142	3.86
Haricot beans	955	3.23
Horse beans	944	3.19
Kale	856	2.89
Bananas	847	2.86
Godere	776	2.62
Millet	505	1.71
Field peas	481	1.63
Avocados	478	1.62
Sweet potato	473	1.6
Mangos	464	1.57
Gesho	457	1.55
Garlic	425	1.44
Pumpkins	410	1.39
Red pepper	350	1.18
Sugarcane	299	1.01
Potatoes	285	0.96
Nueg	282	0.95
Sesame	252	0.85
Rape seed	251	0.85
Papaya	241	0.81
Groundnuts	204	0.69
Chick peas	198	0.67
Linseed	198	0.67
Cassava	192	0.65
Green pepper	182	0.62
Lentils	180	0.61
Vetch	153	0.52
Oranges	131	0.44
Fenugreek	130	0.44
Onion	124	0.42
Guava	111	0.38
Lemons	101	0.34
Rue	95	0.32

	Freq.	Percentage
Cabbage	70	0.24
Tomatoes	69	0.23
Cardamon	61	0.21
Beer root	60	0.2
Sunflower	57	0.19
Oats	53	0.18
Tobacco	53	0.18
Peach	49	0.17
Coriander	43	0.15
Cactus	40	0.14
Ginger	39	0.13
Soyabeans	32	0.11
Gibto	28	0.09
Cotton	23	0.08
Tumeric	21	0.07
Boye/yam	20	0.07
Spinach	16	0.05
Rice	14	0.05
Carrot	14	0.05
Roman	12	0.04
Pineapples	12	0.04
Black cumin	9	0.03
Mandarins	9	0.03
Cauliflower	9	0.03
Citron	8	0.03
Lettuce	6	0.02
White cumin	5	0.02
Apples	4	0.01
Amboshika	4	0.01
Fennel	3	0.01
Comtatie	3	0.01
Gishita	2	0.01
Black pepper	1	0
Timiz kimem	1	0
Kazmir	1	0
Other roots	164	0.55
Other fruits	233	0.79
Other spices	141	0.48
Other pulses	23	0.08
Other oil seed	5	0.02
Other cereal	49	0.17
Other case crops	1	0
Other vegetable	75	0.25
Total	29,575	100

Appendix B

Caloric intake in sub-Saharan African countries

Table 13B1 Household Consumption and the Contribution of Own Production to Consumption from Past Studies/Statistics

Indicator	Unit	Chad	Côte d'Ivoire	Ghana	Kenya	Malawi
		2009	2002	1998	2005	2004
Dietary energy consumption	kcal/capita/day	2461.3	2104.6	2302.3	1798.7	2236.5
Protein consumption	g/capita/day	82.1	63.7	52.5	52.4	74.3
Fat consumption	g/capita/day	52.3	46.7	48	41.9	47.5
Dietary energy (%)						
Share of *purchased* food in total food consumption		57	73.4	55.4	58.7	43.7
Share of *own produced food* in total food consumption		37.9	23.8	36.4	14.1	46.1
Share of food *from other sources* in total food consumption		1.4	0	0	24.7	9.4
Share of food *consumed away from home* in total food consumption		3.8	2.8	8.2	2.5	0.8
Share of own produced food (calories)		933	502	837	253	1031

Indicator	Unit	Mali	Mozambique	Niger	Papua New Guinea	Sudan (former)
		2001	2002	2007	1996	2009
Dietary energy consumption	kcal/capita/day	2276.3	1955.3	1937.6	2002.6	2237.7
Protein consumption	g/capita/day	63.5	51.6	56.7	47.6	69.4
Fat consumption	g/capita/day	44.7	41.8	34	43.5	54.3

Indicator	Unit	Mali	Mozambique	Niger	Papua New Guinea	Sudan (former)
		2001	2002	2007	1996	2009
Dietary energy (%)						
Share of *purchased* food in total food consumption		45.5	35.8	18.2	30.4	79.4
Share of *own produced food* in total food consumption		49.3	62.8	79.2	54.5	7.6
Share of food *from other sources* in total food consumption		2.3	0.9	2.1	10.6	9.2
Share of food *consumed away from home* in total food consumption		2.9	0.5	0.5	4.5	3.8
Share of own produced food (calories)		1121	1227	1535	1091	171

Indicator	Unit	Togo	Uganda	Uganda	Zambia	Ethiopia
		2006	2002	2005	2002	2015
Dietary energy consumption	kcal/capita/day	2159	2158.9	2006.2	1967.5	2313
Protein consumption	g/capita/day	66.4	50.1	47.5	77.8	
Fat consumption	g/capita/day	37.6	28.8	25.1	45.5	
Dietary energy (%)						
Share of *purchased* food in total food consumption		45.5	40.6	40.2	58	87.8
Share of *own produced food* in total food consumption		28.2	52.3	48.1	34.6	11.8
Share of food *from other sources* in total food consumption		0	3.2	6.6	7.1	0.4
Share of food *consumed away from home* in total food consumption		26.3	3.9	5.2	0.3	
Share of own produced food (calories)		610	1130	964	681	273

Source: FAOSTAT (2017).

Source: The statistics from Ethiopia are from Lam et al. (2017).

Note: Share of own produced food (calories) is derived by multiplying the 'Share of own produced food in total food consumption' with 'Dietary energy consumption' so as to obtain the share of calories that is derived from own production.

Note

1 Nkwani is a C. maxima species which falls under the scientific name of cucurbitacease (i.e., cucurbita species). It is somewhat similar to the squash, pumpkin and marrow family, and is produced for consumption of leaves, fruits and seeds (FAO, 1989).

References

Abdalla, S. 2014. Assessment of Caloric and Protein Intake in Sudan. International Working Paper Series No. 14–04. Pavia, Italy: Natural Resources, Agricultural Development and Food Security – International Research Project.

Abdalla, S., I.U. Leonhäuser, S. Bauer and E. Elamin. 2013. Households Coping Strategies and Food Consumption Gap in Sudan. International Working Paper Series No. 13–06. Pavia, Italy: Natural Resources, Agricultural Development and Food Security – International Research Project.

Abele, S., E. Twine and C. Legg. 2007. Food Security in Eastern Africa and the Great Lakes. Crop Crisis Control Project Final Report. Ibadan, Nigeria: International Institute of Tropical Agriculture.

Adepoju, A.O., and K.A. Adejare. 2013. Food Insecurity Status of Rural Households during the Post-Planting Season in Nigeria. *Journal of Agriculture and Sustainability* 4(1): 16–35.

Adeyefa, S. A. (2016). Determinants of Food Insecurity among the Urban Poor in the City of Tshwane, South Africa. *Journal of Economics*, 4(2), 101–114.

Asfaw, S., and G. Maggio. 2017. Gender, Weather Shocks and Welfare: Evidence from Malawi. *The Journal of Development Studies* 1–21. http://dx.doi.org/10.1080/00220388.2017.1283016

Babatunde, R.O., A.O. Adejobi and S.B. Fakayode. 2010. Income and Calorie Intake among Farming Households in Rural Nigeria: Results of Parametric and Nonparametric Analysis. *Journal of Agricultural Science* 2(2): 135–146.

Babatunde, R.O., O.A. Omotesho, E.O. Olorunsanya, and G.M. Owotoki. 2008. Determinants of Vulnerability to Food Insecurity: A Gender-Based Analysis of Farming Households in Nigeria. *Indian Journal of Agricultural Economics* 63(1): 116.

Benfica, R., and T. Kilic. 2015. The Effects of Smallholder Agricultural Involvement on Household Food Consumption and Dietary Diversity: Evidence from Malawi. International Association of Agricultural Economists Conference. August 9–14, Milan, Italy (No. 211218).

Bezabih, M., and S. Holden. 2010. The Role of Land Certification in Reducing Gender Gaps in Productivity in Rural Ethiopia. Environment for Development Discussion Paper 10–23. Gothenburg, Sweden: Environment for Development and Resources for the Future.

Bongaarts, J. 2001. Household Size and Composition in the Developing World in the 1990s. *Population Studies* 55(3): 263–279.

Carr, E.R. 2008. Men's Crops and Women's Crops: The Importance of Gender to the Understanding of Agricultural and Development Outcomes in Ghana's Central Region. *World Development* 36(5): 900–915.

Carter, I. (1997). Rural Women and Food Insecurity. *Foodsteps*, 35(3).

Debela, M., and W. Abebe. 2017. Determinants of Rural Female-Headed Households' Food Security in Ambo District, West Shewa Zone, Oromia Regional State, Ethiopia. *Journal of Science and Sustainable Development* 5(1): 73–87.

Di Falco, S., M. Bezabih and M. Yesuf. 2010. Seeds for Livelihood: Crop Biodiversity and Food Production in Ethiopia. *Ecological Economics* 69(8): 1695–1702.

Di Falco, S., M. Veronesi and M. Yesuf. 2011. Does Adaptation to Climate Change Provide Food Security? A Micro-Perspective from Ethiopia. *American Journal of Agricultural Economics* 93(3): 829–846.

FAO. 1989. *The State of Food and Agriculture*, 37. Food and Agriculture Organization of the UN (FAO).

FAO. 1998. Vitamin and Mineral Requirements in Human Nutrition: Report of a Joint FAO/WHO Expert Consultation, Bangkok, Thailand, September 21–30.

FAO. 2001. The State of Food and Agriculture 2001, No. 33. Rome: Food and Agriculture Organization.

FAO. 2010. Gender Dimensions of Agricultural and Rural Employment: Differentiated Pathways Out of Poverty – Status, Trends and Gaps. FAO, IFAD and ILO.

FAO, IFAD and WFP. 2014. The State of Food Insecurity in the World 2014. Strengthening the Enabling Environment for Food Security and Nutrition. Rome: Food and Agriculture Organization.

FAO, IFAD and WFP. 2015. The State of Food Insecurity in the World 2015. Meeting the 2015 International Hunger Targets: Taking Stock of Uneven Progress. Rome: Food and Agriculture Organization.

FAOSTAT. 2017. Database Collections. Rome: Agriculture Organization of the United Nations.

Flatø, M., R. Muttarak and A. Pelser. 2017. Women, Weather, and Woes: The Triangular Dynamics of Female-Headed Households, Economic Vulnerability, and Climate Variability in South Africa. *World Development* 90: 41–62.

Gürkan, C., and I. Sanogo. 2011. Structural Differences in Rural Food Poverty between Female and Male-Headed Households. *Zambia Social Science Journal* 2(1): Article 5. http://scholarship.law.cornell.edu/zssj/vol2/iss1/5

Habyarimana, J.B. 2015. Determinants of Household Food Insecurity in Developing Countries Evidences from a Probit Model for the Case of Rural Households in Rwanda. *Sustainable Agriculture Research* 4(2): 78. http://dx.doi.org/10.5539/sar.v4n2p78

Haddad, L., Kennedy, E., and Sullivan, J. (1994). Choice of Indicators for Food Security and Nutrition Monitoring. *Food Policy*, 19(3), 329–343.

Hjelm, L., and W. Dasori. 2012. Ghana Comprehensive Food Security and Vulnerability Analysis, 2012: Focus on Northern Ghana. Rome: UN World Food Programme.

Horrell, S., and P. Krishnan. 2007. Poverty and Productivity in Female-Headed Households in Zimbabwe. *The Journal of Development Studies* 43(8): 1351–1380.

International Fund for Agricultural Development. 2015. The State of Food Insecurity in the World 2015. Rome: FAO.

Kabubo-Mariara, J., R. Mulwa and S. Di Falco. 2016. The Impact of Climate Change on Food Calorie Production and Nutritional Poverty: Evidence from Kenya. Environment for Development Discussion Paper No. 16–26. Gothenburg, Sweden: Environment for Development and Resources for the Future.

Kakota, T., D. Nyariki, D. Mkwambisi, and W. Kogi-Makau. (2011). Gender Vulnerability to Climate Variability and Household Food Insecurity. *Climate and Development*, 3(4), 298–309.Kassie, M., S. Wagura Ndiritu and J. Stage. 2014. What Determines Gender Inequality in Household Food Security in Kenya? Application of Exogenous Switching Treatment Regression. *World Development* 56: 153–171.

Kassie, M., J. Stage, H. Teklewold, & O. Erenstein. 2015. Gendered Food Security in Rural Malawi: Why Is Women's Food Security Status Lower? *Food Security*, 7(6), 1299–1320.

Kebede, M. 2009. The Gender Perspective of Household Food Security in Meskan District of the Gurage Zone, Southern Ethiopia. *African Research Review* 3(4): 31–47.

Lam, R.D., Y.A. Boafo, S. Degefa, A. Gasparatos and O. Saito. 2017. Assessing the Food Security Outcomes of Industrial Crop Expansion in Smallholder Settings: Insights from Cotton Production in Northern Ghana and Sugarcane Production in Central Ethiopia.

Sustainability Science 1–17. https://link.springer.com/article/10.1007/s11625-017-0449-x

Mallick, D., and M. Rafi. 2010. Are Female-Headed Households More Food-Insecure? Evidence from Bangladesh. *World Development* 38(4): 593–605.

Modirwa, S., and O.I. Oladele. 2012. Food Security among Male and Female-Headed Households in Eden District Municipality of the Western Cape, South Africa. *Journal of Human Ecology* 37(1): 29–35.

Ndobo, F. 2013. Determining the Food Security Status of Households in a South African Township. Doctoral Dissertation. North-West University.

Ndobo, F., and T.J. Sekhampu. 2013. Determinants of Vulnerability to Food Insecurity in a South African Township: A Gender Analysis. *Mediterranean Journal of Social Sciences* 4(14): 311.

Neuman, S., and Oaxaca, R.L. (2004). Wage Decompositions with Selectivity-Corrected Wage Equations: A Methodological Note. *Journal of Economic Inequality* 2(1), 3–10.

Ngigi, M.W., U. Mueller and R. Birner. 2017. Gender Differences in Climate Change Adaptation Strategies and Participation in Group-Based Approaches: An Intra-Household Analysis from Rural Kenya. *Ecological Economics* 138: 99–108.

Nyariki, D.M., S.L. Wiggins and J.K. Imungi. 2002. Levels and Causes of Household Food and Nutrition Insecurity in Dryland Kenya. *Ecology of Food and Nutrition* 41(2): 155–176.

Oaxaca, R.L., and Ransom, M.R. (1994). On Discrimination and the Decomposition of Wage Differentials. *Journal of Econometrics* 61(1), 5–21.

Oaxaca, R.L., and Ransom, M.R. (1999). Identification in Detailed Wage Decompositions. *The Review of Economics and Statistics* 81(1), 154–157.

Owusu, V., A. Abdulai and S. Abdul-Rahman. 2011. Non-Farm Work and Food Security among Farm Households in Northern Ghana. *Food Policy* 36(2): 108–118.

Plants, F.T.F. 1988. *Traditional Food Plants: A Resource Book for Promoting the Exploitation and Consumption of Food Plants in Arid, Semi-Arid, and Sub-Humid Lands of Eastern Africa.* Rome: FAO.

Silvestri, S., S. Douxchamps, P. Kristjanson, W. Förch, M. Radeny, I. Mutie, C.F. Quiros, M. Herrero, A. Ndungu, N. Nicolas, J. Mango, L. Claessens and M.C. Rufino. 2015. Households and Food Security: Lessons from Food-Secure Households in East Africa. *Agriculture and Food Security* 4(1): 23. https://doi.org/10.1186/s40066-015-0042-4

Tefera, G. 2014. Women's Participation in Ensuring Food Security at Household Level. LAP LAMBERT Academic Publishing. http://www.efdinitiative.org/publications/womens-participation-ensuring-food-security-household-level-evidence-ebinat-district.

Theriault, V., M. Smale and H. Haider. 2017. How Does Gender Affect Sustainable Intensification of Cereal Production in the West African Sahel? Evidence from Burkina Faso. *World Development* 92: 177–191.

Tibesigwa, B., and M. Visser. 2016. Assessing Gender Inequality in Food Security among Smallholder Farm Households in Urban and Rural South Africa. *World Development* 88: 33–49.

Tibesigwa, B., M. Visser, M. Collinson and W. Twine. 2016. Investigating the Sensitivity of Household Food Security to Agriculture-Related Shocks and the Implication of Social and Natural Capital. *Sustainability Science* 11(2): 193–214.

Tibesigwa, B., M. Visser, L. Hunter, M. Collinson and W. Twine. 2015. Gender Differences in Climate Change Risk, Food Security, and Adaptation: A Study of Rural Households' Reliance on Agriculture and Natural Resources to Sustain Livelihoods. Environment for Development Discussion Paper Series No. 15–20. Gothenburg, Sweden: Environment for Development and Resources for the Future.

Tibesigwa, B., M. Visser and J. Turpie. 2015. The Impact of Climate Change on Net Revenue and Food Adequacy of Subsistence Farming Households in South Africa. *Environment and Development Economics* 20(3): 327–353.

World Bank. 2009. *Gender in Agriculture Sourcebook*. Washington, DC: World Bank, Food and Agriculture Organisation and International Fund for Agricultural Development.

Zakari, S., L. Ying and B. Song. 2014. Factors Influencing Household Food Security in West Africa: The Case of Southern Niger. *Sustainability* 6(3): 1191–1202.

14 Gender-differentiated impacts of climate variability in Ethiopia

A micro-simulation approach

Tesfamicheal Wossen

Introduction

Climate variability, manifested by changes in rainfall amount, intensity and timing, as well as through changes in temperature, often causes serious agricultural production losses and exacerbates food insecurity in sub-Saharan Africa (SSA). Given that the direct impacts of climate variability are transmitted through the agricultural sector, improving farm households' capacity to adapt to the adverse effects of climate-related shocks through effective adaptation and policy interventions is imperative (Milman and Arsano, 2013; Arndt et al., 2011; Deressa et al., 2009). Previous studies by Deressa et al. (2009), Block et al. (2008), Arndt et al. (2011), Robinson et al. (2012) and Di Falco et al. (2011) documented that farm households in Ethiopia are vulnerable to the impacts of climate variability. Although a great deal of progress has been made in disentangling the effects of climate variability, uncertainties still remain. For example, it is acknowledged that climate variability matters; however, the exact magnitude of the effect is not yet clear (Milman and Arsano, 2013; Di Falco et al., 2011; Di Falco and Veronesi, 2014; Deressa et al., 2009; De Pinto et al., 2013; Kandulu et al., 2012; Alauddin and Sarker, 2013; Wossen et al., 2015). Studies show that the estimated climate change effects range in the order of 7%–10% decline in GDP compared to a scenario of no climate change (Arndt et al., 2011; Robinson et al., 2012). Cognizant of this fact, the Ethiopian government developed a National Adaptation Program of Action (NAPA) in 2007 (National Meteorological Agency, 2007). The NAPA sets out potential adaptation options suited for small-scale and subsistence farm households.

Because successful implementation of policy interventions in response to climate variability depends on the magnitude and direction of expected effects of variability at a disaggregated level, examining the distributional effects of climate variability as well as the current roles of adaptation strategies will be crucial (Juana et al., 2013). Ideally, such analysis should also include gender dimensions. However, the evidence on the gender-specific effects of climate variability is rather scant. Capturing the gender-specific effects of climate variability is crucial because climate variability may have differential impacts on male-headed households (MHHs) and female-headed households (FHHs) because of differences in the perception of climate variability, adaptive capacity (Bryan et al., 2009), physical assets and social capital and hence adaptive and coping capacity, risk perception and choice of crop portfolios.[1]

In addition, women and men may have different levels of access to extension and climate information. For example, Asfaw and Admassie (2004) found that MHHs are more likely than FHHs to get information about new technologies and to undertake risky businesses. The Ethiopian Rural Household Survey (ERHS) 2009 also shows that only 15% of MHHs are eligible for safety nets, such as work-for-food programs, compared to 26% of FHHs. This result further underscores that FHHs are poorer than MHHs, as access to safety nets is granted based on the initial poverty status of households. Similarly, Tenge and Hella (2004) found that FHHs have limited access to information, land and other resources due to traditional social barriers. Empirical evidence in many developing countries further shows that FHHs own less land and fewer assets and also use fewer improved seed varieties (World Bank, 2013). In line with this, Kilic et al. (2014) documented that female-managed plots are on average 25% less productive and 91% of this difference is explained by the endowment effect. In particular for Ethiopia, Dercon et al. (2005) found that drought shocks have disproportionately higher impacts on FHHs compared to MHHs.

This study examines to what extent FHHs are vulnerable to the impacts of current climate variability compared to MHHs.[2] In particular, the study aims at examining how current climate variability may disproportionately affect FHHs compared to MHHs and to what extent adaptation options such as adoption of new crop varieties may reduce the vulnerability of FHHs. The study also assesses the responsiveness of FHHs to policy interventions compared to MHHs when exposed to the same policy treatment after climate shocks.[3] The study employs a micro-simulation approach that captures farm-level impacts of climate variability while taking into account a wide range of adaptation options. This is quite novel compared to the existing climate variability research which focuses on macro-level impacts. In particular, the micro-simulation approach employs a scenario-based analysis to examine the possible impacts of climate variability on income and food security levels of FHHs and MHHs. The model captures uncertainty in production and consumption decision-making processes, captures causes and outcomes of adaptation processes due to its recursive nature, and assesses trade-offs and synergies among food production, consumption (and hence food security) and environmental impacts resulting from the use of adaptation options. Furthermore, the model captures heterogeneity among households in terms of resource and wealth dynamics, adaptive capacity, production and consumption preference, knowledge and learning ability. Because farm-level costs and returns are explicitly captured, adaptation to climate variability occurs endogenously.

The remainder of the chapter is organized as follows. The next section introduces the conceptual framework developed for evaluating gender-specific roles in adaptation. The following section presents the data sources and the micro-simulation model. The fourth section discusses the findings. The concluding section presents a list of open questions and an outlook on next research steps.

Conceptual framework

In principle, exposure to climate variability should, *ceteris paribus*, have the same effect irrespective of the gender dimension. However, due to differences in initial endowments, climate variability will have differential effects on MHHs and FHHs. In this section, we first show how climate variability may affect productivity, using a conceptual framework developed by Antle and Capalbo (2010). We then show how adaptation strategies in response to climate variability may become gender-biased. Figure 14.1 is a generic representation of how the effectiveness of adaptation options may differ under different weather realizations without taking into account gender dimensions. Y represents expected outcome variables measured to evaluate the impacts of climate variability (in our case, mainly that of expected household income and food security).

A_i, [1 n] represents the different set of adaptation options available to a given household and (C_0 & C_1) are the different weather realizations. (Y, $A_0 C_0$) represents the production function without climate variability. Point *b* represents the corresponding income or food security level of farm households at the level of adaptation (A_0) under no climate variability.[4] With the same level of adaptation (A_0), point *d* then represents the level of income or food security that a household achieves under climate variability (Y,A_0,C_1). The impact of climate variability is represented by the vertical distance (*b* − *d*). In order to reduce the impacts of variability, households may respond by increasing the scale of their adaptation through the use of more credit, off-farm income or adoption of new and improved seed varieties, which is represented by (A_1). Under the new level of adaptation, the level of income or food security achieved by a household is given by point *g* and the corresponding effect of climate variability is given by (*f* − *g*). The vertical difference (*g* − *h*) captures the role of adaptation strategies. In the extreme

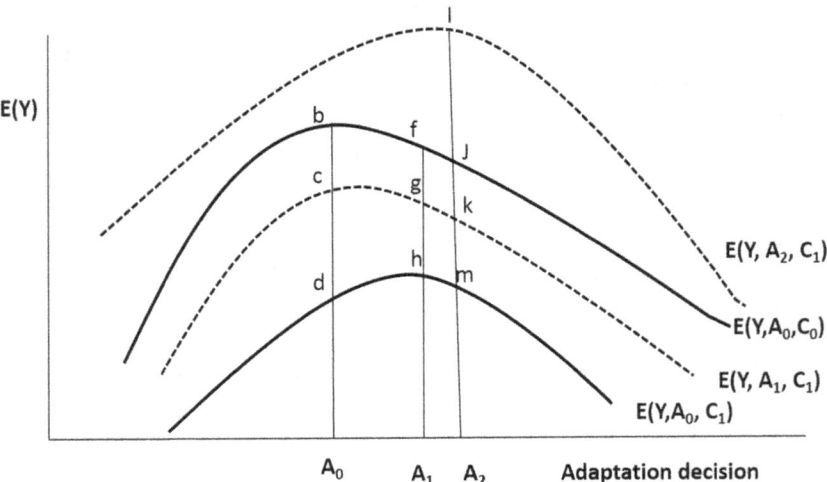

Figure 14.1 Evaluation of the Effectiveness of Adaptation Options

scenario, when the scale of adaptation reaches (A_2), adaptation not only success-fully reduces the impacts of variability but also improves food security and income beyond the initial condition.

However, Figure 14.1 does not take into account gender differences in vulner-ability. As mentioned in the introduction, FHHs may be more vulnerable than MHHs due to the endowment effect. Figure 14.2 further shows how adaptation options may have differential impacts between MHHs and FHHs. As shown in Figure 14.1, adaptation practices through policy interventions can reduce vulner-ability. This leads to the question of what constitutes a successful adaptation strategy. We argue that successful policy interventions aimed at increasing adap-tive capacity should improve the livelihoods of the most disadvantaged and poor groups (irrespective of households being MHHs or FHHs).[5] Adaptation can be successful but still gender-biased. Gender-biased adaptation may produce unin-tended consequences by exacerbating the existing inequality between FHHs and MHHs. We show how successful adaptation might lead, on average, to gender-biased outcomes in the following conceptual framework. In Figure 14.2, Y repre-sents the income level of a given household in the situation of no climate variability, while Y_{mn} and Y_{an} show income levels of MHHs and FHHs, respectively, under climate variability. Y_{ma} and Y_{fn} represent the respective income levels of MHHs and FHHs after adaptation to climate variability has been undertaken. Finally, Y_{aa} represents the average outcome for the whole community (average outcome irrespective of gender). R_d, R_h and R_a refer to the different possible weather realizations (from bad to good).

Because FHHs own fewer assets, they apply less fertilizer, improved seed and other inputs of production.[6] Given that both MHHs and FHHs are exposed to the

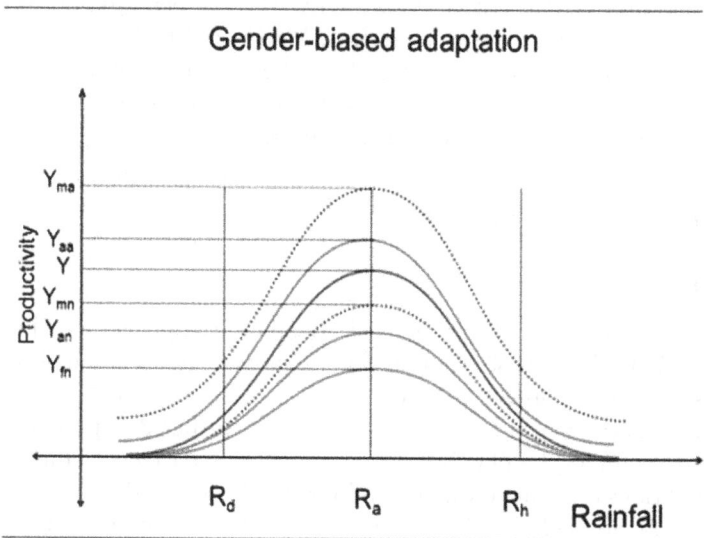

Gender-biased adaptation

Figure 14.2 Example of a Gender-Biased Policy Intervention

same types of climate/weather shocks, we expect MHHs to be less vulnerable than FHHs due to higher use of agricultural inputs. Here, it is clear that FHHs operate at the lower production frontier due to the endowment effects. The difference $(Y_{mn} - Y_{fn})$ is therefore regarded as the endowment effect without climate variability. $(Y - Y_{mn})$ is the average effect of climate variability on MHHs, while $(Y - Y_{fn})$ is the average effect of climate variability on FHHs. The magnitude of the difference between the two then provides the gender-specific effects of climate variability $((Y - Y_{mn}) - (Y - Y_{fn}))$.

Now, let us consider a new adaptation intervention through the promotion of new crop varieties. Such an intervention definitely improves productivity under the same weather exposure level but also requires more investment. As a result, we observe two effects on households' productivity: the initial endowment effect that affects adaptive capacity and the climate variability effect. The endowment effect is always positive – that is, better endowment leads to higher use of inputs and hence higher adaptive capacity. The climate variability effect is, however, negative because it erodes households' ability to adapt. The net effect on household productivity is then the sum of the two effects plus the (positive) new technology effect.[7] Figure 14.2 shows that the new intervention yields a higher outcome level for the community on average $(Y_{aa} - Y_{an})$. However, it has no effect on the income level of FHHs. The average effect is influenced by the higher gains of MHHs $(Y_{ma} - Y_{mn})$. Such an intervention is clearly successful on average but is also gender-biased[8] and leads to higher inequality between MHHs and FHHs. It is unlikely that the objective of a policy intervention is to produce gender-biased outcomes. However, due to the initial levels of inequality, an intervention may yield a higher average outcome but at the expense of higher inequality.

The other important aspect of vulnerability, which is perhaps not well documented, is vulnerability to extreme events. MHHs and FHHs may be equally sensitive to adverse events on average but differ in their vulnerability when extreme events occur. In such a case, adaptation is successful on average but becomes gender-biased when an extreme event occurs.

Data sources and methodology

Data sources

The analysis in this chapter uses the 2009 round of Ethiopian Rural Household Survey (ERHS). This data set contains information about farm household characteristics, crop and livestock production, and food consumption, among other factors, in rural Ethiopia for both MHHs and FHHs. Further, data provided by the National Meteorology Agency (NMA) of Ethiopia is used to specify the meteorological conditions for climate variability. In particular, we used historical records over the last 60 years (1951–2010) and grouped the years into normal, dry, wet, extremely dry and extremely wet categories, using the standardized annual rainfall anomaly index against the 1971–2000 period.[9] The years were grouped into

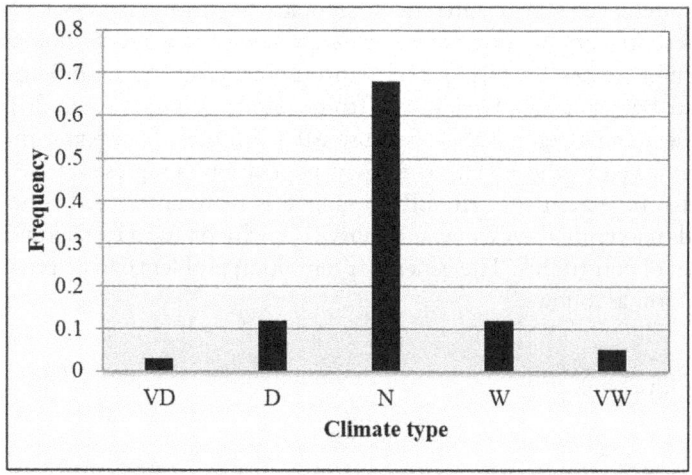

Figure 14.3 Observations and Frequencies of Current Climate Variability

five categories using the Standardized Anomaly Index (SAI) and the distribution of each category is presented in Figure 14.3.

Selling and buying prices on output and input markets were also extracted for each peasant association (PA)[10] from the ERHS and FAO. In the price data set, we found considerable variation of prices across PAs and hence decided to use PA-level prices instead of regional or country average prices. As a result, farm households receive different prices for the same product depending on their geographical location. In general, data quality is sufficient for use in bio-economic modeling but crop-specific labor and fertilizer production functions cannot be estimated from this data source. We therefore used IFPRI's Nile Basin survey (Deressa et al., 2009) as a complementary data source for the estimation of these parameters. Crop data from the Ethiopian Central Statistical Agency (CSA), including yield damage assessments, were used to compute crop yields for very dry, dry, normal, wet and very wet years for each site of the ERHS.

Methodology

We employ the agent-based modeling framework MPMAS, which allows us to simulate farm-level decision making in agricultural systems based on whole-farm mathematical programming (Schreinemachers and Berger, 2011; Schreinemachers et al., 2007; Wossen et al., 2014; Wossen and Berger, 2015). In our MPMAS model, each model agent represents a farm household from the survey (i.e., there is a one-to-one correspondence of agents to their real-world analogues). MPMAS captures the characteristics of each agent household, its demographic composition, land rights, ownership of durable assets and locations within agroecological zones and administrative units based on the ERHS data set. Further, MPMAS captures differences across different households (e.g., MHHs versus FHHs) in terms of resource and wealth dynamics, adaptive capacity, production and consumption preferences, knowledge and learning ability (Wossen and Berger, 2015; Troost and Berger, 2014; Berger and Troost, 2013; Wossen et al., 2014; Schreinemachers and Berger, 2011). Because MPMAS includes every farm household interviewed in the ERHS, the agent population is representative of rural Ethiopia to the extent that the ERHS sample is representative. In the model, households maximize their expected utility (U), which has to be maximized subject to a set of constraints. The general optimization problem can be presented in a generic form as follows:[11]

$$
\begin{cases}
\max U(Z) = \displaystyle\sum_{j=1}^{n} c_j x_j \\[2mm]
\quad s.t. \\[2mm]
\displaystyle\sum_{i=1}^{n} a_i x_i \leq b_i \\[2mm]
\displaystyle\sum_{i=1}^{n} w_i x_i = 0 \\[2mm]
\quad x_j \leq u_j \\[1mm]
\quad x_k \in Z \\[1mm]
\quad x_j \geq 0 \\[1mm]
a,b,c,w,x \in R
\end{cases}
$$

where $U(Z)$ represents the utility that a given agent derives by choosing the optimal combination of crop, livestock and nonfarm activities subject to production and consumption preferences, as well as resource endowment constraints. In the foregoing equation, x_i represents the decision variables (e.g., crop, livestock and nonfarm activities), which can take only non-negative values; c_i is a vector of coefficients of the objective function; a and w are specific constraint coefficients; and b_i captures the resources required to produce one unit of activity x_j. These include resources such as labor, credit, financial capital, land and water. The input requirement a_{ij} of a particular activity x_j can be presented at specific time intervals (monthly, yearly, quarterly or seasonally). For instance, labor requirements are disaggregated on a monthly basis to capture the different growing stages (land preparation, planting, weeding and harvesting). Some activities in the model are subject to upper bounds ($x_j \leq u_j$). For example, households are allowed to take only the maximum allowable credit. As mentioned earlier, the solution to the foregoing maximization problem contains values for x_i, for which $U(Z)$ takes the highest value that can be achieved without violation of specified constraints.

The foregoing maximization problem in MPMAS is implemented in three stages. These include investment, production and consumption decision stages;

Table 14.1 Flow Chart of Household Decision Making in MPMAS

Stage	Investment decision	Production decision	Consumption decision
Timing	Start of the period	Start of the period	End of the period
Yields	Expected	Expected	Actual
Prices	Expected	Expected	Actual
Resource supply	Expected	Expected	Actual

see Table 14.1. Such segmentation of decision making is required to reflect the resource allocation and timing of activities (e.g., liquid assets that a farmer uses for a long-term investment at the start of a cropping season cannot be used in production activities throughout the season). The steps are implemented by recursive solutions of agent mixed integer linear programming (MILP) problems: each decision step involves optimizing a particular MILP and transferring certain parts of the solution vector to the MILP of the next step. Each agent MILP is specified such that, when taking an investment decision, an agent already plans for production and consumption, and, when taking a production decision, an agent plans for consumption. All investment and production decisions are made based on actual resource supply and expected yields and prices. Because production and investments decisions are made based on expected yields and prices, climate and price variability can reduce income due to yield and price prediction errors.

We used the Decision Support System for Agrotechnology Transfer (DSSAT) to estimate the impact of climate variability on crop yields based on weather realizations, as shown in Figure 14.3 (see Jones et al., 2003). These yields are then translated into consumption vulnerability in MPMAS using a parameterized demand system in a three-stage budgeting process (Wossen and Berger, 2015; Wossen et al., 2014; Schreinemachers and Berger, 2011). The budgeting process allocates income into savings and expenditures in the first stage, expenditure into food and nonfood expenditures in the second stage, and finally food expenditure into specific food items, using a parameterized demands system called Almost Ideal Demand System (AIDS). The first stage in the budgeting process allocates income into savings and expenditures using the following simple relationship between total income (Y), savings (S) and total expenditure (TE).

$$Y = S + T \tag{1}$$

For an individual household, savings are specified as a function of income and other household-specific characteristics using the following quadratic specification:

$$S = \alpha_0 + \beta_1 Y + \beta_2 Y^2 + \beta_3 x^{hc} + \sum_{n=1}^{n} \beta_n D + \mu_i \tag{2}$$

where $x^{hc</I>}$ includes a vector of household characteristics such as household size and <I>D is a vector of regional dummies. The next stage uses the following budget share equation to allocate income (after saving) into food and nonfood expenditure:

$$\omega_i = \alpha_0 + \beta_1 \ln(PCE) + \beta_2 x^{hc} + \sum_{n=1}^{n} \beta_n D + \mu_i \qquad (3)$$

where $\omega_{i</I>}$ is the share of food expenditure[12] from the total expenditure and <I>PCE is per capita expenditure. In the final stage of the budgeting process, households allocate their food expenditures to specific food items. At this stage, the food preference of farm households is estimated using the Linear Approximation of the Almost Ideal Demand System (LA-AIDS), which is specified as a function of own price, the price of other goods in the demand system and the real total expenditure on the group of food items, as follows:

$$F_i = \alpha_i + \sum_{j=1}^{j} \gamma_{ij} \ln p_j + \delta_i \left(\frac{x}{\sum_{n=1}^{n} w_n \ln p_n} \right) + \varphi_i x^{hc} + \sum_{n=1}^{n} \beta_n D + \mu_i \qquad (4)$$

where F_i refers to the budget share of food category i, p is a vector of prices, and x refers to the total per capita food expenditure. In MPMAS, the complete household demand system was implemented through piece-wise linear segmentation of the underlying functions according to the size of the expenditure budget. The final income allocation is agent-specific and is defined by the amount of current income and household size and composition of a particular agent. In most cases, households satisfy food requirements through own production and income generating activities. When food production is not enough to satisfy the minimum requirements, households will use other sources of income, such as savings and livestock assets.

Given the foregoing parameterization of production and consumption processes in our model, the relevant question will then be the estimation of welfare outcomes under climate and price variability. Given our objective of examining the vulnerability of households differentiated by headship (i.e., what is the effect of climate variability on household welfare and what would have happened to their welfare without climate variability), a counterfactual analysis would be the obvious choice. A similar counterfactual analysis can also be applied for adaptation to climate variability (what would have happened to the welfare of adopters without adaptation and what actually happened with adaptation). However, constructing a counterfactual for no climate variability is not a trivial matter because the scale of variability differs over time and hence induces behavioral change.[13] The problem in the experimental design is therefore to find a control group which was not exposed to climate variability. In reality, this is impossible as there is no possibility of living in a world without climate variability. We address this problem through the use of a novel simulation approach. In particular, we construct a hypothetical baseline situation without any climate variability based on long-term expected average climate variables. For capturing the effect of climate variability, we then exposed households to random variability based on observed year-to-year variation of weather as obtained from NMA (i.e., for each simulation run, a sequence of specific years was randomly drawn from the climate database, and effects were simulated using the agent-based decision model). As such, this experiment answers the question of what happens in a world of *increased* climate variability without policy intervention. Running the simulation with climate variability but without any form of policy intervention enables us to examine the effect of climate variability on MHHs and FHHs.

Note that the focus of this chapter is to examine the impact of current climate variability differentiated by gender. As such, the no-variability scenario is not a forecast, but instead provides a counterfactual – a reasonable trajectory of income in the absence of climate variability. We choose the baseline as a situation without any climate variability because a lack of an appropriate comparison unit may pose challenges for impact estimation. As a baseline, one can, for example, use current levels of variability as a benchmark. However, without establishing how household income would have evolved without any climate variability, it is impossible to estimate the impact of climate variability on household income. As such, it will not be possible to measure the impact of climate variability by simply assuming an increased percentage relative to current variability; this is because, due to behavioral responses, effects are not additive.

As mentioned in the introduction, one possible intervention in response to climate variability is the introduction of new technologies, which increase agricultural productivity under climate variability. The innovation considered in this study is the promotion of new and improved maize and wheat varieties. In our approach, agents consider adoption of novel adaptation practices only after gaining knowledge and being persuaded by their peer groups (Maertens and Barrett, 2012; Wossen et al., 2013). In order to capture the effect of peer-to-peer communication on an individual's decision to adopt adaptation practices, we implemented a network-threshold model of innovation communication in MPMAS (Berger, 2001; Rogers, 1995). The actual adoption process of adaptation strategies is presented in Figure 14.4. First, the household assesses whether the adoption

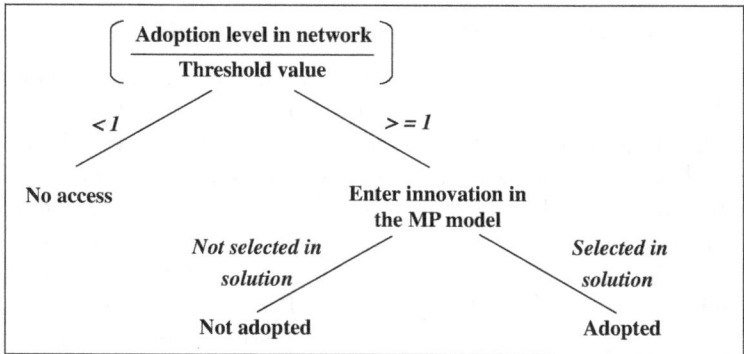

Figure 14.4 Adoption Process as Implemented in MPMAS

level (i.e., exposure) has reached its network threshold. If reached, the second step allows the agent to include the innovation in the decision-making process (through the MILP tableau), allowing an agent to select the innovation if she expects it to be profitable on her specific farm (Berger, 2001). Adoption is subject to various constraints, such as availability of labor, land, cash and other farm assets. Also, the profitability of the innovation is evaluated against that of the cropping options already existing before the farmer had access to the new innovation.

In order to assign network thresholds to households in MPMAS, we use an econometric procedure that reflects the adoption decision-making process. The procedure corresponds to the knowledge and persuasion parts of the adoption process, in which farm households need to reach their social network thresholds before actually considering possible adoption of an innovation. The first key indicator of innovativeness is the time lag between the moment of introduction (technology adopted for the first time) and the individual adoption decision. The shorter the time lag, the higher the innovativeness.[14] However, this ranking based on time lag is incomplete, as many households are associated with identical time lags or unknown time lags. These households were consequently assigned the same rank. We therefore complement the time lag information by the predicted adoption probabilities from a binary adoption model. In particular, we use predicted probabilities of a probit model to assign innovativeness groups conditional on the characteristics of households (e.g., for a household to be in an innovativeness group, it should have a certain amount of land, education level, liquidity, etc., as obtained from the probit model). This procedure leads to an endogenous threshold model of technology diffusion, because innovativeness levels can change over time. Moreover, this approach captures observed differences in socioeconomic characteristics and innovativeness levels of FHHs and MHHs. Finally, we constructed an ideal technical change scenario where all households were given full access to new adaptation practices (new and improved maize and wheat varieties) without incurring any information costs. The result was then compared to the scenario of the network-threshold approach to examine the role of an efficient information delivery system.

Observed gender-specific difference in initial endowments

In order to assess the endowment effect from our data set, we compare MHHs and FHHs in terms of socioeconomic and demographic variables, using ERHS data. According to ERHS data, only 30% of the sample households are FHHs. We hypothesize that differences in endowments in terms of economic and social characteristics can lead to different levels of vulnerability. The descriptive statistics in Table 14.2 show that FHHs have significantly fewer assets compared to MHHs. For instance, MHHs own an average of 1.28 livestock, as measured by tropical livestock units (TLU), higher than FHHs. The difference is statistically significant at the 1% significance level. Similarly, MHHs are more educated than FHHs. This difference is particularly interesting since education is an important

Table 14.2 Comparison of Household Characteristics, by Gender of Household Head

Variable	MHHs	FHHs	Difference
Demographic characteristics			
Household size (family size in numbers)	6.5	4.5	1.95***
Age (age of the household head in years)	54.7	55.6	−0.96
Education (1= household head is literate)	0.65	0.25	0.39***
Assets and resource constraints			
TLU (livestock herd size in tropical livestock units)	3.35	2.07	1.28***
Soil fertility (the level of soil fertility 1=Lem, 2=Lem-Tef, 3=Tef)	1.55	1.68	−0.118***
Land tenure (1= has tenure security)	0.85	0.84	0.018
Access variables			
Access to credit (1= has access to credit)	0.527	0.523	0.037
Access to safety nets	0.15	0.26	−0.105***
Access to extension	0.53	0.38	0.147***
Other variables			
Fertilizer use	0.698	0.565	0.133***
Farm land area	0.41	0.263	0.141***
N	1069	459	

factor for decision making and for the adoption of adaptation mechanisms under climate variability.

In addition, we also found significant differences between MHHs and FHHs in terms of access to safety nets and extension services. According to our data, about 15% of MHHs have access to a production safety net, compared to 26% of FHHs. The difference in safety net access is also significant at the 1% level. As noted earlier, this is additional evidence that FHHs are poorer than MHHs, in that safety net eligibility is based on poverty.

MHHs have better access to extension services compared to FHHs, which is particularly important because extension is an important source of information for the provision of climate information and for acquiring new practices and technologies. We also found that MHHs apply more fertilizer and have larger farm size than FHHs. In terms of credit access, however, we do not find any significant differences between MHHs and FHHs.

Simulation results

In this section, we present the results of our simulation experiment in which we exposed both MHHs and FHHs to similar levels of climate variability. As a reference, we constructed a baseline using constant climate, along with current levels of household characteristics and assets. For measuring vulnerability, we again used the current levels of household characteristics and assets but with variable climate. Because we altered only the level of climate variability, the difference between the two designs will be a result of climate variability. In the previous section, we showed that FHHs own fewer assets and have less access to other services, including social capital. In this section, through the use of our simulation experiment, we intend to show whether such differences are translated into vulnerabilities to climate variability. For simplicity, we divided this section into three main subsections. The first subsection presents the gender-specific effects of climate variability. The second subsection then presents the heterogeneous effects of climate variability. Finally, the third subsection addresses the role of adaptation strategies.

Gender-specific effects of climate variability

In this section, we present gender-specfic impacts of climate variability, focusing on MHHs and FHHs.[15] As shown in Figure 14.5, both MHHs and FHHs are vulnerable to the impacts of climate variability. However, the magnitude of the effect differs. In particular, our result clearly underscores that FHHs are more vulnerable to the impacts of climate variability compared to MHHs, mostly due to the endowment effect. On average, household income in FHHs declined by 12.4% due to climate variability, while income declined by 5.7% in MHHs. Given that we exposed both MHHs and FHHs to the same level of climate shock, the effect is attributed to differences in endowments and adaptive capacity.

Next, we examined whether climate variability has an effect on the income distribution of households. As shown in Table 14.3, climate variability increases

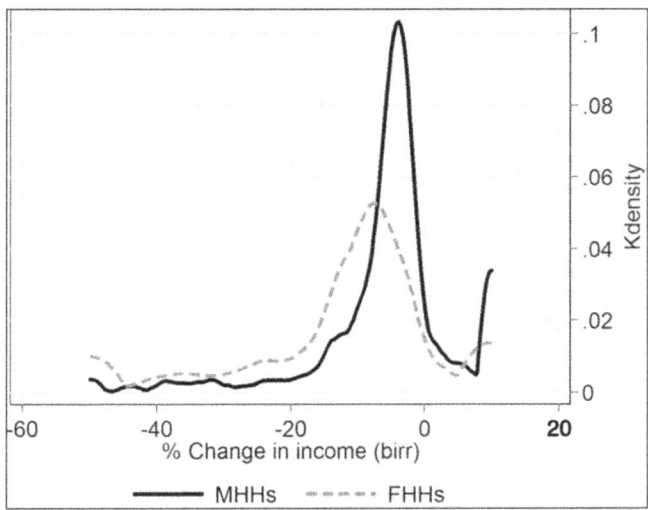

Figure 14.5 Gender-Specific Effects of Climate Variability

Table 14.3 Climate Variability and Income Inequality

Household type	Before climate variability Gini coefficient	After climate variability Gini coefficient
All households	0.47	0.50
MHHs	0.45	0.46
FHHs	0.48	0.54

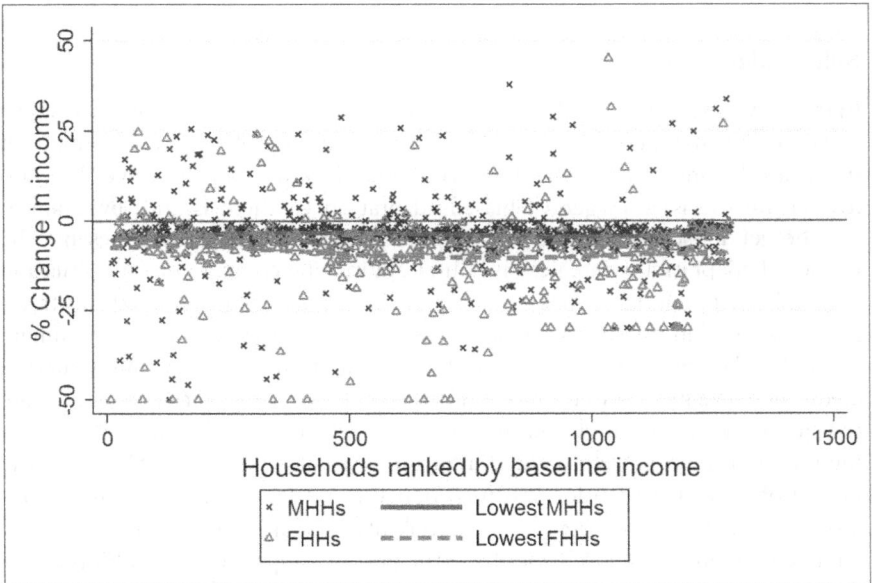

Figure 14.6 Heterogeneous Effects of Climate Variability

overall income inequality, as the Gini coefficient has increased from 0.47 to 0.5 due to climate variability. Because the impact of climate variability is larger among FHHs, the change in income inequality is triggered by FHHs becoming poorer than MHHs as a result of climate variability.

Heterogeneous effects

To further underline the magnitude of the effects, we analyzed the heterogeneous effects of climate variability by considering individual MHHs and FHHs. The result is presented in Figure 14.6. A dot in the scatter plot represents the change of a MHH's or FHH's income under climate variability compared to the income level of the same household under the baseline scenario (without any variability).

The result shows that all households are vulnerable to the impacts of climate variability. However, as shown in the Lowess smoother, the magnitude of the effect is stronger on FHHs compared to MHHs. In addition, the effect of variability is not distributed uniformly across the agent population. To underscore this observed heterogeneity in vulnerability, we examined the different impact pathways of climate variability. We found that FHHs, more so than MHHs, substantially reduce the use of fertilizer and improved seed as a result of climate variability. This should not, however, be interpreted as a behavioral response attributed only to FHHs, as it merely shows vulnerability as a result of lack of adaptive capacity (endowment effects). In particular, we found that FHHs reduce the use of fertilizer by 14.89% while MHHs reduce fertilizer use by 9.2%.

Role of adaptation strategies

In the previous section, we showed that the impact of climate variability is largely negative but also heterogeneous. Here, we discuss to what extent adaptation strategies designed to improve the livelihood of farmers are effective. We also investigate the issue of gender-biased adaptation (if there is any) by examining the gender-specific impacts of adaptation options. Table 14.4 presents the impact of adaptation strategies on household income compared to the situation of no adaptation under climate variability. On average, all the adaptation strategies considered in the simulation are effective in reducing the impacts of climate variability. Policy intervention through the promotion of short-term production credit increases income of MHHs and FHHs by 1.44% and 2% respectively. Note that in our model we implemented a strict repayment rule. As such, the foregoing reported impacts were realized after full repayment of credit. However, the impact of credit intervention was not enough to lift farmers to their initial condition (the condition before climate variability) because the negative impact of climate variability is much larger than the positive impact of credit. Similarly, a 25% fertilizer subsidy has a higher impact than credit but still falls short in compensating the adverse impact of climate variability. The third adaptation strategy considered in the simulation experiment is relaxing information constraints for adoption of improved wheat and maize varieties. In simulating this effect, we relaxed the information constraints that farmers face in accessing information

Table 14.4 Effect of Adaptation Strategies

	FHHs		MHHs	
	Median	Mean	Median	Mean
Climate variability effects (%)	−9.1	−12.4	−4.4	−5.7
Effect of adaptation strategies				
Access to credit (%)	0.51	2	0.31	1.44
Fertilizer subsidy (%)	1.7	3.1	1.5	2.9
Access to information (%)	0.58	3.6	0.29	3.5
All policy packages (%)	3.5	7.4	2.8	7.1

about new technologies (here, we assume ideal technical change in which both FHHs and MHHs access adaptation practices equally and without delays because of imperfect information). As shown in Table 14.4, relaxing information constraints improves income compared to the situation of no adaptation. Moreover, the benefits are slightly higher for FHHs.

The final adaptation strategy, referred as "All policy packages," is a combination of credit access, a 25% fertilizer subsidy and access to improved wheat and maize varieties. Because adoption of new maize and crop varieties is rather expensive, FHHs may not adopt because of the endowment effect. As a result, the difference in the effectiveness of the policy intervention captures both the endowment and the technology-specific effect. Because our objective is mainly to examine the gender dimensions of adaptation (the technology effect), we designed a strategy to offset the gender-specific endowment effect by granting all households credit access irrespective of gender.[16] As such, the values reported in the row "All policy packages" show the maximum possible effect of policy intervention including adoption of improved wheat and maize varieties. The results show that adaptation through a combination of these policy actions offsets the adverse effect of climate variability for MHHs. Impacts on FHHs are also high compared to other adaptation options; the income level of FHHs increased by 7.4%. However, the median effects of all the adaptation strategies considered in this chapter are much lower than the mean effects, suggesting heterogeneity in the effects of adaptation options.

Conclusion

In this chapter, we addressed the question of whether climate variability affects MHHs and FHHs differently and whether adaptation to climate variability exacerbates income inequality and results in gender-biased outcomes. To address the gender-specific effects, we developed a conceptual framework for evaluating climate variability effects and the effectiveness of adaptation strategies. In particular, we stressed that successful policy interventions aimed at increasing adaptive capacity should improve the livelihood of the most disadvantaged and poor groups (irrespective of households being MHHs or FHHs). The results of our descriptive analysis reveal that FHHs own significantly fewer assets, particularly land and livestock. In addition, FHHs are less educated and have less access to extension services than do MHHs. These existing differences in endowments make FHHs more vulnerable to the impacts of climate variability.

The main findings of this study can be summarized as follows. First, both MHHs and FHHs are vulnerable to climate variability. However, the magnitude of the effect differs. In particular, FHHs are more vulnerable to the impacts of climate variability compared to MHHs, mostly due to differences in initial endowments. Second, climate variability not only affects income adversely but also increases income inequality. Third, the effect of climate variability is not distributed uniformly among MHHs and FHHs and between MHHs and FHHs. Fourth, policy interventions through the promotion of new crop varieties, which are adapted to the local climate conditions, yield gender-unbiased outcomes and were largely successful in offsetting the impacts of climate variability. Overall, our analysis suggests that climate variability is a major threat but its impact can be reduced significantly if carefully designed adaptation options are implemented.

Finally, this chapter examined gender-specific impacts of climate variability focusing on MHHs and FHHs. However, the impact of climate variability on women can be much larger because the majority of adult women live in male-headed households, and intra-household allocation decisions may mean that women in male-headed households are also hurt more than men by increased climate variability. As such, considering intra-household decision making while examining the impact of climate variability would be an important future research area. In addition, due to a lack of future climate projections at the level of disaggregation required in this study, we did not consider future climate variability in our simulation experiment. Given the importance of future climate variability, it will be interesting to examine the gender-specific effects of future climate variability.

Notes

1 The focus of previous research has been rather on how gender-related differences (mainly differences in empowerment between men and women in a male-headed household) affect welfare outcomes, instead of examining how vulnerable FHHs are compared to MHHs. In line with this, Alkire et al. (2013) and Sraboni et al. (2014) reported a positive relationship between women's empowerment and productivity and food security outcomes. Moreover, Fafchamps et al. (2009) documented that the relative nutrition of spouses is associated with bargaining power. Wiig (2013) also reported that joint property rights have a strong effect on the decision to make large investments in agriculture. In addition, previous studies captured the gender-specific effects of climate variability, using regression-based approaches where gender effects were captured through a gender dummy. This approach, however, does not take into account the existence of interaction effects between gender and other socioeconomic variables (i.e., each individual socioeconomic variable has the same effect and only the intercept differs between MHHs and FFHs).

2 Examining the gender-specific effects of climate variability is not a trivial matter due to the problem of overcontrolling and endogeneity bias (Dell et al., 2014). In particular, some of the socioeconomic variables that affect intra-household decision-making and bargaining power are also directly affected by climate variability. In this case, controlling for household-specific characteristics can have the effect of partially eliminating the explanatory power of climate even if climate is the underlying fundamental cause (Dell et al., 2014). A second key methodological issue is the endogeneity of female headship status for some types of FHHs.

3 Our approach does not differentiate between the de facto FHHs and de jure FHHs due to lack of data.

4 We assume that climate variability (C1) will have a higher adverse effect than the situation of no variability (C0).

5 In this regard, while considering adaptation options, both economic efficiency and equity objectives should be taken into account.

6 Note that this assumption also can be made for male-headed households depending on the context and hence our assumption does not change the implications of our conceptual framework.

7 The technology effect is positive because the adoption decision is based on profitability.
8 Gender-biased refers to an outcome that exacerbates inequality between men and women. Note that a gender-biased intervention could improve the welfare of FHHs but at the same time result in deterioration of the welfare of MHHs.
9 Normal (N): $-0.5 < SAI < 0.5$; Very Dry (VD): $SAI <= -1.0$; Dry (D): $-1.0 < SAI < -0.5$; Wet (W): $1.0 < SAI < 0.5$; Very Wet (VW): $SAI >= 1.0$.
10 A PA is the lowest administrative unit in Ethiopia.
11 Note that our agent-based model has 8,175 activities, 769 constraints and 133 integers.
12 Household food consumption comprises monetary expenditures on food, quantity of consumption from own harvest, and gifts. The quantity of own consumption was converted into imputed values using PA-level price information for food items.
13 Constructing a counterfactual for climate variability through experimental methods is unfortunately impossible.
14 However, a shorter time lag may not necessarily imply higher innovativeness levels, because differences in economic conditions, farm size and asset endowment may be the major reasons for differences in time lag. Identifying the determinants of the time lag is therefore a key step in order to use it as an indicator of innovativeness levels. To this end, we analyzed the determinants of time lag using a regression model.
15 Note that this study does not consider female members of MHHs.
16 Note that, in principle, credit access will not remove the full effects of the endowment effect. However, because we considered a technology in which the endowment effects operate through the liquidity constraints, controlling for credit will provide a robust comparison unit.

References

Alauddin, M., and A.R. Sarker. 2013. Climate Change and Farm-Level Adaptation Decisions and Strategies in Drought-Prone and Groundwater-Depleted Areas of Bangladesh: An Empirical Investigation. *Ecological Economics* 106: 204–213.

Alkire, S., R. Meinzen-Dick, A. Peterman, A. Quisumbing, G. Seymour and A. Vas. 2013. The Women's Empowerment in Agriculture Index. *World Development* 52: 71–91.

Antle, J., and S. Capalbo. 2010. Adaptation of Agricultural and Food Systems to Climate Change: An Economic and Policy Perspective. *Applied Economic Perspectives and Policy* 32(3): 386–416.

Arndt, C., S. Robinson and D. Willenbockel. 2011. Ethiopia's Growth Prospects in a Changing Climate: A Stochastic General Equilibrium Approach. *Global Environmental Change* 21: 701–710.

Asfaw, A., and A. Admassie. 2004. The Role of Education on the Adoption of Chemical Fertilizer under Different Socioeconomic Environments in Ethiopia. *Agricultural Economics* 30: 215–228.

Berger, T. 2001. Agent-Based Spatial Models Applied to Agriculture: A Simulation Tool for Technology Diffusion, Resource Use Changes and Policy Analysis. *Agricultural Economics* 25(2–3): 245–260.

Berger, T., and C. Troost. 2013. Agent-Based Modelling of Climate Adaptation and Mitigation Options in Agriculture. *Journal of Agricultural Economics* 65: 323–348. doi:10.1111/1477-9552.12045.

Block, P., K. Strzepek and X. Diao. 2008. Impacts of Considering Climate Variability on Investment Decisions in Ethiopia. *Agricultural Economics* 39: 171–181.

Bryan, E., T. Deressa, G. Gbetibouo and C. Ringler. 2009. Adaptation to Climate Change in Ethiopia and South Africa: Options and Constraints. *Environmental Science and Policy* 12(4): 413–426.

Dell, M., F.M. Jones and A.M. Olken. 2014. What Do We Learn from the Weather? The New Climate-Economy Literature. *Journal of Economic Literature* 52(3): 740–798.

De Pinto, A., R. Robertson and B.D. Obiri. 2013. Adoption of Climate Change Mitigation Practices by Risk-Averse Farmers in the Ashanti Region, Ghana. *Ecological Economics* 86: 47–54.

Dercon, S., J. Hoddinott and W. Tassew. 2005. Shocks and Consumption in 15 Ethiopian Villages, 1999–2004. *Journal of African Economies* 14: 559–585.

Deressa, T., M. Hassan and C. Ringler. 2009. Determinants of Farmer's Choice of Adaptation Methods to Climate Change in the Nile Basin of Ethiopia. *Global Environmental Change* 19: 248–255.

Di Falco, S., and M. Veronesi. 2014. Managing Environmental Risk in the Presence of Climate Change: The Role of Adaptation in the Nile Basin of Ethiopia. *Environmental and Resource Economics* 57(4): 553–577.

Di Falco, S., M. Yesuf, G. Köhlin and C. Ringler. 2011. Estimating the Impact of Climate Change on Agriculture in Low-Income Countries: Household-Level Evidence from the Nile Basin, Ethiopia. *Environmental and Resource Economics* 52: 457–478.

Fafchamps, M., B. Kebede and A. Quisumbing. 2009. Intra-Household Welfare in Rural Ethiopia. *Oxford Bulletin of Economics and Statistics* 71(4): 567–599.

Jones, J.W., G. Hoogenboom, C.H. Porter, K.J. Boote, W.D. Batchelor, L.A. Hunt, P.W. Wilkens, U. Singh, A.J. Gijsman and J.T. Ritchie. 2003. DSSAT Cropping System Model. *European Journal of Agronomy* 18: 235–265.

Juana, S., Z. Kahaka and N. Okurut. 2013. Farmers' Perceptions and Adaptations to Climate Change in Sub-Sahara Africa: A Synthesis of Empirical Studies and Implications for Public Policy in African Agriculture. *Journal of Agricultural Science* 5(4): 1–15. http://www.ccsenet.org/journal/index.php/jas/article/viewFile/23791/15889

Kandulu, J., B. Bryan, D. King and J. Connor. 2012. Mitigating Economic Risk from Climate Variability in Rain-Fed Agriculture through Enterprise Mix Diversification. *Ecological Economics* 79: 105–112.

Kilic, T., P.A. Lopez and M. Goldstein. 2014. Caught in a Productivity Trap: A Distributional Perspective on Gender Differences in Malawian Agriculture. *World Development* 70: 416–463. http://dx.doi.org/10.1016/j.worlddev.2014.06.017

Maertens, A., and C. Barrett. 2012. Measuring Social Networks Effects on Agricultural Technology Adoption. *American Journal of Agricultural Economics* 95: 353–359.

Milman, A., and J. Arsano. 2013. Climate Adaptation and Development: Contradictions for Human Security in Gambella, Ethiopia. *Global Environmental Change* 29: 349–359. doi:10.1016/j.gloenvcha.2013.11.017.

National Meteorological Agency. 2007. National Adaptation Program of Action. Available online at https://unfccc.int/resource/docs/napa/eth01.pdf

Robinson, S., D. Willenbockel and K. Strzepek. 2012. A Dynamic General Equilibrium Analysis of Adaptation to Climate Change in Ethiopia. *Review of Development Economics* 16: 489–502.

Rogers, E. 1995. *Diffusion of Innovations*, 4th Edition. New York: The Free Press.

Schreinemachers, P., and T. Berger. 2011. An Agent-Based Simulation Model of Human Environment Interactions in Agricultural Systems. *Environmental Modelling and Software* 26: 845–859.

Schreinemachers, P., T. Berger and J. Aune. 2007. Simulating Soil Fertility and Poverty Dynamics in Uganda: A Bio-Economic Multi-Agent Systems Approach. *Ecological Economics* 64(2): 387–401.

Sraboni, E., H.J. Malapit, A. Quisumbing and A. Ahmed. 2014. Women's Empowerment in Agriculture: What Role for Food Security in Bangladesh? *World Development* 61: 11–52.

Tenge, J.D., and J.P. Hella. 2004. Social and Economic Factors Affecting the Adoption of Soil and Water Conservation in West Usambara Highlands, Tanzania. *Land Degradation and Development* 15(2): 99–114.

Troost, C., and T. Berger. 2014. Dealing with Uncertainty in Agent-Based Simulation: Farm-Level Modelling of Adaptation to Climate Change in Southwest Germany. *American Journal of Agricultural Economics* 97: 833–854. doi:10.1093/ajae/aau076.

Wiig, H. 2013. Joint Titling in Rural Peru: Impact on Women's Participation in Household Decision-Making. *World Development* 52: 104–119.

World Bank. 2013. Levelling the Field: Improving Opportunities for Women Farmers in Africa. Washington, DC: The World Bank.

Wossen, T., and T. Berger. 2015. Climate Variability, Food Security and Poverty: Agent-Based Assessment of Policy Options in Northern Ghana. *Environmental Science and Policy* 47: 95–107.

Wossen, T., T. Berger and S. Di Falco. 2015. Social Capital, Risk Preference and Adoption of Improved Farm Land Management Practices in Ethiopia. *Agricultural Economics* 46: 1–17.

Wossen, T., T. Berger, T. Mequaninte and B. Alamirew. 2013. Social Network Effects on the Adoption of Sustainable Natural Resource Management Practices in Ethiopia. *International Journal of Sustainable Development and World Ecology* 20: 477–483.

Wossen, T., T. Berger, N. Swamikannu and T. Ramilan. 2014. Climate Variability, Consumption Risk and Poverty in Semi-Arid Northern Ghana: Adaptation Options for Poor Farm Households. *Environmental Development* 12: 2–15.

Part V

The broader development context

15 The land certification program in Ethiopia

A review of achievements, constraints and opportunities

Mintewab Bezabih Ayele and Hailu Elias

Introduction

Given its heavy dependence on agriculture, the performance of the Ethiopian economy is highly intertwined with access to and productivity of agricultural land. Indeed, there is a wealth of evidence indicating that land use and land tenure systems are at the core of the well-being of rural households living on and dependent on land (Deininger, Ali, Holden et al., 2008; Holden and Yohannes, 2002). Further, land tenure systems that encompass land ownership and transfers have been central features of governance throughout Ethiopia's long history of changing political systems and leadership.

Much like the case of Ethiopia, the design and characteristics of land tenure systems in the rest of the developing world are critical to the functioning of the economy as a whole. In recognition of this, a number of international organizations have put the issue of land tenure high on their international policy agenda. Examples include the Commission for Legal Empowerment of the Poor, World Bank, and Global Land Tool Network (GLTN); see Augustinus, 2005; World Bank, 2006). Similarly, the issue of land tenure systems has been a national policy priority in many countries (Mekonnen et al., 2013; Deininger, Ali and Yamano, 2008) and a number of reforms and laws have been passed to that effect, particularly since the 1990s (Deininger et al., 2011). However, the programs have been plagued by both design and implementation issues (Deininger, Ali and Yamano, 2008). Indeed, the degree to which these reforms have brought about the desired economic effects remains debatable. Hence, it is important to rethink relatively rigid approaches and replace them with fresh and innovative methods that take into account the shortcomings of the previous programs (Deininger, Ali, Holden et al., 2008).

The Ethiopian land certification program is widely viewed as one such example in which there has been a move toward new approaches, in terms of both implementation and learning from past mistakes. In particular, the program is a low-cost, bottom-up program, which ascribes farmers written user rights to demarcated pieces of land, and puts a greater emphasis on gender equity. These features are considered as significant departures from traditional land titling interventions (Deininger et al., 2011).

Previous studies on the efficiency and equity impacts of traditional land reforms have yielded decidedly mixed results. For instance, Deininger, Ali, Holden et al. (2008) argued that enhanced tenure security associated with reforms could lead to increased incentives for land-related investment, enhanced gender equality and bargaining power by women, improved governance, reduced potential for conflict, and lower transaction costs for productivity-enhancing land transfers (either rental or sale). In line with this, Holden, Xu et al. (2009) showed that forest tenure reforms in China, which instituted written documentation of forest land rights (in the form of forest land certificates for a specific time period), enhanced tenure security beyond the effect of other perceived use rights to land. Similarly, in their study of the impact of low-cost land certification in the Amhara region of Ethiopia, Deininger et al. (2011) found that the land reform increased soil conservation investment and participation in the land rental market. However, empirical evidence of the impact of land reform does not always ensure that land reforms live up to intended objectives. For instance, land titling in Kenya brought little increase in land market activity (Place and Migot-Adholla, 1998); Jacoby and Minten (2006) found the same results in a similar study in Madagascar. In some instances, land titling programs benefit the wealthy and powerful at the expense of the poor and marginalized (Besley and Burgess, 2000; Cotula et al., 2004; Deininger et al., 2003). Even where legislation does strengthen women's property rights, lack of legal knowledge and weak implementation may act as impediments to the realization of the full benefits of the reform (Deininger, Ali and Yamano, 2008).

Given the mixed results observed in previous studies, there is a need to scientifically investigate the contribution of new land reforms on their own merits; hence, the study of the impacts of the land certification program. The program has been undertaken in the four major regions of Ethiopia. Due to the specific features discussed earlier, the program has attracted a huge interest in evaluating the different aspects of its impact on the welfare of the rural poor.

The purpose of this chapter is to look critically into studies that have examined the program with respect to several economic and behavioral variables, with case studies from the four regions in Ethiopia. The findings generally show that the program has addressed the gaps in tenure security and has increased women's participation in the land rental market, as well as productivity and investment in land. However, for the reduction in tenure security to have permanent impacts, sustained efforts need to be carried out in the realm of public policy. We further argue that there are important lessons to be learned from the program for duplication in other countries.

The study is organized as follows. The next section describes the land tenure system and the certification program in the different parts of Ethiopia. The following section is an overview of theoretical and empirical findings on tenure security. The fourth section presents a brief synthesis of the findings of different certification studies. The fifth section discusses areas for further research. The concluding section notes caveats related to policy making. Issues of methodology are discussed in the appendix.

The land tenure system in Ethiopia and the land certification program

General history of the land tenure system in Ethiopia

Ethiopia's contemporary land tenure system is largely shaped by the land tenure reform in 1975, characterized by expropriation of all land for state ownership and the "land to the tiller" policy. Farmers were granted usufruct rights that were egalitarian in three major ways: (1) all resident households in the community had the constitutional right to land; (2) the allocation depended on household size; and (3) each household was assigned a fair share of each major land quality class in the community. Size and quality were assured through sampling each major land quality class (Kebede, 2002, 2008). The farmers' membership in the peasant associations made them claimants, endowed with rights such as access, some management rights and limited exclusion rights (Crewett et al., 2008). Land sales, mortgaging and rentals were illegal and so was the hiring of labour. As a result, the only viable means of access to land for newly formed households was land redistribution, which maintained the egalitarian features of the system by taking land from the most land-rich households and giving it to the new households when other communal land was no longer available. Frequent land redistribution and the ban on land market activities – the two main features of the system – have long been blamed for creating a zero-sum game of enhanced tenure insecurity with possible negative investment and productivity effects (Deininger and Jin, 2006; Holden and Yohannes, 2002).

The EPRDF[1]-led government that took power in 1991 followed suit in keeping all rural and urban land under public (government) ownership, with two notable changes: formal confirmation that land rights are to be granted to men and women and the right to lease out land, albeit with restrictions on the period of the lease and the share of land to be leased out (Crewett et al., 2008).

The formalization of women's right to land could be considered a major milestone in the history of land reforms in Ethiopia, as rural women in Ethiopia have historically held an inferior position in relation to men in terms of property rights.[2] Prior to 1975, Ethiopia's long feudalistic system of land tenure rarely recognized independent land ownership by women, except through marriage and inheritance. Per the 1975 legislation, spouses enjoyed joint ownership of the land, implying that, on paper, men and women were entitled to the same land rights. However, women's rights to land depended on marriage and were, in most cases, not registered separately (Crewett et al., 2008). The program's aims to explicitly incorporate women's participation can be considered an additional stride in strengthening women's land rights (Bezabih et al., 2016).

Description of the certification program

The land certification program, considered one of the major milestones in the history of Ethiopian land reforms, was implemented in four of the major regions in the country, where more than 20 million plots and 6 million households had received land certificates by 2006. The program essentially provided perpetual user rights to the land (Holden and Tefera, 2008). The launch and implementation of the program in the different regions were backed by land proclamations to the respective regions and a federal land administration and utilization proclamation. The program was first implemented in the Tigray region, and the region was also the first to have its own regional land proclamation (TRLAUP, 1997). This was followed by the Amhara region in 2003, and the accompanying land administration, utilization and proclamation (ARNLAB, 2003). The Oromia and Southern Nations Nationalities and Peoples Regions (SNNPR) began these processes in 2004 with the accompanying regional proclamations (Oromiya Region, 2007; SNNPR, 2007).

The program is considered one of the largest land registration programs in the world, comparable to the 11 million certificates awarded in Vietnam in 1993–2000 and the 8.7 million titles issued in Thailand in 1980–2005. The program also cost much less than similar programs, considering that it has been the largest land registration effort in the last decade in Africa and possibly in the world (Deininger, Ali and Yamano, 2008).

The certification is initiated by a team of experts from the *woreda* (district)[3] who are responsible for overseeing a series of activities. The first step in the process is an awareness meeting between the woreda and *kebele* (village) administration and farmers, about the purpose and organization of land registration and certification, followed by the election of Land Administration and Use Committees (LACs), and provision of training for the elected LAC members (Palm, 2010).

The LAC then assumes responsibility for systematic field-based adjudication of rights. This second step starts with the identification of individual household plots and plot borders, which involves the neighbouring farm households walking the farm fields, together with LAC members (Adnew and Abdi, 2005). A form consisting of the plot location (by name), plot size (using local measurement methods and units) and land quality class is filled out for each plot. The information is registered in land registry books, where each household has a number. The adjudication process culminates with the production of preliminary registration certificates that identify size and neighbours for each of a holder's plots; these are displayed in public. There is an attempt to resolve conflicts by discussion among neighbours and help from elders. Any outstanding conflict is passed to the courts and the result of the land adjudication is presented to the public for a month-long verification in order to allow for corrections. After a period for raising complaints, the third step of entering the recordings into registry books commences. Each household is then provided a certificate which contains information for each of their farm plots, names (and in many cases pictures of the household head and spouse) and space for maps and spatial information expected to be added in a second stage. Copies of the certificates are kept at the kebele and woreda level. The woreda has responsibility for approving the legal status of the holding (Olsson and Magnérus, 2007; Palm, 2010).

Theoretical relationships and major empirical findings

In this section, we discuss how tenure security, the major parameter directly related to certification, affects various economic and social variables of interest. The major socioeconomic variables believed to be impacted by the program (which are covered by the studies reviewed in this report) include tenure security and land-related conflict; land-related investment and tree planting; land productivity; gender and land market participation; off-farm employment; wealth-related measures; and overall (subjective) welfare and levels of trust.

Land certification, tenure insecurity and land conflict

The certification program, tenure security and conflict resolution

One of the most direct outcomes of the program is an increased sense of secure ownership and reduction in conflict that was prevalent pre-certification (with lack of properly registered and mutually agreed boundaries on adjacent plots). In line with this, Deininger et al. (2011) find that certification significantly reduced fear of land loss, although it failed to eliminate a sense of insecurity about land tenure.[4] Similar patterns of increased perceptions of tenure security were registered in southern Ethiopia for women and men. Particularly for women, this increased sense of security is reported to have been the result of having their names and pictures appearing on the land certificates, which they believed would strengthen their position in case of divorce or death of their husbands. There was, however, evidence of concern by the later wives of polygamous households about how much land they would keep upon divorce (Holden and Tefera, 2008).

The role of the program in conflict reduction has been investigated through interviews with local conflict mediators and households, to determine their perceptions; see Table 15.1. Holden and Tefera (2008) showed that the program has helped reduce the number of border disputes and inheritance disputes, particularly with better plot demarcation with neighbours as witnesses. Local conflict

Table 15.1 Summary of Certification Studies Related to Tenure Security and Land Conflict

Empirical study	Data source	Variable of interest affected by certification
Deininger et al. (2011)	Amhara	Reduction in sense of insecurity and fear of land loss
Holden et al. (2011)	Tigray	Changes in the frequency of land border disputes
Holden and Tefera (2008)	SNNP	– Sense of security differentiated by gender and by monogamous vs polygamous households – Local conflict mediation, effectiveness of courts

mediators also considered it beneficial to have joint certification and to have the names and pictures of wives on the land certificates. However, there were widespread reports of suspicion over fair judgments and the perception that the courts benefited the wealthy and influential (Holden and Tefera, 2008). Similarly, Holden et al. (2011) assess changes in the frequency of land border disputes as a result of the program. Their findings show that implementation aspects of the program, such as the quality of land demarcation and measurement and involvement of local elders, were strong determinants of the certification outcome.

The certification program, tenure security and soil conservation investment

A number of studies show that investment in land improvement can be enhanced by increased tenure security (e.g., Besley, 1995). Conversely, incomplete, inconsistent or non-enforced property rights can be characterized as the causes of many environmental problems (Bromley and Cernea, 1989). These studies elsewhere in the world have empirical support from studies on Ethiopia; see Table 15.2. In line with this, perceived security by farmers or length of time the farmer has worked on the land is shown to have positive impacts on soil and water conservation investments (e.g., Gebremedhin and Swinton, 2003; Deininger and Jin, 2006). Similarly, studies that have looked at tree growing indicate that tree-planting decisions hinge on private tenure security (Holden and Yohannes, 2002; Deininger and Jin, 2006; Mekonnen, 2009).

In a pioneering study assessing the impact of the certification program in Ethiopia, Holden, Deininger et al. (2009) show that stone terrace activities are enhanced by certification, while the certificate variables were never significant for soil bunds, with existing public conservation programs essentially functioning as complements. Deininger et al. (2011) find stronger results, showing that the propensity to invest in soil and water conservation measures increases by between 20 and 30 percentage points and the number of hours spent on such activities increases significantly with program participation. Using household plot-level data in the Amhara and Tigray regions, Mekonnen et al. (2013) investigate the impacts of land certification on the number of trees grown. They find a significant effect of the program on tree-growing behaviour. However, public investment appeared to have hampered tree growing on private plots; this may be related to

Table 15.2 Summary of Certification Studies Related to Soil Conservation Investment

Empirical study	Data source	Variable of interest affected by certification
Holden et al. (2009b)	Tigray	Soil bund and terrace structure Tree planting in general, eucalyptus
Mekonnen et al. (2013)	Tigray, Amhara	Tree planting on private land
Deininger et al. (2011)	Amhara	Soil and stone bund (both modern and traditional), furrows, other soil conservation activities

the restrictions on tree planting on arable land, especially for eucalyptus. The tree-planting activities are found to be confined to non-fertile plots, the reason being that tree production is much more profitable than crop production on very marginal arable land, where production of annual crops is likely to be less sustainable than growing of trees. Allowing more tree planting on private land could also provide an alternative coping strategy in times of shock and distress. This finding is also supported by Holden, Deininger et al. (2009), who find that, in the Tigray region, the program stimulated tree planting, including planting of eucalyptus, even with restrictions on tree planting on arable land.

Tenure security and land productivity

Previous studies have shown that tenure insecurity has major implications for agricultural productivity via several channels; see Table 15.3. First, by ensuring investors' ability to benefit from long-term investments, tenure security enhances long-term agricultural investments, such as the planting of multi-year (perennial) crops and the adoption of irrigation technologies (Dercon and Ayalew; Do and Iyer, 2008). Tenure security also fosters land-related investments in soil and water conservation, fertilizer and other agricultural inputs, which are expected to increase agricultural productivity (Holden et al., 2009b; Deininger et al., 2011). Second, well-functioning land markets foster allocative efficiency in the agricultural sector by transferring land to more productive but land-poor households, a concept commonly referred to as the factor equalization role of tenure security.

Holden, Deininger et al. (2009) measure the productivity increase from certification to be as high as 45%. They examine the productivity impacts with a series of parametric regressions with alternative specifications, including with and without plots prior to certification (1998 plots), with and without time-variant household variables, and with and without the public investment in conservation variables. In the parametric estimations, they found no significant productivity effects from conservation investment variables, which they attributed to possible correlation between conservation structures and plot characteristics. To remedy this correlation, non-parametric methods that match observations based on observable characteristics were used to measure the impacts of investments in stone terraces on land productivity. Their findings showed a significant and positive effect of such investments on land productivity, implying that the investment impacts may partially explain the productivity

Table 15.3 Summary of Certification Studies Related to Productivity

Empirical study	Data source	Variable of interest affected by certification
Holden et al. (2009b)	Tigray	Plot-level productivity
Deininger et al. (2011)	Amhara	Plot-level productivity
Bezabih et al. (2016)	Amhara	Gender differentials in productivity

impacts. Deininger et al. (2011) find a large magnitude of increases in the propensity to invest as a result of certification. Further, production was significantly responsive to the presence of a functioning conservation structure. Similarly, Bezabih et al. (2016) analyze the productivity impacts of the program using plot-level panel data from the East Gojjam and South Wollo Zones in the Amhara region of Ethiopia. The general productivity impacts are positive and significant. The program is also shown to have a strong gender component; there are different impacts on male and female productivity, with female-headed households gaining significantly more. In sum, it can be argued that improved tenure security, in terms of formal land registration, has improved the productivity of farmers in general and that of female farmers in particular.

This possibility of increased productivity advantages for women is associated with the fact that female-headed households, in the absence of land certification, have a lower level of tenure security than male-headed households. Given that the land certification program improves tenure security for all certificate holders equally, the program would then allow for higher overall tenure security for female-headed households. Furthermore, as enhanced tenure security generally leads to increased productivity, the land certification program is likely to enhance farm-level productivity. It therefore follows that the larger the increase in tenure security enjoyed by female-headed households due to the certification program, the greater the expected increase in the productivity of female-owned farms.

The certification program, tenure security and benefits from land market participation

Secure tenure in the context of land markets may facilitate the propensity to rent out land, thereby enhancing the allocative efficiency of land rental markets (Besley, 1995; Otsuka, 2007). The essence of allocative efficiency of land rental markets in enhancing productivity and reducing poverty has been examined by a number of studies (Bliss and Stern, 1982; Bell and Sussangkarn, 1988; Skoufias, 1995). Indeed, there is strong evidence suggesting that tenure insecurity is a major constraint in land rental market participation (Deininger, Ali and Alemu, 2008; Ghebru and Holden, 2008; Lunduka et al., 2008; Holden and Bezabih, 2008; Holden et al., 2011; Rozelle et al., 2002).

To the extent that land legislation programs bring about increased tenure security and, thereby, an improvement in the functioning of the land rental market, gains in efficiency are likely to follow (Teklu and Lemi, 2004; Deininger and Jin, 2006; Deininger and Jin, 2006; Holden et al., 2011). Despite such perceived importance in allocative efficiency, however, the findings on the impact of land titling on the allocative efficiency of land sales and land rental markets remain mixed (e.g., Place and Migot-Adholla, 1998; Jacoby and Minten, 2006).

Land market participation also has particular significance for the welfare of rural women, particularly in Ethiopia. Due to the taboo against women undertaking major farming activities (e.g., plowing with oxen), female-headed households are heavily reliant on the land rental market for production (Gebreselassie, 2005;

Table 15.4 Summary of Certification Studies Related to Land Market Participation

Empirical study	Data source	Variable of interest affected by certification
Holden et al. (2011)	Tigray	Gender differentials in land market participation
Deininger et al. (2011)	Amhara	Land market participation: static and dynamic
Bezabih et al. (2016)	Amhara	Land market participation: general and by gender
Holden and Tefera (2008)	SNNP	Gender-focused land market participation, particularly in polygamous households

Bezabih et al., 2016). However, due to the low degree of tenure security, land market participation by female-headed households is likely to be lower than what is optimal, given their labour needs. Table 15.4 summarizes recent research on gender and other dynamics of land markets.

Given that the current tenure system allows for no land sales, the program's objective of enhancing tenure security could be primarily reflected in its aiding land market activities. To the extent that certification-induced rental market effects allow productive use of plots that had been left uncultivated, or allow greater freedom in the choice of transaction partner (so that land can be transferred to those with greater farming ability), such effects could enhance the productivity of land use (Deininger et al., 2011). In one of the earlier works assessing the allocative efficiency of the land rental market following the certification program in Tigray, Holden et al. (2011) find that land certification initially enhanced land rental market participation of (potential) tenant and landlord households, especially those that are headed by women. In addition, Holden et al. (2011) has shown reduction in transaction costs in the land rental market by making poor female (potential) landlord households more willing to rent out their land. It has, therefore, also become easier for (potential) tenants to access land to rent. However, Holden and Tefera (2008) argue that the formal reporting requirements associated with land renting, including the consent of the whole family, may have reduced the amount of land market activities since the land reform. In addition, Deininger et al. (2011) estimated that the amount of land rented out increased by about one-tenth of a hectare at the mean, and the propensity to rent out increased by 13 percentage points. In addition, participation decisions and the amount of land transacted on both sides of the rental market are found to be strongly state-dependent, indicating that policy interventions affecting market participation at any given point in time will affect households' long-term trajectories.

The program's impact on the land rental market and particularly on women's participation in the land rental market was also examined using data from the Amhara region. Bezabih et al. (2016) argue that the benefits for female-headed households come about because the increased tenure security enables female-headed households to enter into longer-term contracts or to select more productive tenants beyond their immediate social network. As virtually all land is rented under sharecropping contracts, any productivity effects would translate directly into improved welfare for female-headed households.

The certification program, tenure security and benefits
from off-farm employment

Other parameters of interest that are direct outcomes of the program and that are not related to productivity/investment include off-farm employment, consumption, welfare and overall levels of wealth.

In settings where land is state-owned property, farmers' ability to temporarily transfer land constitutes an important ingredient for enabling a structural transformation of the rural economy. Most importantly, temporary land transfers enable relatively unproductive but land-rich farmers to lease out their land to more productive farmers while the owners engage in off-farm activities (Do and Iyer, 2008). This engagement, in turn, has the potential to relax farm households' credit constraints on the purchase of agricultural inputs and adoption of agricultural technologies (Lanjouw and Lanjouw, 2001; Ellis, 2000; Barrett et al., 2001). However, in the absence of private ownership of land, any income diversification effort of households that involves off-farm employment or migration signals excess holding, thus implying a high cost because the land holding may be subject to redistribution. Rupelle et al. (2009), for instance, found that the temporary nature of China's rural migration is associated with tenure insecurity for fear of land redistribution following permanent migration to cities. Studies that have assessed the role of the program on off-farm employment in Ethiopia are based on these premises; see Table 15.5.

In line with this, the analysis by Bezabih et al. (2013) focuses on household-level off-farm employment decisions and activity choices, using panel logit estimation. The results show that the coefficient of certification is positive and significant at the 1% level, indicating that the program has considerable impact on off-farm participation. The findings also show a strong effect of certification on activity choices. In addition, there is a negative impact of farm size on off-farm employment, given certification. This could be for two reasons. On the one hand, certification and the resulting increased tenure security could lead to increased confidence to leave the farm and engage in off-farm activities. On the other hand, given the persistent insecurity regarding ownership of farmland, certification could increase own labour investment on the farm, thus reducing the supply of labour to off-farm activities.

Turning to measures of wealth, using data from the Tigray region, Holden and Ghebru (2013) show that real household consumption expenditure per adult equivalent has increased with the duration of land certificate ownership, most significantly for female owners of land certificates. However, Holden and Tefera

Table 15.5 Summary of Certification Studies Related to Behavioural Parameters

Empirical study	*Data source*	*Variable of interest affected by certification*
Bezabih et al. (2011)	Amhara	Trust
Akay et al. (2013)	Amhara	Happiness, subjective well-being

(2008) find a very limited impact of the program on women's ability to influence farm management and hence on their wealth generation ability.

In terms of the effect on poverty, Holden and Tefera (2008) showed that there is no evidence of anti-poor aspects in the implementation of the land registration and certification because poorer households have had the same probability of receiving land certificates as less poor households. In particular, both certification programs and the accompanying federal and regional proclamations are geared toward strengthening the land rights of women, who are among the poorest. In addition, there is an attempt to extend egalitarianism into land market participation; for instance, the consent of the family is required before the head of the household can rent out land, and priority in inheritance should be given to family members depending on the land for livelihood.

Tenure security and behavioural indicators

Behavioural parameters that have been largely overlooked in classical economic theory are increasingly becoming the focus of economists' attention as keys to explaining economic behaviour. Issues related to trust and happiness have been shown to be linked to both tenure security and healthy economic performance; see Table 15.6. In line with this, Putnam (2000) and Teraji (2008) argue that, because almost all economic transactions have embedded within them a certain level of trust, the outcomes of efficiency-enhancing market interactions rely on the level of trust, while a lack of trust has negative economic consequences. The issue of property rights has come to the forefront as one of the determinants of trust (Fehr et al., 1997; Cox et al., 2009). Similarly, the last decade has witnessed increasing interest in the scientific study of happiness or subjective well-being (SWB). SWB is considered to be a sufficient measure for welfare and is used to evaluate the impact of various life domains and events (Easterlin, 1995; Frey and Stutzer, 2002; Clark et al., 2009).

Bezabih et al. (2011) assess the impact of a land certification program in Ethiopia on the level of interpersonal and institutional trust among households. Data used for this study is obtained from the Amhara region. The findings show that the acquisition of a land certificate is associated with a 14.2% increase in the likelihood of perceiving the regional government to be completely trustworthy, and a 16.4% increase in the probability of perceiving the federal government as completely trustworthy. For the police authority and the Swedish International Development Agency (Sida), the corresponding percentage increase in perceived degree of trust is 3.9% and 4.9%, respectively. Taken together, these results may point to the effect of increased interaction with authorities and thus better knowledge of whether to trust them.

Overall, trust is found to be enhanced by the certification program, with trust toward formal institutions being more responsive to the program than interpersonal trust. The greater responsiveness of institutional trust to exogenous policy measures, as opposed to interpersonal trust, implies that interpersonal trust takes time to form. Hence, policy measures that aim toward building stronger interpersonal trust should focus on more long-term measures.

Synthesis and evaluation of empirical findings

The essence of this chapter is to examine the achievements of the certification program in light of whether it has hit its intended objectives of increasing tenure security, and in a broader sense, whether it has contributed to improving the welfare of Ethiopia's smallholder farmers. In addition, as a pioneer case in low-cost land certification intervention, the Ethiopian case provides lessons to other poor countries characterized by high land pressure, tenure insecurity, severe rural poverty and land degradation (Deininger, Ali, Holden et al., 2008).

Given that the main objective of the land certification program in Ethiopia is to reduce the inherent insecurity of land holdings associated with state ownership of land, the studies that evaluated the impact of the program assessed features that are likely affected by tenure security. Hence, if efficient, the Ethiopian Land Certification program has the potential to ease constraints related to tenure security, such as limited land market participation, and to promote better land management and investment, reduce constraints on productivity associated with female land ownership, and encourage better production decisions due to a more secure sense of ownership.

The initiation of the program is coincidental with the Federal Rural Land Administration Proclamation 1997, revised in 2005, and has been implemented in the four most populous regions of the country: Tigray, Amhara, Oromiya, and the Southern Nations and Nationalities (SNNP) (Adnew and Abdi, 2005). Using community and household-level data collected recently from these four major regions, several empirical analyses have been conducted to assess the program's effectiveness against various social and economic parameters.

As would be expected, a larger share of the studies comes from the Tigray region, where the program was implemented first, followed by Amhara, SNNP and Oromiya. A pioneering study is that of Holden, Deininger et al. (2009), who show that the certification program enhanced the maintenance of soil conservation structures, investment in trees, and land productivity, using data from the Tigray region. Other studies based on data from the Tigray region include Holden et al. (2011), who investigate the impact of the program on households' participation in the land rental market and show that the program significantly enhanced women's participation in the land rental market. Furthermore, Holden et al. (2010) find a significant reduction in land border disputes. Using data from the Amhara region of Ethiopia, Deininger et al. (2011) find the impacts of the program to be significant for tenure security, investment decisions and land productivity. Similarly, Deininger et al. (2011) find a significant enhancement of land rental market activity. In line with the program's objective to be gender-sensitive, Bezabih et al. (2016) explicitly look into the program's impacts on productivity, differentiated by gender. Mekonnen et al. (2013) use household and plot-level panel data collected from the Amhara region of Ethiopia to assess the tree-planting impacts of the program and find significant positive impacts despite policy constraints. More behaviour-focused studies on the impacts of land certification in Ethiopia have identified significant positive trust and happiness effects (Bezabih et al., 2013; Akay et al., 2013; Bezabih et al., 2011).

In sum, the program can be said to have enhanced tenure security for households/villages participating in the program. Looking at particularities, certification was found to have had positive effects on productivity and appears to boost investment in soil conservation and tree planting. The program has been particularly important for female-headed households in terms of enabling them to achieve a relatively higher productivity gain from the intervention and an increased tendency to participate in the land rental market, showing the potential of such reforms in reaching rural women. The major policy implication of the study is a confirmation that effective land reforms do improve welfare (in terms of productivity) of rural households in general and female-headed households in particular. The studies also make several methodological contributions in line with identifying program impacts in the face of potential endogeneity.

Gaps in previous studies

The program can be seen as an important example of attempts to formalize land rights at low cost while also addressing land market participation and gender issues, largely left out in previous similar reforms. However, there are several caveats identified in the studies, regarding the nature of the programs and their implementation, as well as the analytical and data limitations of the studies.

Regarding analytical and data issues, one severe limitation of the analyses is the complete focus on measuring the post-certification changes in the variables of interest, potentially neglecting pre-existing factors that could shape post-certification outcomes. For instance, there could be potential differences in the pre-certificate level of insecurity among female-headed households, depending on whether they acquired the land independently from the government (through redistribution) or through inheritance from family or husband. The degree and source of insecurity, as well as their impact on land market participation, are likely to be different in the two cases. This heterogeneity is interesting from both academic and policy perspectives, and is worth investigating as a separate future research question. Future studies that relate imperfections in the process of land reforms with intended economic outcomes can disentangle constraints associated with preconditioning factors from those associated with the design and implementation of land reforms.

Second, almost all the studies focused on accurate quantification of the impacts of the program in terms of gains associated with a particular variable of interest. While these quantitative findings can be understood within the specific context of each study, comparison of the studies in terms of their quantified findings could be misleading. This is for two reasons: the studies are undertaken by several individual scientists using different methods of study, implying potential methodological discrepancies and the associated differences in measurement errors. In addition, given differences in data coverage (spatially and temporally), differences in program implementation strategies across regions, and unobserved non-program related differences, the resulting impacts of the program may not be comparable across studies, at least in terms of magnitude.

Third, a full cost-benefit analysis of the program that enables comparison with other land reform programs, particularly classical programs, is also needed. A comprehensive analysis of the costs of the program in terms of time, knowledge and resources required to implement the program, vis-à-vis aggregate quantification of the welfare gains, is required to gain a full picture of the effects of the program.

Fourth, given that the program is touted for being unique in its genuinely democratic and grassroots participation, some caveats need to be addressed. In particular, in-depth investigation of the nature and process of participation is needed. For such a program to be truly a success, its implementation needs to be inclusive of the participants at a grassroots level. However, given the authoritarian nature of many of these settings, questions may arise as to what extent the formation of the Land Administration Committee is indeed truly inclusive, particularly of vulnerable household groups, such as the poor and female-headed households.

Potential policy caveats

In terms of program design and implementation, one of the caveats of such a rapid approach is that demarcation of borders may not be scientific and conflicts may arise. In addition, while important strides have been made in highlighting the land ownership status of women, given the sociocultural constraints, it is worth investigating with longer panel data the extent to which women are able to exercise their full land rights and gain access to needed legal support, as well as how often disputes end with positive outcomes for women a few years into certification (see Holden and Tefera, 2008).

Another caveat is that some rigid aspects of the program's specifications need to be relaxed. For instance, a system of voluntary registration of land rental contracts may be better than the compulsory registration of all land rental contracts that the law now attempts to impose. According to the new regional and federal land laws, consent of the family is required for land to be rented out and land rental contracts should be reported to the village. Although this may strengthen the rights of women, it may also increase the transaction costs in the land rental market and cause such rental arrangements to go unrecorded (Holden and Tefera, 2008).

Furthermore, while the current program has not accommodated a fast-growing landless population, such programs need to look into ways of including such groups in the processes in the future.

Appendix

Data and methodology

Description of data employed in the empirical analyses

Data used in the empirical analyses of the impacts of the program are obtained from all four regions. The analyses from the Tigray region are based on a household panel data set covering 400 households in 16 communities in the Tigray region. The first-round survey was carried out in 1997/98 and considered the pre-registration and certification period, while the three follow-up survey rounds were conducted in 2000, 2005 and 2008. For the Amhara region, the data used is largely from the Sustainable Land Management Survey, conducted in 2000, 2002, 2005 and 2007 in the zones of East Gojjam and South Wollo of Amhara National Regional State. In each round, the data on more than 1,500 randomly selected households and over 7,500 plots was collected. The sample covers 14 kebeles from five different woredas in the two regional zones. About half of the sampled kebeles in each zone received certification at least 12 months before the beginning of the last survey.

Data for the analysis in the SNNP and Oromia regions was collected using stratified random samples, based on agro-ecosystem variation, market access, population density, and regional differences in land laws and implementation of land registration and certification. The survey was conducted in 2007 with 615 households, with the sample including three major ethnic groups from 16 peasant associations in five districts in Southern Ethiopia: three in the Oromia region and two in the Southern Nations, Nationalities and Peoples (SNNP) region (Holden and Tefera, 2008).

Empirical methodology

Methodologically, all the analyses reviewed here are program impact assessment studies that strive to appropriately quantify the impact of the program on given variables of interest. Accurate identification of the impact of a project would require that the ex ante chance of participating in the program is identical between participants and nonparticipants. This implies that observed or unobserved attributes prior to the introduction of the program have no impact on the likelihood of actual participation. Thus, a study evaluating impact needs to address selection bias stemming from inadequate controls for observable heterogeneity plus bias stemming from unobservable heterogeneity (Ravallion, 2007). The different studies reviewed in this report used one or more of the methods we discuss ahead.

The propensity score matching method

Balancing the observed distribution of covariates across households in the certified and treatment groups is the main purpose of the propensity score estimation. This is done by estimation of the propensity score, which is the probability of participation in the program conditional on observable variables but independent of the program outcomes (Rosenbaum and Rubin, 1983). The first step of computing a propensity score is to estimate a standard probit or logit participation model with control variables. This is followed by computation of the standardized before and after matching, which is essentially a covariate balancing indicator. The purpose of this process is that, after matching (and discarding non-matching observations), there should be no systematic differences in the distribution of covariates between the treatment and control groups (Kassie et al., 2010).

Difference in differences approach

The approach compares the change in the outcome of interest in the certified group with the corresponding change in the control group. By controlling for both observed and unobserved differences between the control and treatment groups (Wooldridge, 2005), the difference-in-difference method captures the causal effect of the program on the outcome variable of interest.

Controlling for unobserved heterogeneity

In due consideration of the endogeneity of the policy variable, another approach is to generate a random policy variable that enables unbiased impact assessment, using three different instrumental variables estimation approaches (Holden, Deininger et al., 2009). In the first approach, administrative processes that lead to non-random selection of the treatment units are instrumented for using a "years since certification" variable in a village fixed effects specification. In the second specification, observable household characteristics are added into the village effects regression to test for the validity of the instrument used. The importance of such characteristics in the specification would be indicative of endogeneity bias. In the third and final specification, household fixed effects are estimated in a linear probability framework. Predictions from all three specifications and the actual certificate variable were added for further robustness testing.

Identification of treatment and control villages and households

For the studies based on the Amhara region data, the timing of the implementation of the program is a critical parameter used in the categorization of the households/villages as belonging to the certified (treated) or non-certified (control) categories. (It should be noted that the program has covered all farm land in the regions studied; hence, there are no natural control groups except for gaps in implementation time.) Because the program was gradually rolled out to the villages, villages that had been reached by the program at the time of the data collection could be considered as treated and those enrolled in the program at later stages as control villages.

The geographical gap in the implementation of the program was the result of shortages in financial and manpower resources at both at kebele and woreda levels. In addition, significant gaps in the timing of the implementation between the treated and control woredas assured the distinction of the kebeles/woredas into two groups (Deininger et al., 2011; Bezabih et al., 2016). Deininger et al. (2011) argue that defining treatment at the village level is likely to underestimate impact, owing to inclusion of non-certified households, whereas defining it at the household level would overestimate impacts.

Notes

1 EPRDF (Ethiopian People's Revolutionary Democratic Front) is the ruling political coalition in Ethiopia.
2 Gender-biased disadvantages in land-related property rights are not unique to Ethiopia. For examples of studies from Asia, see Agarwal (1994), Quisumbing et al. (1996), Lastarria-Cornhiel (1997) and Kevane and Gray (1999). For Africa, see Dey (1981), Okali (1983), Crummy (2000) and Yngstrom (2002).
3 *Kebele* is the smallest administrative unit in Ethiopia, while *woreda* is the next largest, formed of a collection of kebeles.
4 Tenure security, in most cases, is measured in terms of expectation of changes in land holdings over the next five years.

References

Adnew, B., and F. Abdi. 2005. *Land Registration in the Amhara Region, Ethiopia: Research Securing Land Rights in Africa, Report 3*. London: International Institute for Environment and Development.

Agarwal, B. 1994. *A Field of One's Own: Gender and Land Rights in South Asia*. Cambridge: Cambridge University Press.

Agarwal, B. 2003. Gender and Land Rights Revisited: Exploring New Prospects via the State, Family, and Market. *Journal of Agrarian Change* 3(2): 184–224.

Akay, A., M. Bezabih and P. Martinsson. 2013. Happiness and the Land Certification Program in Ethiopia. Mimeo. The Grantham Institute for Climate Change and the Environment.

ARNLAB (Amhara Region Natural Resource and Land Use Administration Bureau). 2003. Land Use Policy 46/96 Implementation Guideline.

Augustinus, C. 2005. Global Network for Pro-Poor Land Tools. Nairobi, UN-Habitat.

Barrett, C.B., T. Reardon and P. Webb. 2001. Nonfarm Income Diversification and Household Livelihood Strategies in Rural Africa: Concepts, Dynamics, and Policy Implications. *Food Policy* 26(4): 315–331.

Bell, C., and C. Sussangkarn. 1988. Rationing and Adjustment in the Market for Tenancies: The Behaviour of Landowning Households in Thanjavur District. *American Journal of Agricultural Economics* 70(4): 779–789.

Besley, T. 1995. Property Rights and Investment Incentives: Theory and Evidence from Ghana. *Journal of Political Economy* 103(5): 903–937.

Besley, T., and R. Burgess. 2000. Land Reform, Poverty Reduction and Growth: Evidence from India. *Quarterly Journal of Economics* 115(2): 389–430.

Bezabih, M., S. Holden and A. Mannberg. 2016. The Role of Land Certification in Reducing Gaps in Productivity between Male and Female-Owned Farms in Rural Ethiopia. *Journal of Development Studies* 52(3): 360–376.

Bezabih, M., G. Köhlin and A. Mannberg. 2011. Trust, Tenure Security and the Land Certification Program in Ethiopia. *Journal of Socioeconomics* 52(3): 360–376.

Bezabih, M., A. Mannberg and E. Siba. 2013. Off-Farm Employment, Farm Size and the Land Certification Program in Ethiopia. Mimeo. The Grantham Institute for Climate Change and the Environment.

Bliss, C.J., and N.H. Stern. 1982. *Palanpur: The Economy of an Indian Village.* Clarendon: Oxford University Press.

Bromley, D., and M. Cernea. 1989. The Management of Common Property Natural Resources. World Bank Discussion Paper No. 57. Washington, DC: World Bank.

Clark, A.E., P. Frijters and M.A. Shields. 2008. Relative Income, Happiness, and Utility: An Explanation for the Easterlin Paradox and Other Puzzles. *Journal of Economic Literature* 46: 95–144.

Cotula, L., C. Toulmin and C. Hesse. 2004. *Land Tenure and Administration in Africa: Lessons of Experience and Emerging Issues.* London: International Institute for Environment and Development.

Cox, J., E. Ostrom, J.M. Walker, A. Castillo, E. Coleman, R. Holahan, M. Schoon and B. Steed. 2009. Trust in Private and Common Property Experiments. *Southern Economic Journal* 75(4): 957–975.

Crewett, W., A. Bogale and B. Korf. 2008. Land Tenure in Ethiopia: Continuity and Change, Shifting Rulers, and the Quest for State Control. CAPRi Working Paper No. 91. Washington, DC: International Food Policy Research Institute.

Crummy, D. 2000. Land and Society in the Christian Kingdom of Ethiopia. Champaign: University of Illinois Press.

Deininger, K., D. Ali and T. Alemu. 2011. Impacts of Land Certification on Tenure Security, Investment, and Land Markets. *Land Economics* 87(2): 312–334.

Deininger, K., D. Ali and T. Alemu. 2008. Assessing the Functioning of Land Rental Markets in Ethiopia. *Economic Development and Cultural Change* 57(1): 67–100.

Deininger, K., D. Ali, S.T. Holden and J. Zevenbergen. 2008. Rural Land Certification in Ethiopia: Process, Initial Impact, and Implications for Other African Countries. *World Development* 36(10): 1786–1812.

Deininger, K., D. Ali and T. Yamano. 2008. Legal Knowledge and Economic Development: The Case of Land Rights in Uganda. *Land Economics* 84(4): 593–619.

Deininger, K., G. Feder, G. Gordillo de Anda and P. Munro-Faure. 2003. Land Policy to Facilitate Growth and Poverty Reduction. In *Land Reform, Land Settlement, and Cooperatives,* Special Edition. Edited by P. Groppo. Washington, DC: World Bank and FAO.

Deininger, K., and S. Jin. 2006. Tenure Security and Land-Related Investment: Evidence from Ethiopia. *European Economic Review* 50(5): 1245–1277.

De Janvry, A., G. Gordillo, J.P. Platteau and E. Sadoulet. 2001. *Access to Land, Rural Poverty, and Public Action*. Oxford: Oxford University Press.

Dercon, S., and D. Ayalew. 2007. Land Rights, Power, and Trees in Rural Ethiopia. Working Paper, Oxford: University of Oxford.

Dey, J. 1981. Gambian Women: Unequal Partners in Rice Development Projects? *Journal of Development Studies* 17(3): 109–122.

Do, Q.-T., and Iyer, L. 2008. Land Titling and Rural Transition in Vietnam. *Economic Development and Cultural Change* 56(3): 531–579.

Easterlin, R. 1995. Will Raising the Incomes of All Increase the Happiness of All? *Journal of Economic Behaviour and Organization* 27: 35–47.

Ellis, F. 2000. The Determinants of Rural Livelihood Diversification in Developing Countries. *Journal of Agricultural Economics* 51(2): 289–302.

Fehr, E., S. Yachter and G. Kirchsteiger. 1997. Reciprocity as a Contract Enforcement Device: Experimental Evidence. *Econometrica* 65(4): 833–860.

Frey, B.S., and A. Stutzer. 2002. What Can Economists Learn from Happiness Research? *Journal of Economic Literature* 40: 402–435.

Gebremedhin, B., and S.M. Swinton. 2003. Investment in Soil Conservation in Northern Ethiopia: The Role of Land Tenure Security and Public Programs. *Agricultural Economics* 29: 69–84.

Gebreselassie, M. 2005. Women and Land Rights in Ethiopia. Unpublished. Mekelle: Relief Society of Tigray and the Development Fund.

Ghebru, H., and S. Holden. 2008. Factor Market Imperfections and Rural Land Rental Markets in Northern Ethiopian Highlands. In *The Emergence of Land Markets in Africa: Impacts on Poverty and Efficiency*, edited by S. Holden, K. Otsuka and F. Place. Washington, DC: Resources for the Future.

Holden, S., S. Benin, B. Shiferaw and J. Pender. 2003. Tree Planting for Poverty Reduction in Less-Favoured Areas of the Ethiopian Highlands. *Small-Scale Forest Economics, Management and Policy* 2(1): 63–80.

Holden, S., and M. Bezabih. 2008. Gender and Land Productivity on Rented Land in Ethiopia. In *The Emergence of Land Markets in Africa: Impacts on Poverty and Efficiency*, edited by S. Holden, K. Otsuka and F. Place. Washington, DC: Resources for the Future.

Holden, S., K. Deininger and H. Ghebru. 2009. Impacts of Low-Cost Land Certification on Investment and Productivity. *American Journal of Agricultural Economics* 91(2): 359–373.

Holden, S., K. Deininger and H. Ghebru. 2009b. Tenure Insecurity, Gender, Low-Cost Land Certification and Land Rental Market Participation. *Journal of Development Studies* 47(1): 31–47.

Holden, S., and H. Ghebru. 2013. Welfare Impacts of Land Certification in Tigray, Ethiopia. In *Land Tenure Reform in Asia and Africa: Assessing Impacts on Poverty and Natural Resource Management*, edited by S. Holden, K. Otsuka and K. Deininger. London: Palgrave Macmillan.

Holden, S., and T. Tefera. 2008. Early Impacts of Land Certification on Women in Ethiopia. UNHABITAT.

Holden, S., J. Xu and X. Jiang. 2009. Tenure Security and Forest Tenure Reform in China. Photocopy. Department of Economics and Resource Management. Norwegian University of Life Sciences.

Holden, S., and H. Yohannes. 2002. Land Redistribution, Tenure Insecurity and Intensity of Production: A Study of Farm Households in Southern Ethiopia. *Land Economics* 78(4): 573–594.

Jacoby, H.G., and B. Minten. 2006. *Land Titles, Investment, and Agricultural Productivity in Madagascar: A Poverty and Social Impact Analysis*. Washington, DC: The World Bank.

Kassie, M., P. Zikhali, J. Pender and G. Köhlin. 2010. The Economics of Sustainable Land Management Practices in the Ethiopian Highlands. *Journal of Agricultural Economics* 61(3): 605–627.

Kebede, B. 2002. Land Tenure and Common Pool Resources in Rural Ethiopia: A Study Based on Fifteen Sites. *African Development Review* 14(1): 113–149.

Kebede, B. 2008. Land Reform, Distribution of Land and Institutions in Rural Ethiopia: Analysis of Inequality with Dirty Data. *Journal of African Economies* 17(4): 550–577.

Kevane, M., and L. Gray. 1999. A Woman's Field Is made at Night: Gendered Land Rights and Norms in Burkina Faso. *Feminist Economics* 5(3): 1–26.

Lanjouw, J.O., and P. Lanjouw. 2001. The Rural Non-Farm Sector: Issues and Evidence from Developing Countries. *Agricultural Economics* 26(1): 1–23.

Lastarria-Cornhiel, S. 1997. Impact of Privatization on Gender and Property Rights in Africa. *World Development* 25(8): 1317–1333.

Lunduka, R., S. Holden and R. Øygard. 2008. Land Rental Market Participation and Tenure Security in Malawi. In *The Emergence of Land Markets in Africa: Impacts on Poverty and Efficiency*, edited by S. Holden, K. Otsuka and F. Place. Washington, DC: Resources for the Future.

Mekonnen, A. 2009. Tenure Security, Resource Endowments and Tree Growing: Evidence from the Amhara Region of Ethiopia. *Land Economics* 85(2): 292–307.

Mekonnen, A., H. Ghebru, S. Holden and M. Kassie. 2013. The Impact of Land Certification on Tree Growing on Private Plots of Rural Households: Evidence from Ethiopia. In *Land Tenure Reform in Asia and Africa: Assessing Impacts on Poverty and Natural Resource Management*, edited by S. Holden, K. Otsuka and K. Deininger. London: Palgrave Macmillan, pp. 308–332.

Okali, C. 1983. *Cocoa and Kinship in Ghana: The Matrilineal Akan of Ghana*. London: Kegan Paul International.

Olsson, M., and K. Magnérus. 2007. Transfer of Land Rights in Rural Areas. A Minor Field Study in the Amhara Region, Ethiopia. Stockholm: The Royal Institute of Technology.

Oromiya Region. 2007. Proclamation to Amend the Proclamation No. 56/2002, 70/2003, 103/2005 of Oromiya Rural Land Use and Administration.

Otsuka, K. 2007. Efficiency and Equity Effects of Land Markets. In *Handbook of Agricultural Economics*, Volume 3, edited by R.E. Evenson and P. Pingali. Amsterdam: Elsevier, pp. 2671–2703.

Palm, L. 2010. Quick and Cheap Mass Land Registration and Computerisation in Ethiopia FIG Congress. Facing the Challenges – Building the Capacity. April 11–16, Sydney, Australia.

Place, F., and S. Migot-Adholla. 1998. The Economic Effects of Land Registration on Smallholder Farms in Kenya: Evidence from Nyeri and Kakamega Districts. *Land Economics* 74(3): 360–373.

Putnam, R.D. 2000. *Bowling Alone: The Collapse and Revival of American Community*. New York: Simon and Schuster.

Quisumbing, A. 1996. Male-Female Differences in Agricultural Productivity: Methodological Issues and Empirical Evidence. *World Development* 24(10): 1579–1595.

Ravallion, M. 2007. Evaluating Anti-Poverty Programs. In *Handbook of Development Economics*, Volume 4. Amsterdam, The Netherlands: Elsevier, pp. 3787–3846.

Rosenbaum, P., and D. Rubin. 1983. The Central Role of the Propensity Score in Observational Studies for Causal Effects. *Biometrika* 70: 41–55.

Rozelle, S., L. Brandt, L. Guo and J. Huang. 2002. Land Rights in China: Facts, Fictions, Issues. *China Journal* 47(1): 67–97.

Rupelle, M. de la, D. Quheng, S. Li and T. Vendryes. 2009. Land Rights Insecurity and Temporary Migration in Rural China. IZA Discussion Papers No. 4668. Bonn, Germany: Institute for the Study of Labor.

Skoufias, E. 1995. Household Resources, Transactions Costs, and Adjustment through Land Tenancy. *Land Economics* 71(1): 42–56.

SNNPR. 2007. SNNPR Rural Land Administration and Utilization Proclamation. 110/2007.

Teklu, T., and A. Lemi. 2004. Factors Affecting Entry and Intensity in Informal Rental Land Markets in Southern Ethiopian Highlands. *Agricultural Economics* 30(2): 117–128.

Teraji, S. 2008. Property Rights, Trust, and Economic Performance. *The Journal of Socio-Economics* 37(4): 1584–1596.

TRLAUP. 1997. Rural Land Administration and Utilization Proclamation. Ethiopia: Tigray Regional State.

Wooldridge, J. 2005. Simple Solutions to the Initial Conditions Problem in Dynamic, Nonlinear Panel Data Models with Unobserved Heterogeneity. *Journal of Applied Econometrics* 20: 39–54.

World Bank. 2006. *Land Policy and Administration*. Washington DC: The World Bank.

Yngstrom, I. 2002. Women, Wives, and Land Rights in Africa: Situating Gender Beyond the Household in the Debate over Land Policy and Changing Tenure Systems. *Oxford Development Studies* 30(1): 22–40.

16 Migration as an adaptation strategy to weather variability

An instrumental variables probit analysis

Yonas Alem, Mathilde Maurel and Katrin Millock

Introduction

Global environmental threats, in particular climate change, have been identified as potential causes of large migration flows in the future (UNFPA, 2009; IPCC, 2014). Although the climate is a global good, available projections indicate that climate change damages will differ significantly according to region and will affect developing countries more than developed countries (IPCC, 2014). There is now a large degree of certainty that parts of the developing world will suffer a greater incidence of extreme weather events, which may increase the necessity of displacement migration (Hulme et al., 2001). There are several reasons for this: vulnerability of coastal zones, greater dependence on agriculture and less adaptive capacity (Tol et al., 2004; Mendelsohn et al., 2006). The term "environmental refugees" was popularized by Myers (see, e.g., Myers, 1997; Bates, 2002). The term has since been criticized, in particular because what matters in predicting future migration flows is ultimately identifying vulnerable populations, which is a result of the joint influence of household characteristics, social networks, access to infrastructure, political environment and so forth (Raleigh et al., 2010; Black et al., 2011). More recent credible reports, such as the Stern Report (2007), warn that, by 2050, 200 million people could be at risk of climate events that *may* induce migration.

Certainly, environmental conditions have always influenced habitat, and increased stress on natural resources constitutes an important factor in migration. Some of the most prevalent examples are drought and increased water scarcity and soil erosion, which lead farmers to move away from arid land. Specific types of migration form an integral part of human adaptation to changes in the environment: seasonal or circular migration have very different drivers than displacement migration following a disaster (Adger, 2000; Perch-Nielsen et al., 2008). The most recent literature studies climate change and uses weather factors that are strictly exogenous. In this chapter, we investigate both the ex ante and ex post impacts of climate variables on the decision to engage in migration by smallholder farm households in Ethiopia. We use the coefficient of variation of rainfall and objectively defined rainfall shock variables to test our hypotheses on the relationship between weather and migration.

Rural Ethiopia provides a valuable setting to investigate our hypotheses. About 82% of its population depends on rain-fed agriculture, access to irrigation is insignificant (CIA, 2015) and the country has been affected by frequent climatic shocks in recent decades (Dercon, 2004). Future projections show a significant impact on African migration flows, in particular, following climate change–induced water stress (Le Blanc and Perez, 2008). According to the IPCC's 5th Assessment Report, sub-Saharan Africa is the region with the highest exposure to drought in the world, in terms of the share of the exposed population (Niang et al., 2014). In this regard, our analysis provides important insights about the possible impacts of climate change on migration flows.

There is a small but growing literature that investigates the link between climate change and migration. One of the first studies measured the impact of rainfall on the distribution of population among rural and urban areas in Africa, and found a significant effect of rainfall patterns on the population distribution in sub-Saharan Africa (Barrios et al., 2006). Some recent studies analyze macro-level data on international migration flows to examine the effect of rainfall and temperature on out-migration from sub-Saharan Africa (Marchiori et al., 2012), and on in-migration to OECD countries from developing countries (Coniglio and Pesce, 2015). Using yearly migration data from the OECD, Coniglio and Pesce (2015) find that interannual rainfall variability increases out-migration from developing countries to OECD countries, especially from countries with large agricultural sectors. Beine and Parsons (2015) include both climate factors and natural disasters in an analysis of international migration flows derived from ten-year interval migrant stock data from the World Bank from 1960 to 2000. From that standardized data, which also include South-South migration, they find no statistically significant long-run average effects of either climate factors or natural disasters on international migration flows. They find a statistically significant effect only of natural disasters on internal migration flows, proxied by the rate of urbanization. This result seems compatible with the argument of Marchiori et al. (2012), who find that rainfall and temperature anomalies first increased migration from rural to urban areas, and then, in a second step, also increased international out-migration from sub-Saharan Africa over the period 1960–2000, depending on the size of urban agglomeration externalities. Recently, Maurel and Tuccio (2016) also find a similar effect on international bilateral migration, using the same standardized data as in Beine and Parsons (2015).

Our study uses household data to analyze the migration decision in the "new economics of migration" framework (Stark and Bloom, 1985; Stark, 1993). While a study from one country hardly can be used to infer more general conclusions, compared to the studies on international migration flows, the use of detailed household data enables us to disentangle the many factors that interact to explain migration. Early evidence of environmentally induced migration is found in geographical studies from Burkina Faso (Henry, Piche et al., 2004; Henry, Schoumaker et al., 2004), Ethiopia (Ezra and Kiros, 2001; Meze-Hausken, 2000) and Mali (Findley, 1994). Most of these studies use proxies for climate and its consequences – vulnerability to famines, for example, according to an NGO assessment in the Ezra and Kiros (2001) study on Tigray, Northern Ethiopia. Very few of them can therefore separate the impact of household and community characteristics from the possibility of exogenous environmental push factors, such as rainfall variability and rainfall shocks. Our key contribution is thus to assess both the ex ante and the ex post impacts of one of the most relevant climate variables for sub-Saharan Africa – rainfall – in addition to household socioeconomic characteristics, based on detailed and representative household panel data from Ethiopia.

Household data have been used in studies on migration and climate factors in Ecuador (Gray, 2009), Nigeria (Dillon et al., 2011), Malawi (Lewin et al., 2012), Indonesia (Bohra-Mishra et al., 2014), Bangladesh (Gray and Mueller, 2012b) and Ethiopia (Gray and Mueller, 2012a), among others (see the survey in Millock, 2015). Among the few of these studies to use actual rainfall data, Bohra-Mishra et al. (2014) find a significant positive but small effect from rainfall on household permanent migration in Indonesia (and a larger significant effect of temperature); both effects are nonlinear. Below a certain threshold, an increase in rainfall reduces out-migration, whereas an increase in rainfall above that threshold increases out-migration. In an analysis of Malawi, a very poor country, Lewin et al. (2012) find that rainfall shocks lead to a lower probability of migration. This result is consistent with the hypothesis that severe weather shocks reduce a household's income and stock of capital so much that the household does not have the resources necessary to migrate.

The study that is the most closely related to our work is Gray and Mueller (2012a). Using the Ethiopian Rural Household Survey (ERHS), which we use in the current study, these authors investigate the effects of drought on farm households' mobility and document that men migrate more than women following drought. Gray and Mueller (2012a) model migration as an individual decision and use subjective indicators of exposure to drought reported by households. In this chapter, we follow the new economics of migration (Stark, 1993) and model migration as a household decision. We also use objectively measured rainfall data to define our key variables of interest: weather variability, as measured by the coefficient of variation of rainfall, and rainfall shocks, constructed as one standard deviation from the long-term mean two years before migration. This is the second contribution of the current study, since it is plausible to expect that reported shocks would be correlated with household characteristics that determine the migration decision. Our study also extends the analysis of Gray and Mueller (2012a) by taking into account endogeneity of income – an important determinant of migration – using an instrumental variables estimator and the level of rainfall as an instrument.

Instrumental variables probit regression results suggest that the likelihood of sending a migrant increases with rainfall variability, measured by the coefficient of variation. This can be interpreted as an ex ante measure of the riskiness of a household's environment. Results also provide suggestive evidence that households respond to rainfall shocks by sending a migrant, but this effect is not statistically significant at the conventional levels. We also show that controlling for endogeneity of income is important to clearly identify its role in migration. Our findings have important implications for possible future impacts of climate change related to migration and households' adaptation strategies.

The remainder of the chapter is organized as follows. The next section describes the relevance of studying the case of Ethiopia. The conceptual framework that defines our hypotheses is outlined in the following section. The data and the econometric strategy are described in the fourth section and the variables in the fifth section. The estimation results are presented and discussed in the sixth section, followed by the concluding section.

Rainfall and agricultural income in Ethiopia

We test the hypothesis that migration is induced by exogenous environmental factors, such as rainfall variability, in a predominantly agrarian country. It is particularly relevant to focus on Africa and especially on Ethiopia because the uneven impact of climate change has been shown to affect the less developed countries more, in particular those that are exposed to severe water stress, and where the low level of adaptive capacity may lead to environmentally induced migration (Black et al., 2008). Ethiopia is one sub-Saharan African country that has been classified as extremely vulnerable to drought and other natural disasters, such as floods, heavy rains, frost and heat waves. These extreme weather events cause loss of lives and property and disrupt livelihoods. A large proportion of the population of Ethiopia is heavily dependent on rain-fed agriculture, which is affected by climate change (Kurukulasuriya et al., 2006). According to the World Bank development indicators, less than 0.5% of agricultural lands in Ethiopia were irrigated in the period 2001–2005. According to the CIA World Factbook (2015), the livelihood of 85% of the population depends on this sector. Agriculture also constitutes approximately 40% of the GDP, and 60% of export earnings.

Using a Ricardian approach of measuring the impacts of climate change, Deressa and Hassan (2009) show significant negative marginal impacts of increases in temperature and decreases in rainfall on crop revenue per hectare. Their projections of damages up to 2050 and 2100 call into question the very survival of the Ethiopian agricultural sector. In the face of such serious impacts, it is natural to expect that farmers use migration as one coping strategy, especially as alternative means of insurance are limited. Agricultural adaptation measures, such as planting trees, have been shown to reduce damages to some extent (Di Falco et al., 2012), but may not be sufficient to compensate for the expected losses. Also, the most vulnerable households are the least likely to undertake adaptation measures (Di Falco et al., 2011). Although higher rainfall levels in the preceding season increase yields and income, and thus relax credit constraints that may limit farmers' use of productivity-enhancing inputs, the effect of rainfall variability as such is different. Alem et al. (2010) show that rainfall variability decreases both the probability of applying fertilizer and the amount applied by farmers in their sample from the Ethiopian highlands, because of the increase in risk. The risk induced by rainfall variability has even been shown to directly affect farmers' well-being (Alem and

Colmer, 2016). The risk of a negative rainfall shock and its impact on consumption discourages households' adoption of a risky input such as fertilizer and can leave households in poverty traps (Dercon and Christiaensen, 2013). Poverty and market imperfections also reduce households' investments in soil quality (Shiferaw and Holden, 1999). The existing evidence thus underscores the importance of rainfall variability in Ethiopia and the fact that most vulnerable smallholder households do not have the means to adapt to climate change in the most effective manner. These households may thus be trapped in poverty, unless they have some means to send a household member as a migrant.

Conceptual framework

We view migration as a risk-coping strategy of the household, again following the new economics of migration (Stark and Bloom, 1985; Stark, 1993). A household bases its migration choice on the expected utility of consumption with migration compared to no migration. The impact of rainfall variability is an indirect effect, in that it affects agricultural production, and hence consumption, should the household not be able to smooth its consumption through other strategies, such as drawing down on assets, or searching for employment in the nonagricultural sector, for example. Drought is indeed the most commonly cited shock in rural Ethiopia and harvest failure is the most cited cause of hardship (Dercon and Krishnan, 2000). Using a subsample from the ERHS data from 1989 to 1997, Dercon (2004) finds that a 10% decline in rainfall reduces food consumption by about 5% and that the effect lingers on several years afterwards. In a subsequent analysis, Porter (2012) analyzes the complete ERHS survey data beyond 1997 and finds that extreme rainfall realizations drive the results. Rainfall in the lowest quintile can reduce consumption by 10% to 20%. Alternative strategies, such as diversification from nonagricultural income, do not succeed in reducing the losses from particularly bad rainfall shocks, whereas idiosyncratic shocks can be insured through informal mechanisms.

The migration decision depends upon household characteristics (family size, highest education level in the household, asset holdings) and the costs of migration (in terms of distance to major destinations, or road access). Agricultural income is endogenous in the year of migration and we will use an instrumental variables approach to instrument it in the migration equation.

Following Rose (2001), we separate rainfall variability as a determinant of ex ante migration decisions and rainfall levels – and hence shocks – as a determinant of ex post migration decisions. Prior to observing the current year's (season's) rainfall, household i in PA j may take an ex ante decision to send a migrant as a function of household characteristics F_{ij}, costs of migration from the PA (C_j), and rainfall variability based on past observations of the rainfall distribution. A relevant measure for rainfall variability is the rainfall coefficient of variation measured at the village level (RCV_j). This ex ante decision can be expressed as

$$M^{ante}_{ij,t-1} = f(F_{ij,t-1}, C_{j,t-1}, RCV_{j,t-1}) \tag{1}$$

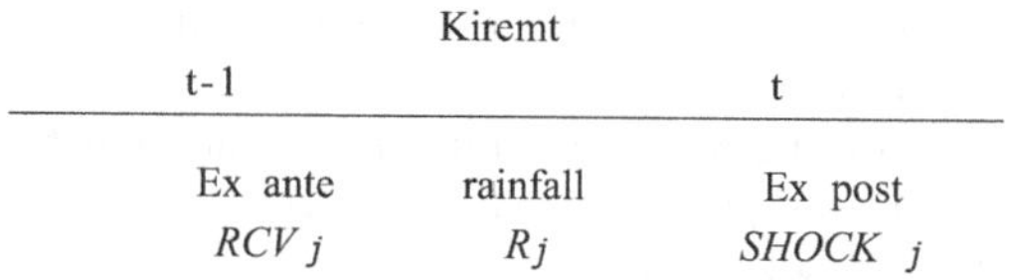

	Kiremt	
t-1		t

| Ex ante | rainfall | Ex post |
| RCV_j | R_j | $SHOCK_j$ |

Figure 16.1 Timeline

Ex post, after observing the current year's (season's) rainfall outcome and its impact on the harvest, and thus on agricultural income (Y^{AG}), the household may decide to send a member away, but this ex post decision will be based on the actual rainfall level and the shock it represents compared to the expected (mean) rainfall:

$$M_{ij,t}^{post} = f(F_{ij,t}, Y_{ij,t}^{AG}, C_{j,t}, SHOCK_{j,t}) \tag{2}$$

Taken together, the final migration observed in household i in PA j in year t is a function of both rainfall variability and the rainfall shock:

$$M_{ij,t} = f(F_{ij,t}, Y_{ij,t}^{AG}, C_{j,t}, RCV_{j,t-1}, SHOCK_{j,t}) \tag{3}$$

The relationship between the ex post expectation and the ex ante shock is shown in Figure 16.1. Note that the final specification includes controls for other coping strategies that are available to the household. After a rainfall shock, households may draw down on assets (Fafchamps et al., 1998; Hoddinott, 2006; Kazianga and Udry, 2006; Porter, 2012) or may increase their activities in the nonagricultural sector (Bezabih et al., 2010). In particular, the vector of household characteristics F_{ij} controls for household assets, represented by land owning and livestock, and human capital, which increases employability in the nonagricultural sector. We also use membership in an *iddir* – an informal risk-sharing institution in Ethiopia – to control for access to informal insurance against shocks.

Data and empirical strategy

Data

We use data from the Ethiopia Rural Household Survey (ERHS), a longitudinal survey implemented by the Department of Economics of Addis Ababa University Ethiopia, in collaboration with the International Food Policy Research Institute (IFPRI) and the Centre for the Study of African Economies (CSAE) at Oxford University. Data were collected for the first time in 1989 from six villages that suffered drought in 1984. The survey was further expanded in 1994 to encompass 15 peasant associations (PA) across the four major regions of Ethiopia (Tigray, Amhara, Oromia and Southern Nations and Nationalities and People's region), constituting 1,477 households, and subsequent rounds took place in 1995, 1997, 1999, 2004 and 2009. The sample of villages and households was chosen randomly to represent the major agroecological zones of the country, excluding the nomadic population. The data thus give a representative sample of Ethiopia apart from nomadic pastoral lands.

In order to investigate the role of migration as an adaptation to weather variability, we use the 1999 and 2004 waves of the ERHS. Attrition in the panel has been low, at 1%–2% of households per year (Dercon and Hoddinott, 2011). The ERHS documents detailed information on individual and household characteristics, assets, expenditures, consumption, health, agricultural production, and use of agricultural inputs.

In addition to the ERHS panel data, annual rainfall data have been collected from the Ethiopian meteorology agency for the weather stations nearest to each village. The rainfall data have been matched with the ERHS villages using the geo codes of the villages to compute our key weather variables.

Empirical strategy

We use a random effects probit framework to model the decision by households in rural Ethiopia to send at least one household member as a migrant in response to weather variability. Let the latent model of migration be specified as

$$m^*_{it} = x'_{it}\beta + \varepsilon_{it} \qquad i = 1, 2, \dots, N; \quad t = 1, \dots, T \tag{4}$$

$$\varepsilon_{it} = c_i + u_{it} \tag{5}$$

where m^*_{it} is a latent dependent variable; m_{it} is the observed binary outcome variable defined as

$$m_{it} = \begin{cases} 1 & \text{if } m^*_{it} > 0; \\ 0 & \text{otherwise.} \end{cases} \tag{6}$$

x_{it} represents a vector of time-varying and time-invariant variables which influence m^*; b represents a vector of parameters to be estimated; and ε_{it} is a composite error term which can be decomposed into c_i, a term capturing unobserved individual (household in our case) heterogeneity, and $u_{it} \sim IN(0, \sigma^2)$, a random error term. The subscripts i and t refer to households and time periods respectively. One can marginalize the likelihood function by assuming that, conditional on the x_{it}, the unobserved individual heterogeneity term $c_i \sim IN(0, \sigma^2)$ is independent of x_{it} and u_{it}.

Assuming that the distribution of the latent variable m^*, conditioned on c_i, is independent normal (Heckman, 1981), the vector of parameters – that is, b – can be estimated easily. Thus,

$$\Pr(m_{it} = 1 | c_i, x_{it}) = \Pr\left(\frac{u_{it}}{\sigma_u} > \frac{-x'_{it}\beta - c_i}{\sigma_u}\right) = \Phi(v_{it}) \tag{7}$$

where

$$v_{it} = -(x'_{it}\beta + c_i)/\sigma_u \qquad (8)$$

and Φ is the distribution function of the standard normal variable. Consequently, the likelihood function to be maximized (which is marginalized with respect to c) is given by

$$\prod_i \left(\int_{-\infty}^{\infty} \prod_{t=1}^{T} \left[1 - \Phi\left(x'_{it}\beta^* + \sqrt{\frac{\lambda}{1-\lambda}} c^* \right) \right]^{1-m_{it}} \times \left[\Phi\left(x'_{it}\beta^* + \sqrt{\frac{\lambda}{1-\lambda}} c^* \right) \right]^{m_{it}} \phi(c^*) dc^* \right)$$

where $\beta^* = \beta/\sigma_u$ and $c^* = c/\sigma_c$. A standard software can be used to estimate β^* and λ, which are normalized on σ_u.

The time-varying and time-invariant variables captured in the vector x_{it} include weather variability (as measured by the coefficient of variation), income from agriculture, wealth (as measured by the value of livestock owned and the size of land), human capital, membership in informal risk-sharing institutions, and village-level variables, such as access to a good road and to town. A detailed description and motivation of these variables are provided in the next section.

It is plausible to argue that some of the explanatory variables (e.g., income) would be endogenous. We attempt to take care of endogeneity of income using the level of rainfall in the previous production year as a credible instrumental variable (IV), and estimate a binary instrumental variables model of migration. We specifically estimate the instrumental variables probit model using the ivprobit command in Stata. This estimator identifies the parameters of a model with a binary dependent variable and an endogenous explanatory variable (or variables). As in the linear instrumental variables estimator, the instrumental variables probit model is estimated in a two-stage process. Consistent estimation is based on the assumption that the error terms of the two equations (in both the first and second stage) are independently and identically distributed multivariate normal. If this assumption is not fulfilled, one could use clustered standard errors to control for the lack of independence (Maddala, 1983).

Explanatory variables

We discuss the ex ante hypotheses on each explanatory variable used in the migration equation and in the agricultural income equation. The rainfall variability and shock measures are discussed separately in the next section.

The migration decision

Agricultural income

Migration depends first upon agricultural income (in logarithms). The standard hypothesis is that droughts decrease agricultural income and push households to migrate (Munshi, 2003). The higher the agricultural income of the household, the less likely it is to have to resort to migration to cope with climatic shocks. However, studies on Burkina Faso (Henry, Piche et al., 2004; Henry, Schoumaker et al., 2004) and Ecuador (Gray, 2009) found the opposite result. A positive correlation is consistent with the so-called hump-shaped pattern of migration (Hatton and Williamson, 2002), which refers to the cost of migration that is too much for households to afford below a critical income threshold. The probability of migration by poor households may thus vary positively with agricultural income. The measure of agricultural income in the ERHS comprises total agricultural income, from both crop and livestock production. Income is given in Ethiopian birr,[1] adjusted for spatial price differences using carefully constructed price deflators.

Household assets

The more assets the household possesses, the less likely it is to have to rely upon migration as a risk-coping strategy. For example, the landless have a higher propensity to migrate as a response to shocks, as demonstrated by Jayachandran (2006). Asset ownership is often proxied by the number of livestock units owned (Kazianga and Udry, 2006). Livestock sales can be used as one strategy to smooth consumption when there is an environmental shock (Hoddinott, 2006). Rogg (2005) provides an in-depth study of the asset portfolio responses of the ERHS households to adverse shocks during the first four rounds of the survey, and also of portfolio responses to ex ante uncertainty. He finds that households in more risky environments hold significantly less livestock. He also shows that households use buffer-stock assets, such as crop/food stocks and some types of livestock, to smooth consumption when income fluctuates. In order to reduce endogeneity problems, we use the value of livestock in t-1 as a control variable for migration in t.

Household size

We control for household size before migration, which would enter as a factor of production in working the field to produce agricultural income. Household size also takes into account our assumption that the migration decision is part of a household-level optimization strategy. More populous households are more likely to send more migrants away. Ideally, we would have preferred to use the dependency ratio, defined as the number of dependent persons (under 15 and above 65 years of age) over the number of working persons. The rationale is that a higher dependency ratio may be an obstacle to migration, as migration would result in fewer adults available to take care of the youngest and oldest. This control variable has not been retained, however, for two main reasons. First, the number of working persons over nonworking persons is endogenous to migration, as the latter has been shown to reduce child and female work (see Acosta, 2011, for an application to El Salvador). Second, age data is incomplete on former members, so the variable would very likely be biased.

Education

Education is generally considered to have a positive influence on internal migration (Lucas, 1997; Taylor and Martin, 2001) and positive evidence of this is found in many studies (Henry, Schoumaker et al., 2004; Konseiga, 2006; Tsegai, 2007; Beegle et al., 2011). On the one hand, it should increase the employment possibilities of a member who migrates. On the other hand, it also increases the opportunities to find nonagricultural work within the same PA, thus decreasing the probability of sending a member away. The first effect seems to dominate in Beegle et al. (2011), a study that highlights a strong positive and nonlinear effect of education on migration. We measure human capital as the highest education level in the household, according to the following categories: some primary education or adult literacy program, completed primary education, secondary education or some university education. In the analysis presented here, we use a dummy variable indicating whether the household has some education at all compared to the base case of no education.

Membership in iddir

The literature suggests a range of mechanisms that households living in risky environments have developed to shield their consumption from risk, including social insurance arrangements, particularly important in the absence of formal insurance or credit markets. An iddir is a kind of mutual insurance association that pays for funerals. The variation is not that large in this variable; in 2004, for example, 80% of the households were members of an iddir. Nevertheless, we use it to control for access to credit or support if the household were to suffer a reduction in its income.

Distance to town and good road

Accessibility is measured by the distance between the PA in which the household resides and the nearest town. We also use a measure of road accessibility from the PA. Following Dercon et al. (2009), we define road accessibility as a dummy variable equal to 1 if the PA has at least one road that is accessible to trucks and buses in both rainy and dry weather, and equal to 0 otherwise. The influence of distance to town and accessibility by road is a priori ambiguous. On the one hand, closer proximity to markets makes it easier to purchase inputs and sell crops, and it should also increase the possibilities of seeking alternative wage employment in the origin location as a diversification strategy. These factors all speak for a negative influence of the accessibility variables on the migration decision. On the other hand, proximity to a nearby village and road accessibility reduce the cost of migration and should thus positively influence migration, as in Henry, Schoumaker et al. (2004), for example. These last two variables are measured at the level of the PA, and do not vary across households in the same PA.

The agricultural income equation

Household agricultural income is a function of the household's number of laborers, proxied by the total number of household members before migration. This variable should be refined to include only working-age people, but, since the sample shows that people start working in the field as early as six years old, there is no clear cutoff for the definition of useful labor. The value of livestock holdings in the current year is included as a production factor for agricultural income. Following a Ricardian production factor approach (Mendelsohn and Dinar, 2009), we control for the size of the household's landholdings, measured in standardized units (ha).

Weather variables

Rainfall variability has been shown to be a relevant climate parameter for agricultural yields in other parts of Africa (Sultan et al., 2010). The household survey data is matched with data on rainfall over the period 1967–2004 from the weather station closest to the PA. The rainfall data used in the study thus vary only across PAs and over time. We use the rainfall in the 12 months preceding the survey to account for the impact of rainfall on agricultural income. The rainfall shock variable is defined as equal to one if the annual rainfall in the 12 months preceding the survey was below one standard deviation from the long-term mean of 30 years of rainfall data, representing a negative rainfall shock. Apart from average rainfall based on past observations, and deviations from this rainfall level, another relevant measure is the variability of rainfall. We use the coefficient of variation to measure the riskiness in the household's environment. The coefficient of variation is defined as the standard deviation over the mean, multiplied by 100. We will thus use the coefficient of variation of the 30-year rainfall distribution as a measure of the riskiness of the household's environment with respect to rainfall. This should be the basis for a household's ex ante decision to send a member as a migrant, before the actual rainfall level and its impact on the harvest have been observed. Ex post, once the rainfall level of the particular year (harvest season) has been observed, it is the actual rainfall shock that will determine a migration decision. Because the survey does not enable us to separate the two types of decisions (ex ante and ex post), it is important to include both measures in the equation. A test of the hypothesis of a nonzero coefficient on any of the variables should test for its relevance in the final migration decision, which includes both ex ante and ex post considerations.

Table 16.1 shows the descriptive statistics of the variables used in the study.

Table 16.1 Descriptive Statistics of Variables over Time

	1999		2004	
	Mean	SD	Mean	SD
Household sent a migrant	0.02	0.15	0.09	0.28
Agricultural income (birr)	2014.38	2651.72	2410.73	3167.17
Value of livestock (birr)	1800.73	2087.54	2578.60	3388.25
Land size (ha)	1.20	1.02	1.61	2.10
Household size adjusted for migrants	6.79	3.05	5.88	2.59
Household has primary or secondary educ.	0.78	0.42	0.75	0.43
Member of at least one iddir (dummy)	0.72	0.45	0.80	0.40
Coefficient of variation of rainfall	28.35	13.45	28.35	13.45
Village experienced rainfall shock	0.20	0.40	0.12	0.32
Distance to nearest town (km)	8.55	5.91	8.55	5.91
Accessible road (dummy)	0.44	0.50	0.69	0.46
Annual rainfall (mm)	956.79	295.15	945.44	313.88
Observations	1347		1347	

Source: Authors' computation from ERHS 1999 and 2004.

Results

Table 16.2 shows random effects probit regression results and the corresponding marginal effects for the household model of migration presented in Equation 4. The results show that household size and household education positively affect the decision to send a member as a migrant, while distance to the nearest town reduces the likelihood of sending a migrant. The results also show that more households sent a migrant in 2004 relative to 1999, which is the base year. We do not find any statistically significant effect of our key variables of interest: household income, weather variability and rainfall shocks. The random effects probit treats all the right-side variables as exogenous. It is, however, plausible to expect that most of the household variables are endogenous. We use the level of rainfall in the main agricultural season as an instrument to take care of endogeneity of agricultural income. Other household variables, such as value of livestock, land size, household size and education, are also likely to be endogenous. However, we do not find credible IVs to instrument them. Thus, we do not make a causal inference between migration and these variables, but we control for them in the estimations.

The first-stage and second-stage regression results from the instrumental variables probit estimator are reported in Tables 16.3 and 16.4. We begin with the first-stage regression results. Table 16.3 shows that the first-stage relationship between rainfall and agricultural income is strongly positive: the level of annual rainfall is significantly related to agricultural income at the 1% significance level. This relationship is robust to exclusion of the other variables and the

Table 16.2 A Household Model of Migration: Random Effects Probit Regression Results

	RE-Probit		Marginal effects	
	Coeff.	SE	Coeff.	SE
Agricultural income (log)	−0.023	0.031	−0.003	0.004
Value of livestock (log)	−0.015	0.020	−0.002	0.002
Land size	0.024	0.019	0.003	0.002
Household size adjusted for migrants	0.087***	0.017	0.010***	0.002
Household has primary or secondary educ.	0.222*	0.119	0.025*	0.014
Member of at least one iddir	0.250	0.153	0.029	0.018
Coefficient of variation of rainfall	0.007	0.004	0.001	0.001
Village experienced rainfall shock	−0.040	0.127	−0.005	0.015
Distance to town in km	−0.021**	0.008	−0.002**	0.001
Accessible road	−0.063	0.099	−0.007	0.011
Year 2004	0.611***	0.102	0.065***	0.010
Intercept	−2.647***	0.303		
Log likelihood	−494.08			
Observations	2297			

Note: *** p < 0.01,
** p < 0.05,
* p < 0.1.

Table 16.3 A Household Model of Migration: IV Probit – First-Stage Regression

Agricultural income	First stage	
	Coeff.	SE
Value of livestock (log)	0.184***	0.014
Land size	0.094***	0.021
Household size adjusted for migrants	0.093***	0.014
Household has primary or secondary educ.	0.399***	0.084
Member of at least one iddir	0.506***	0.106
Coefficient of variation of rainfall	−0.013***	0.004
Village experienced rainfall shock	−0.201**	0.099
Distance to town in km	0.017***	0.007
Accessible road	0.288***	0.083
Year 2004	0.708***	0.074
Level of rainfall	0.001***	0.000
Intercept	3.481***	0.273
Observations	2297	

Note: *** p < 0.01,
** p < 0.05,
* p < 0.1.

Table 16.4 A Household Model of Migration: IV Probit – Second-Stage Regression

	[1] IV probit		[2] Marginal effects	
	Coeff.	SE	Coeff.	SE
Agricultural income (log)	0.459***	0.106	0.116**	0.057
Value of livestock (log)	−0.098***	0.021	−0.025**	0.011
Land size	−0.027	0.021	−0.007	0.007
Household size adjusted for migrants	0.012	0.029	0.003	0.007
Household has primary or secondary educ.	−0.047	0.119	−0.012	0.032
Member of at least one iddir	−0.095	0.147	−0.024	0.042
Coefficient of variation of rainfall	0.014***	0.003	0.003**	0.001
Village experienced rainfall shock	0.132	0.099	0.033	0.029
Distance to town in km	−0.024***	0.006	−0.006***	0.002
Accessible road	−0.143*	0.073	−0.036*	0.022
Year 2004	0.053	0.208	0.013	0.049
Intercept	−3.644***	0.243	–	–
athrho	−1.050**	0.420		
lnsigma	0.498***	0.015		
Log likelihood	−4894.483			
Observations	2297			

Note: *** p < 0.01,
** p < 0.05,
* p < 0.1.

Wald test rejects the null hypothesis of no endogeneity (p-value=0.001). In a country like Ethiopia where more than 95% of smallholder farmers are rain-dependent for their livelihood, positive rainfall typically leads to better agricultural production. The first-stage regression results also show that all the other correlates of agricultural income have the theoretically expected signs and are statistically significant at the 1% level, except the rainfall shocks variable, which is significant at the 5% level.

Regression results from the second stage of the instrumental variables probit model, presented in Table 16.4, demonstrate the importance of controlling for endogeneity of agricultural income in the migration equation. Column 1 presents the parameter estimates and Column 2 presents the corresponding marginal effects. The log of agricultural income is now statistically significant at the 1% level in Column 1 and positively and strongly determines rural Ethiopian households' decisions to send a member as a migrant. From the marginal effects reported in Column 2, we observe that a 1% increase in agricultural income leads to about an 11.6% increase in the probability of sending a member as a migrant. This is consistent with findings in earlier studies of migration (Henry, Schoumaker et al.,

2004; Gray, 2009), where the liquidity constraints argument seems to be the dominant mechanism in explaining the effect of income on migration. On the one hand, migration is costly and higher income provides more resources for households to cover the cost of migration. The value of livestock, on the other hand, is negatively correlated with the probability of sending a household member as a migrant and the effect is statistically significant at the 1% level. A 1% increase in the value of livestock is associated with a 2.5% decrease in the probability of sending a migrant (Column 2). Livestock are important as wealth and factors of production in a smallholder farming setup like rural Ethiopia. Consequently, other factors being constant, greater livestock wealth gives the household the opportunity to be productive in farming and get better access to other productivity-enhancing modern agricultural inputs, such as chemical fertilizer (Alem and Broussard, 2016).

We will now analyze how climate factors affect households' decisions to send a migrant. Instrumental variables probit regression results presented in Table 16.4 suggest that the decision to send a household member as a migrant is positively influenced by weather variability, as measured by the coefficient of variation of rainfall. A one-unit increase in the coefficient of variation increases the probability of sending a migrant by about 0.3%. The effect is statistically significant at the 1% level. This provides strong and important evidence supporting our hypothesis that households in areas with highly variable weather are more likely to adapt through sending a migrant. This is an *ex ante* effect of weather variability. Because climate change is expected to result in extreme weather events and the sub-Saharan African region is going to experience even greater variability in rainfall (IPCC, 2014), our findings imply that there will be increased migration by household members in rural communities in the future. On the data studied here, no evidence is found that migration decisions are driven by negative rainfall shocks over the short run (24 months); rather, such decisions are taken due to assessments based on the long-term risk of variability in rainfall as given by the coefficient of variation.

We finally note that distance to the nearest town negatively affects the decision to send a migrant. This may sound counterintuitive, as proximity to towns should make migration easier. However, it is also plausible that households close to towns have the option of participating in off-farm activities without the need to leave home. This last result thus yields support to the proximity to markets argument: better access to markets for inputs and outputs enables the household to be more productive in agriculture and also to find alternative employment in urban centers (Bezabih et al., 2010).

Conclusions

Climate change is predicted to impact society and ecosystems by resulting in extreme weather events, changing precipitation and declining agricultural productivity in large parts of the world (IPCC, 2014). As a result of its dependence on climate variables, agriculture is more vulnerable than other sectors to the effects of climate change. Sub-Saharan Africa, whose agricultural sector is known for its low productivity, is one of the most vulnerable regions. This chapter attempts to shed light on whether households in rural areas of Ethiopia use migration as a strategy to adapt to weather variability. We use Ethiopian Rural Household Survey panel data combined with village-level rainfall data to investigate both the ex ante and ex post impact of climate variables on households' likelihood of sending a migrant. Compared to earlier studies, the key contributions of the chapter are in using objective measures of rainfall variability and rainfall shocks, which are exogenous to households, and in teasing out the ex ante and ex post effects. We also control for endogeneity of income – an important determinant of migration – using a credible instrument.

Instrumental variables probit regressions suggest that households in rural Ethiopia adapt to weather variability by sending a migrant. Smallholder farm households that live in places with higher rainfall variability are likely to send a household member as a migrant. This effect is economically important and statistically significant at the 1% level. We also find suggestive evidence on the ex post impacts of weather, proxied by the prevalence of negative rainfall shocks between one and two standard deviations below the long-run mean, in the year preceding migration. The variable positively affects the decision to send a migrant, but it is not statistically significant at the conventional levels, most probably due to large standard errors. We also find that income (instrumented by the level of rainfall) has a significant positive impact on the decision to migrate, while wealth (proxied by the value of livestock owned by the household) is negatively associated with the likelihood of engaging in migration. Regression results also show that living close to a town reduces the likelihood of sending a migrant.

Extensions to this analysis could study the role of irrigation, not included in our study because the measurement errors in the variable are large in the ERHS data. Improving water access would be a useful policy objective, given the strong impact that we find from rainfall variability. Developing rainfall insurance products that could cushion the impact of rainfall variability also seems an important policy implication of the results, in line with previous research on the negative effects of rainfall variability on smallholder farm households. Because several factors enter into the migration decision, it is also clear that some migration is unavoidable – and even desirable from a household viewpoint – and policy measures should also be directed toward supporting development in the urban sectors of the country that will receive the out-migrants from the rural areas that are subject to high rainfall variability.

Note

1 One Ethiopian birr=USD .043 as of August 2017.

References

Acosta, P. 2011. School Attendance, Child Labour, and Remittances from International Migration in El Salvador. *The Journal of Development Studies* 47(6): 913–936.

Adger, N. 2000. Social and Ecological Resilience: Are They Related? *Progress in Human Geography* 24(3): 347–364.

Alem, Y., M. Bezabih, M. Kassie and P. Zikhali. 2010. Does Fertilizer Use Respond to Rainfall Variability? Panel Data Evidence from Ethiopia. *Agricultural Economics* 41: 165–175.

Alem, Y., and N.H. Broussard. 2016. The Impact of Safety Nets on Technology Adoption: A Difference-in-Differences Analysis. Environment for Development Discussion Paper Series No. 16–13. Gothenburg, Sweden: Environment for Development and Resources for the Future.

Alem, Y., and J. Colmer. 2016. Consumption Smoothing and the Welfare Cost of Uncertainty. https://sites.google.com/site/jonathancolmer/

Barrios, S., L. Bertinelli and E. Strobl. 2006. Climatic Change and Rural-Urban Migration: The Case of Sub-Saharan Africa. *Journal of Urban Economics* 60(3): 357–371.

Bates, D. 2002. Environmental Refugees? Classifying Human Migrations Caused by Climate Change. *Population and Environment* 23(5): 465–477.

Beegle, K., J. De Weerdt and S. Dercon. 2011. Migration and Economic Mobility in Tanzania: Evidence from a Tracking Survey. *The Review of Economics and Statistics* 93(3): 1010–1033.

Beine, M., and C. Parsons. 2015. Climatic Factors as Determinants of International Migration. *Scandinavian Journal of Economics* 117(2): 723–767.

Bezabih, M., Z. Gebreegziabher, L. GebreMedhin and G. Köhlin. 2010. Participation in Off-Farm Employment, Rainfall Patterns and Rate of Time Preferences: The Case of Ethiopia. Environment for Development Discussion Paper Series No. 10–21. Gothenburg, Sweden: Environment for Development and Resources for the Future.

Black, R., S.R.G. Bennett, S.M. Thomas and J.R. Beddington. 2011. Migration as Adaptation. *Nature* 478: 447–449.

Black, R., D. Knighton, R. Skeldon, D. Coppard, A. Murata and K. Schmidt-Verkerk. 2008. Demographics and Climate Change: Future Trends and Their Policy Implications for Migration. Working Paper No. T27, Development Research Centre on Migration, Globalisation and Poverty. Falmer, UK: University of Sussex.

Bohra-Mishra, P., M. Oppenheimer and S. Hsiang. 2014. Nonlinear Permanent Migration Response to Climatic Variations but Minimal Response to Disasters. *Proceedings of the National Academy of Sciences of the USA* 111(27): 9780–9785.

CIA. 2015. The World Factbook 2013–14. Washington, DC: Central Intelligence Agency.

Coniglio, N., and G. Pesce. 2015. Climate Variability and International Migration: An Empirical Analysis. *Environment and Development Economics* 20(4): 434–468.

Dercon, S. 2004. Growth and Shocks: Evidence from Rural Ethiopia. *Journal of Development Economics* 74(2): 309–329.

Dercon, S., and L. Christiaensen. 2013. Consumption Risk, Technology Adoption and Poverty Traps: Evidence from Ethiopia. *Journal of Development Economics* 96: 159–173.

Dercon, S., and J. Hoddinott. 2011. The Ethiopian Rural Household Surveys 1989–2009: Introduction. Mimeo.

Dercon, S., and P. Krishnan. 2000. Vulnerability, Seasonability and Poverty in Ethiopia. *Journal of Development Studies* 36(6): 25–53.

Dercon, S., D. O'Gilligan, J. Hoddinott and T. Woldehanna. 2009. The Impact of Agricultural Extension and Roads on Poverty and Consumption Growth in Fifteen Ethiopian Villages. *American Journal of Agricultural Economics* 91(4): 1007–1021.

Deressa, T.T., and R.M. Hassan. 2009. Economic Impact of Climate Change on Crop Production in Ethiopia: Evidence from Cross-Section Measures. *Journal of African Economies* 18(4): 529–554.

Di Falco, S., M. Veronesi and M. Yesuf. 2011. Does Adaptation to Climate Change Provide Food Security? A Micro-Perspective from Ethiopia. *American Journal of Agricultural Economics* 93(3): 829–846.

Di Falco, S., M. Yesuf, G. Köhlin and C. Ringler. 2012. Estimating the Impact of Climate Change on Agriculture in Low-Income Countries: Household-Level Evidence from the Nile Basin. *Environmental and Resource Economics* 52(4): 457–478.

Dillon, A., V. Mueller and S. Salau. 2011. Migratory Responses to Agricultural Risk in Northern Nigeria. *American Journal of Agricultural Economics* 93(4): 1048–1061.

Ezra, M., and G.-E. Kiros. 2001. Rural Out-Migration in the Drought-Prone Areas of Ethiopia: A Multi-Level Analysis. *International Migration Review* 35(3): 749–771.

Fafchamps, M., C. Udry and K. Czukas. 1998. Drought and Saving in West Africa: Are Livestock a Buffer Stock? *Journal of Development Economics* 55(2): 273–305.

Findley, S. 1994. Does Drought Increase Migration? A Study of Migration from Rural Mali during the 1983–1985 Drought. *International Migration Review* 28(3): 539–553.

Gray, C. 2009. Environment, Land, and Rural Out-Migration in the Southern Ecuadorian Andes. *World Development* 37(2): 457–468.

Gray, C., and V. Mueller. 2012a. Drought and Population Mobility in Rural Ethiopia. *World Development* 40(1): 134–145.

Gray, C., and V. Mueller. 2012b. Natural Disasters and Population Mobility in Bangladesh. *Proceedings of the National Academy of Sciences of the USA* 109(16): 6000–6005.

Hatton, T.J., and J.G. Williamson. 2002. What Fundamentals Drive World Migration? NBER Working Paper No. 9159. Washington, DC: National Bureau of Economic Research.

Heckman, J.J. 1981. The Incidental Parameters Problem and the Problem of Initial Conditions in Estimating a Discrete Time-Discrete Data Stochastic Process. In *Structural Analysis of Discrete Data with Econometric Applications*, edited by C.F. Manski and D. McFadden. Cambridge, MA: MIT Press, pp. 114–178.

Henry, S., V. Piche, D. Ouedraogo and E. Lambin. 2004. Descriptive Analysis of the Individual Migratory Pathways. *Population and Environment* 25(5): 397–422.

Henry, S., B. Schoumaker and C. Beauchemin. 2004. The Impact of Rainfall on the First Out-Migration: A Multi-Level Event History Analysis in Burkina Faso. *Population and Environment* 25(5): 423–460.

Hoddinott, J. 2006. Shocks and Their Consequences across and within Households in Rural Zimbabwe. *Journal of Development Studies* 42(2): 301–321.

Hulme, M., R. Doherty, T. Ngara, M. New and D. Lister. 2001. African Climate Change: 1900–2100. *Climate Research* 17(2): 145–168.

IPCC. 2014. Climate Change 2014: Impacts, Adaptation, and Vulnerability. The Working Group II. Contribution to the IPCC Fifth Assessment Report (WGII AR5). Geneva: Intergovernmental Panel on Climate Change.

Jayachandran, S. 2006. Selling Labour Low: Wage Responses to Productivity Shocks in Developing Countries. *Journal of Political Economy* 114(3): 538–575.

Kazianga, H., and C. Udry. 2006. Consumption Smoothing? Livestock, Insurance and Drought in Rural Burkina Faso. *Journal of Development Economics* 79: 413–446.

Konseiga, A. 2006. Household Migration Decisions as Survival Strategy: The Case of Burkina Faso. *Journal of African Economies* 16(2): 198–233.

Kurukulasuriya, P., R. Mendelsohn, R. Hassan, J. Benhin, T. Deressa, M. Diop, H.M. Eid, K. Yerfi Fosu, G. Gbetibouo, S. Jain, A. Mahamadou, R. Mano, J. Kabubo-Mariara, S. El-Marsafawy, E. Molua, S. Ouda, M. Ouedraogo, I. Sène, D. Maddison, S.N. Seo and A. Dinar. 2006. Will African Agriculture Survive Climate Change? *The World Bank Economic Review* 20(3): 367–388.

Le Blanc, D., and R. Perez. 2008. The Relationship between Rainfall and Human Density and Its Implications for Future Water Stress in Sub-Saharan Africa. *Ecological Economics* 66: 319–336.

Lewin, P.A., M. Fisher and B. Weber. 2012. Do Rainfall Conditions Push or Pull Rural Migrants? Evidence from Malawi. *Agricultural Economics* 43(2): 191–204.

Lucas, R. 1997. Internal Migration in Developing Countries. In *Handbook of Population and Family Economics*, edited by M.R. Rosenzweig and O. Stark. Amsterdam, Netherlands: Elsevier Science B.V., pp. 721–798.

Maddala, G.S. 1983. *Limited Dependent and Qualitative Variables in Econometrics*. Cambridge: Cambridge University Press.

Marchiori, L., J.-F. Maystadt and I. Schumacher. 2012. The Impact of Weather Anomalies on Migration in Sub-Saharan Africa. *Journal of Environmental Economics and Management* 63: 355–374.

Maurel, M., and M. Tuccio. 2016. Climate Instability, Urbanisation and International Migration. *Journal of Development Studies* 52(5): 735–752.

Mendelsohn, R., and A. Dinar. 2009. *Climate Change and Agriculture: An Economic Analysis of Global Impacts, Adaptation and Distributional Effects*. Cheltenham, UK and Northampton, MA: Edward Elgar, New Horizons in Environmental Economics.

Mendelsohn, R., A. Dinar and L. Williams. 2006. The Distributional Impact of Climate Change on Rich and Poor Countries. *Environment and Development Economics* 11(2): 159–178.

Meze-Hausken, E. 2000. Migration Caused by Climate Change: How Vulnerable Are People in Dryland Areas? *Mitigation and Adaptation Strategies for Global Change* 5(4): 379–406.

Millock, K. 2015. Migration and Environment. *Annual Review of Resource Economics* 7: 35–60.

Munshi, K. 2003. Networks in the Modern Economy: Mexican Migrants in the US Labor Market. *Quarterly Journal of Economics* 118(2): 549–599.

Myers, N. 1997. Environmental Refugees. *Population and Environment* 19(2): 167–182.

Niang, I., O.C. Ruppel, M.A. Abdrabo, A. Essel, C. Lennard, J. Padgham and P. Urquhart. 2014. Africa. In *Climate Change 2014: Impacts, Adaptation, and Vulnerability, Part B: Regional Aspects: Contribution of Working Group II to the Fifth Assessment Report of the Intergovernmental Panel on Climate Change*, edited by V.R. Barros, C.B. Field, D.J. Dokken, M.D. Mastrandrea, K.J. Mach, T.E. Bilir, M. Chatterjee, K.L. Ebi, Y.O. Estrada, R.C. Genova, B. Girma, E.S. Kissel, A.N. Levy, S. MacCracken, P.R. Mastrandrea and L.L. White. Cambridge: Cambridge University Press, pp. 1199–1265.

Perch-Nielsen, S.L., M.B. Battig and M. Imboden. 2008. Exploring the Link between Climate Change and Migration. *Climatic Change* 91: 375–393.

Porter, C. 2012. Shocks, Consumption and Income Diversification in Rural Ethiopia. *Journal of Development Studies* 48(9): 1209–1222.

Raleigh, C., L. Jordan and I. Salehyan. 2010. Assessing the Impact of Climate Change on Migration and Conflict. Commissioned for the Social Dimensions of Climate Change Workshop. Washington, DC: The World Bank.

Rogg, C. 2005. Precautionary Saving and Portfolio Management in Uncertain Environments: Evidence from Rural Ethiopia. PhD Thesis. University of Oxford, Oxford, UK.

Rose, E. 2001. Ex Ante and Ex Post Labour Supply Response to Risk in a Low-Income Area. *Journal of Development Economics* 64: 371–388.

Shiferaw, B., and S. Holden. 1999. Soil Erosion and Smallholders' Conservation Decisions in the Highlands of Ethiopia. *World Development* 27(4): 739–752.

Stark, O. 1993. *The Migration of Labour*. Cambridge: Blackwell.

Stark, O., and D.E. Bloom. 1985. The New Economics of Labour Migration. *American Economic Review* 75(2): 173–178.

Stern, N. 2007. *The Economics of Climate Change: The Stern Review*. Cambridge: Cambridge University Press.

Sultan, B., M. Bella-Medjo, A. Berg, P. Quirion and S. Janicot. 2010. Multi-Scales and Multi-Sites of the Role of Rainfall in Cotton Yields in West Africa. *International Journal of Climatology* 30(1): 58–71.

Taylor, J.E., and P.L. Martin. 2001. Human Capital: Migration and Rural Population Change. In *Handbook of Agricultural Economics*, Volume 1, edited by B. Gardner and G. Rausser. Amsterdam: Elsevier Science B.V., pp. 457–511.

Tol, R., T.E. Downing, O.J. Kuik and J.B. Smith. 2004. Distributional Aspects of Climate Change Impacts. *Global Environmental Change* 14: 259–272.

Tsegai, D. 2007. Migration as a Household Decision: What Are the Roles of Income Differences? Insights from the Volta Basin of Ghana. *The European Journal of Development Research* 19(2): 305–326.

UNFPA. 2009. *The State of World Population 2009: Facing a Changing World: Women, Population and Climate*. New York: United Nations Population Fund.

17 A changing climate in a changing land

Peter Berck, Cyndi Spindell Berck,
and Tyler N. Jacobson

Introduction

East Africa is a rapidly changing land, with rapid economic growth and increasing levels of urbanization, education, and road building. East Africa's climate is changing as well. Climate change brings more heat, more rain to some parts and less rain to others, and more variable rain to the whole region. Those changes challenge an agricultural industry in East Africa that is already far from reaching optimum output or profit. There are many interventions that can raise that agricultural performance. Governments, businesses, and NGOs can take steps to improve credit and insurance markets, improve human health and education, and help farmers manage risk. Farmers can plant new seed types, plant trees, choose different crops, increase fertilizer use, and so on. Although these interventions are well known, they are not widely adopted. Increasing adoption of these measures is a path to both coping with climate change and increasing current output. Yet, the solutions for climate change and for increasing current output should diverge because climate change is happening over decades and the concern with increasing output is immediate. Given the slow uptake of new measures and the apparently wide gap between productivity in East Africa and in other parts of the world, climate change will add greatly to the woes of the agriculture-dependent populations.

In this chapter, we suppose that instead of East Africa facing the future challenges of climate change as East Africa is today, that East Africa instead faces those challenges as it grows, educates, and urbanizes at its recent growth rates. We ask, what can be done to adapt to climate change in the agricultural sector in a sub-Saharan Africa that has a 170% greater gross national income and a 65% greater secondary school enrollment rate every 13 years (International Monetary Fund, 2017)?

Our story is as follows. Agricultural adaptation to climate change calls for adoption of modern technologies. The technologies that can increase farm production in response to climate change are, to a great extent, the same technologies that would increase farm productivity even if climate change were not an issue. Adoption of modern technologies, whether in farming or other sectors, is constrained by inadequacies in three categories: human capital, infrastructure, and institutions. When these categories are underdeveloped, farmers, like everyone else in a poor society, remain poor. On the positive side, education, infrastructure, and institutions tend to improve at the same time as income. We concentrate on these because there are policy levers that work directly on them and, in East Africa, there are already actions being taken to do so. To improve human capital, governments can invest in general education and expand agricultural extension services. To improve infrastructure, governments and other parties can invest in roads and other transportation infrastructure, such as pipelines, electrification, and information technology. To improve institutions, governments can expand property rights and financial markets.

In short, we think that agricultural adaptation to climate change will be driven (or held back) by the same factors that promote or retard development more generally. The policy implication is that investing in schools or roads or land titling will reap benefits in adapting to climate change. In some case, such investments may even be a better climate investment than investing directly in modern agricultural techniques – although this chapter doesn't attempt to quantify that intuition.

We begin with a brief overview of how climate change affects different agricultural sectors.

Climate change's impact on agriculture

Chapter 22 in the Intergovernmental Panel on Climate Change's 5th Report provides projections for rainfall and temperature changes through the end of the century. Broadly, in the RCP 8.5 modeling – the extreme scenario for climate change – there is an increase in temperature of between 4°C and 6°C for the late twenty-first century, with the more northern and southern reaches of Africa projected as hotter. Rainfall for East Africa is not forecast to decrease (though it is for both the Maghreb and South Africa). In addition to the overall forecasted trends, there is the possibility that rainfall will happen in larger events with longer dry periods in between.

The general consensus is that high heat is damaging to crop yields. (Although plant growth depends in part on light, climate change is not forecast to increase the amount of light available for plants.) Corn, in particular, would be pushed well over its maximum temperature (29°C in Schlenker and Roberts, 2009), with an average 6°C increase in temperature causing the plant to wilt. An increase in the length of hot and dry periods is also very deleterious for crop growth. An increase in the number of dry days in a row results in dry soil and plant damage or death. So, a change in the pattern of rainfall, even if rainfall increases on average, can be harmful if it leads to these longer periods of dry days.

These effects already have been researched in East Africa. In Ethiopia, farmers typically grow a variety of crops, each on a different plot, with more than one growing season in a year, and with virtually no weather insurance. Bezabih et al. (2014) found negative effects on crop revenue at the extreme ends of the distribution of long-term temperature, pointing to the importance of both short-term weather and long-term climate patterns. Although extreme temperature caused negative effects on revenue for all seasons and crop types, the impact varied drastically across crops and across seasons for each crop, which reminds us that policy design needs to consider the heterogeneous impacts of climate change even within a farm and a given year. Surprisingly for rain-fed agriculture, they found that rainfall had a less important role to play than temperature.

In chapter 4 of this book, Kabubo-Mariara and Kabara have analyzed the effect of climate variables on food security in Kenya. Kenya has high levels of poverty, heavy dependence on rain-fed agriculture, low levels of human and physical capital, and poor infrastructure. Agricultural productivity is already declining, in part because severe droughts have harmed harvests, pastures, and livestock. Attention to agroecological zones is important; Kenya has several climatic zones that differ in terms of moisture, rainfall, and vegetation, and, as a result, diverse farming systems. Their results show that high rainfall at certain times in the growing season is crucial for crop productivity, but excessive rainfall is harmful. Similarly, the relationship between temperature and yield differs depending on the season. The effects of temperature and rainfall also depend on the crop; maize is likely to be hard hit by climate change, with sorghum, beans, and millet more resilient. Overall, a decline in crop yields by up to 69% is projected by 2100, depending on crop and climate scenario. However, increasing population density has a positive effect on food security, and, specifically, on maize and bean yields, possibly due to greater availability of labor. As policy responses, the authors recommend providing farmers with information about both climate conditions (through climate monitoring and intensified early warning systems) and adaptation options. Specific strategies include irrigation (drawing on international success stories in managing limited water supplies) and storage of rain for use during the dry seasons.

Strategies for adapting to climate change

Broadly, there are three types of things farmers can do to adapt to climate change: (1) change how their land is used, whether it be changing the type of crop grown or converting pasture to cropland or vice versa. This can be as small a change as finding a more heat-tolerant variety of the same crop or as large a change as replacing a whole cropping system – for instance, changing maize and green beans to wheat and chickpeas (garbanzo beans/hummus). (2) Change the farming technology – for instance, adopting a low-tillage system or a water harvesting technology. (3) Adding and managing inputs – primarily fertilizer, pesticides, and herbicides. For any of these strategies, there are a few considerations to make them most effective.

Adopting packages rather than individual technologies

Both today's agriculture and the agriculture of the climate change era use packages of technologies – in other words, sets of different practices that work well together. Traditional agriculture in Debre Zeit, Ethiopia, for instance, uses a three-crop rotation of teff (a small grain), wheat, and chickpeas. The teff contributes both a marketable crop and very substantial amounts of hay, which is fed to animals, including draft animals. Those draft animals help control weeds through tillage, going over the land many times. Wheat is also a marketable crop, while the chickpeas fix nitrogen in the soil and provide protein in the diet. Although there are definitely modern touches, such as growing durum wheat to sell for pasta, rather than either traditional varieties or a higher-yielding short-stem variety, the essentials of this system are a set of practices that work together. Some modern practices are simply add-ons. For instance, monitoring for insects on the chickpeas, and spraying once if the monitoring criterion is met, is simply added on. So is digging a borehole well and using the water for supplemental irrigation for the chickpeas (done only if there is a dry spell, particularly early in the season).

More widespread technical change needs to come in packages as well. When land is subject to erosion, either because of soil type or slope or both, low-till systems can preserve soil. This will increase yield in the long run, although without added fertilizer, it will take a few years for the benefits of soil conservation to translate into higher yield. Faced with a future of heavier and less frequent rain, a low-till system would be adaptive. Such a system is a system, however – it needs to be considered as part of a package, not in isolation. In addition to planting with minimum soil disturbance, a low-till system needs a weed control mechanism. The standard US system uses glyphosate for the weed control and therefore uses "Roundup-ready" corn and soybeans. The system is also very dependent on added fertilizer and produces very high yields. Although it is possible to design a low-till system with fewer features than this, such a system must still handle fertilization and weed control, crop rotation, and pest control. For smallholder farmers in Africa, this might mean weeding by hand, which means that time is taken away from education, off-farm employment, and so forth.

Tailoring adaptation to local conditions

Different agroecological zones need different packages matched to them. Today, as one descends from Addis Ababa to the Rift Valley, one goes through at least three different cropping systems: wheat, teff, and chickpeas; maize and green beans; and rice and other crops. In a highland area, one would find flat land and sloping land, each with different prescriptions. Therefore, there is not just one package that works for large areas, but rather many packages, each deployed in its proper place. Kassie et al. (2008) looked at the profitability of stone bunds (low walls that control water runoff and soil erosion) in the Ethiopian highlands and found that the bunds pay only with moderate rainfall. So, for lands with low rainfall, the bunds are not profitable. Farming practices are not homogenous. So, farming practices need to be figured out for each farm. That requires human capital – in other words, education. It also requires roads to trade the materials needed to implement the technology packages.

Additionally, the need to match crops with practices is a part of the explanation as to why single technological breakthroughs, such as improved seeds, do not act as magic bullets, instantly increasing output and being rapidly adopted. The need to adopt packages to increase yield or to adapt to climate change makes adaptation much harder than it would be if it were only a matter of using different seeds or building bunds. The need to adopt packages also creates a premium for education, as the full instructions for how to farm using a low-till system are much more complex than a sentence or two.

Barriers to adoption of adaptive techniques

Recall our three categories of adaptation options: (1) change what is being raised on the land; (2) change technology; and (3) add inputs. The first type of adaptation is observed in East Africa, where the dominant crops are indeed fitted to the weather and soil. It is not too hard to imagine that a slow change in climate will be met with a slow adaptation in crop type. Most of the development attention is, however, on the second and third set of activities. These strategies for adapting to future climate change are exactly the same activities advocated for improving today's agricultural systems. Generally, the development community feels that these measures are under-adopted and runs programs to increase their adoption. What are the barriers to adoption of adaptive techniques, and how might these barriers change with policy and with development?

The main underlying barriers include human capital/education (illiteracy, lack of high school-level education, financial illiteracy, and lack of access to extension services); underdeveloped infrastructure (in particular, roads, and also electricity); and underdeveloped institutions (markets, secure property rights, and access to credit).

The literature on barriers to adoption, however, emphasizes more specific causes of non-adoption, such as risk and credit rationing. Risk is usually measured experimentally by willingness to accept a small gamble instead of a sure payoff. That willingness comes from two things: (1) an attitude toward risk that is part of the subject's individual preferences and (2) the subject's current level of income. In Di Falco et al.'s (2014) risk experiments, a good harvest resulted in a much lower measured risk aversion. So, extreme risk aversion is a consequence of underdevelopment due to lower incomes (as well as a way to perpetuate it, as taking on risk is inherent in investing in new, adaptive technologies).

Credit rationing or astronomical interest rates are a direct consequence of institutional arrangements, such as a lack of transferable land titles and the absence of a country-wide or regional integrated banking system. The future development of credit markets would lower farmers' risk through the addition of crop insurance in case of crop failure.

On the whole, we see the myriad proximate causes of non-adoption as consequences of the underlying underdevelopment and particularly the underdevelopment of human capital, infrastructure, and institutions.

The adoption challenge isn't specific to agriculture

In every part of development economics, there are obviously beneficial practices that are not adopted. Improved stoves provide perhaps the longest-running non-adoption story. Cooking with solid fuel in a simple stove produces intense indoor air pollution and results in pulmonary problems and early death for the cook. Improved stoves also require less fuel, which is a major benefit given the increasing population and the denuding of the forest land. Technically, stoves seem like an easy problem to solve, either by venting or by changing fuels. At least 40 years have passed since improved stoves have been tested. Today there are NGOs devoted to improved stoves and randomized control trials for stoves. Yet, long-term adoption of this technology is disappointing (Gil, 1987). In comparison, the Franklin stove (circa 1770) saw rapid adoption in America. While there is a different story for each non-adopted stove type, there is also a clear pattern that even simple technology does not get easily adopted.

There are two classes of reasons for non-adoption: (1) it doesn't work for the user and (2) there are barriers to adoption. When households obtain a new piece of technology, try it, and then discard it, revealed preference is that they prefer the old. From the user's point of view, it didn't work. Many of the stove models fall into this category. So do abandoned bunds: the user tried it and didn't like it. However, there is another class of cases in which the item does work, but just not for the user. Modern agricultural packages require a certain amount of human capital, or education, to manage them. For instance, nitrogen fertilizer does not work well if the soil is too acidic (Ekbom, 2009). The test for acidity is not all that hard, but one needs to know what an acid is and what a litmus strip is. In practice, one needs to read and, preferably, read an international language (e.g., Arabic, English, French, or Mandarin) because packages are labeled with instructions in international languages. Without literacy, the user may be incapable of making the package work, even if the package itself works fine. This clearly is not the same as a stove meant to feed a family of four when the family size is eight and the stove must also cook feed for the animals (Rehfuess et al., 2014).

When an adoption of a new technology would provide a net benefit and yet it is not adopted, there must be a reason for the non-adoption. Because a new technology is a risky venture, all the usual reasons for not adopting a risky venture should be important: risk aversion, credit constraint, and asymmetric information among them. These usual reasons have underlying causes that can often be traced back to education, infrastructure, and institutions.

Barriers and how growth changes them

Despite weak growth in many sub-Saharan countries in 2016, long-term trends for the region have been promising (World Bank, 2017). Per capita income has doubled for the region since 1990. That economic development has brought along with it improvements to the education, infrastructure, and institutions of the countries in the region. While policy efforts to improve those three areas often do not explicitly target farmers' abilities to adapt in the face of climate change, there are real implications for making farmers' efforts to adapt more likely and more successful.

Education

Because agriculture has been practiced for thousands of years and has employed over half the population – even in developed countries until very recently – the image is that agriculture is a less technical occupation than industry or services. While that is true of subsistence agriculture if compared to software engineering, it is not true of modern agriculture compared to many other occupations. Modern agriculture is based on science and a farm operator needs to have basic technical education to be successful. In the United States, we take it for granted that farm operators have a high school education or more. A modern small farm operator has to understand chemistry and soil science, accounting and taxation, hydrology, welding, machinery operation, finance, environmental laws, pest management, and so on. Problematically, it would be surprising to find many high school graduates on farms in East Africa. It would not be surprising to find college graduates on farms in the United States. Improving education and building a large extension service are two of the ways Ethiopia is responding to the need for educated farmers.

Ethiopia has committed to developing the largest agricultural extension service in sub-Saharan Africa and could serve as a model for other countries in the region. It is estimated that 8,500 farmers training centers (FTCs) have been established at the *kebele* (smallest administrative unit) level, with roughly 2,500 of them reported fully functional (Ministry of Agriculture and Rural Development, 2009). In addition, it was reported that about 45,000 development agents (DAs) currently on duty at the *kebele* level include about 12% to 22% women, depending on the region. The number of frontline extension personnel is expected to increase to roughly 60,000 when all FTCs have been established and are fully functional (Ministry of Agriculture and Rural Development, 2009).

The 60,000 extension agents planned for Ethiopia are nearly as many agents as there would be farmers (86,000) if this were the United States.[1] Another way to view that is that there will be nearly enough extension agents to substitute for an educated farm population in the very near future.

Extension agents will spread important soil conservation techniques, such as low-till systems; establish safe borehole wells; and introduce higher-yielding varieties along with the technology packages needed to sustain them.

To a great degree, a large extension service is like a very specialized school in agriculture, delivered at a very local level. Extension services could help farmers choose the right package of crops and input technologies for their specific situation.

American extension included a home extension segment that focused on food preservation, home gardens, and other work that was traditionally the responsibility of women. Food safety remains a focus of the extension service. In the United

States, this would include education on the dangers of raw milk and raw beef. Although home production has become much less important in the United States, to this day, recipes and instruction for the handling and preserving of food can be found at cooperative extension service websites.

The increasing general level of education will have an interesting interaction with extension. As farmers get closer to a high school education, they will benefit more from extension, as it will be much easier to teach the basics of agriculture to this group. At some point, far off in the development process when the farmers themselves have a better agricultural education than today's extension agents, the service can be pared back.

Infrastructure

Increasing the road network is the most radical thing that is happening in East Africa. Annual external funding for infrastructure tripled between 2004 and 2012 (Gutman and Chattopadhyay, 2015) and there are substantive public efforts as well. Several development agencies recently formulated the Programme for Infrastructure Development in Africa (PIDA). This continental initiative, based on regional projects, is intended to address the infrastructure deficit that hampers Africa's competitiveness in the world market (PIDA, 2012).

In addition to increasing farmers' access to inputs, education, and nonfarm employment opportunities, improving infrastructure stabilizes the markets for agricultural goods. Bad roads impede the law of one price (Brenton et al., 2014). Prices in one place can be much higher than prices in a place nearby if it is difficult to get to that nearby place. In the high-price place, people can be malnourished because they cannot afford the high prices, while in the low-price place there is no further incentive to produce food. Roads even this out, allowing trade from the low- to the high-priced region. Porteous (2016) found that improving SSA roads to an international benchmark would reduce average food prices by 46% and increase GDP by 2.2%.

Roads have consequences well beyond agricultural staple trade. Good roads provide access for manufactured goods to remote villages. For instance, in Mexico in the 1970s, car parts could be shipped by bus, within hours, to small village repair shops hundreds of miles from a major city.[2] Going in the other direction, high-value agricultural products (e.g., watermelons) can be shipped to cities, again in a timeframe of hours, not days. Roads also facilitate input markets, giving farmers greater ability to purchase inputs like fertilizer. In general, the encouragement of linkages both ways between cities and countryside is part of agricultural demand-led industrialization, advocated by Adelman (1984) based on the experience of South Korea.

Climate change increases the variability of weather and thus increases the payoff to the road network. With more places in drought or flood, the value of all-weather roads to smooth out prices across all of Africa will be much larger than the value today.

Institutions

Of the three categories, institutions perhaps have the most room for improvement in East Africa. Both banking (which implicitly assumes ownership of collateral) and insurance are underdeveloped institutions. Opportunities for financial diversification constitute a fundamental feature of advanced agriculture, so we shall group banking among the institutions where growth is possible.

Banking

Banking institutions are not well developed, yet, in East Africa. In contrast, agricultural production credit is a standard and essential feature in developed countries. Farmers borrow for planting time expenses and pay back at harvest. In the United States, the Land Bank was founded in 1916 and the Production Credit Associations date from 1933 (Farm Credit Administration, 2016). These institutions were the policy response to the lack of credit in farm areas and are regarded as an important part of the success of US agriculture. Unlike most credit efforts in East Africa, neither of these American institutions is remotely micro-credit in scale; they were established to move money from money markets (Wall Street) to the farm areas. Ultimately, they involved no subsidies. The United States entered this part of the banking business because there were excellent credit institutions in cities and poor credit institutions in the farm areas. In both places, there were well-established titles to land and machinery and well-established courts that would change the ownership of those titles upon the default of loans.

Micro-credit institutions like the Grameen Bank are neither big enough nor efficient enough to provide for modern agriculture. The size of the US Farm Credit System is much bigger than any micro-credit system. Today there are $194 billion in outstanding loans. The expense ratio is 0.33%. In contrast, the Grameen Bank has outstanding loans of $1.16 billion and an expense ratio of 9.8%. This is another example of how strategies in use today may not be the strategies of tomorrow.

Not being able to borrow in a formal financial market means that a farmer must pay for a new, adaptive technology by gifts or loans from relatives, by selling assets, or through nonfarm earnings. Borehole wells are an example. Borehole wells even out the water supply across the season, so if climate change leads to more variance in rainfall, then drawing on well water during the dry times adapts to the problem. Because there is variability today in weather, borehole wells also can pay off today.[3] However, adoption of this and other technologies is slowed considerably by lack of functioning credit markets.

Mortgages on farmland are also a standard feature of developed country economies. The operator of a farm does not have to finance expansion with retained earnings or money raised from friends and relatives.

Credit markets are an endogenous phenomenon. Village money lenders' services have been in demand for all of written history (see, e.g., the Bible on usury).

Organized banking has had two pieces: the lending end (possibly the money lender, possibly a branch bank) and the intermediation end. The intermediation end is the way money is moved from excess supply regions (e.g., New York City) to excess demand regions (e.g., Iowa in 1910). The Lombardi and Jewish banking syndicates of the Middle Ages were examples of local lenders with an intermediation system that was outside the control of government. These early banking syndicates largely lent to the poor on pawn. However, they also lent to going enterprises on other collateral. One expects that, as sub-Saharan Africa develops, governments will reform their banking sectors. They will provide for intermediation and for stable banking institutions, possibly along the lines of the regional and local land banks of the United States 100 years ago. Credit constraint is less likely to be an issue in 20 years. Certainly banking reform is a policy lever that increases both income and adaptation to climate change.

Crop insurance

One of the current barriers to the adoption of modern practices in SSA is that farmers do not choose to take the risk entailed in the modern practices. Risk arises in two forms. First, some adaptation practices offer higher average yield than current practices, but with greater variance – in other words, the potential rewards are high but so is the risk. Second, spending or borrowing money to try something new is itself risky; there is always a risk that rainfall won't come when needed; then, a farmer would not only have a crop failure but also would have to pay back the money he borrowed to buy the new input.

Crop insurance would reduce these risks. Crop insurance is a policy that is widespread in the developed world and has been tried as a part of packages in the developing world. However, uptake of crop insurance in the developing world is low.

To encourage farmers to adopt a planting time practice (fertilizer, improved seed), they can be offered credit, with repayment not necessary until harvest, and/or crop insurance, in the form of either insuring yield per area planted or ensuring against bad weather. Such an experiment was tried in Malawi by Giné and Yang (2009). Two groups of farmers were offered credit to purchase improved seeds. For one of those groups, the offer was conditional on also buying insurance, at actuarially fair (unsubsidized) rates. The insurance partially or fully forgave the loan in the event of poor rainfall. The unexpected result was that uptake of credit was lower when insurance was a required part of the package. The authors of that study noted that the terms of the loan contract and local enforcement practices limited the lender's recourse to seize assets if a loan was not repaid, and that wealthier farmers, who would have more to lose if assets were seized, were more likely than poorer farmers to accept the package with insurance.

The Malawi study leads to some questions: Does a rational farmer always want insurance if she has to pay its full cost? Is subsidized insurance necessarily the best way to encourage technology adoption? If so, should subsidies continue indefinitely? As with all subsidies, are they reaching the poorer people who need them more? If not, is there a clear decision to subsidize wealthier people in order to achieve a goal such as higher agricultural productivity?

In the developed world, government subsidies for crop insurance are "Depression bred, prosperity fed."[4] Agricultural subsidies in the United States are now politically driven. In the United States, the demand for crop insurance at fair actuarial value is low. It is the government subsidy that makes it sell: the United States pays 60% of the actuarial value and the farmers pay 40% (Congressional Budget Office, 2013). The argument to use such a policy in a low-income country must rest entirely on the policy causing some unrelated desired behavior, such as the adoption of a technology package. Of course, the Congressional Budget Office's conclusion, "Reduce Spending in the Crop Insurance Program" (2013), strongly suggests that as East Africa develops, this should not be a major policy pillar, at least in the long run.

Diversification of income sources

The prime way to reduce risk in agricultural enterprises is diversification. Diversification of crops is discussed elsewhere in this book. Here, we first discuss diversification of nonfarm income sources, and then talk about how overall economic development can stimulate demand for a diversified range of farm products.

In both developed and developing economies, income from off-farm employment is an important way for farmers to reduce risk. Diversification beyond the farm is most effective when the earnings from the nonfarm activity are uncorrelated with those of the farm – for instance, if one member of the family works in town in a non-agriculture-dependent industry. Partnerships with nonfarm family members are also quite effective – for instance, two brothers own the farm, one works it, and the other lives 2,000 miles away and works in an unrelated business.

If one measures risk by standard deviation of income, and assumes that both activities, farm and nonfarm, have the same standard deviation and are independent, then average income has only $1/\sqrt{2}$ times the original standard deviations. In simpler terms, it's less likely for income from two different activities to drop at the same time than it is for income from one activity to drop. Put in a development perspective, the putative borehole well driller can pay back his loan and still survive if his farm does badly one year but his off-farm income does well. The risk of disaster is cushioned by a second activity. All developed economies have much larger nonfarm than farm sectors, most simply because there is more economic opportunity away from the farm. Additionally, better infrastructure would more easily allow members of farm families to travel for other forms of income. In the United States, farm families actually derive more income from off-farm than from on-farm activities.

Development brings higher-income jobs to regions; with further development, farming households will have better access to those jobs. Improving education systems opens up access to higher-income jobs for members of farming households. Infrastructure improvements make it easier to travel farther for economic activity. Improvements to institutions and the expansion of markets increase the number of jobs available.

An income increase in other economic sectors would also lead to more demand for agricultural products. Currently in East Africa, bad roads and a high percentage of national income on farms lead to low trade volumes and only higher-valued products being traded. City demand for varied agricultural goods, combined with good roads, can lead to much-increased demand for lower-valued goods. Today there are many classical grain and livestock farms in East Africa. While livestock buffers year-to-year income variation, and livestock prices tend to move counter to feed prices, these farms are not very diversified endeavors. As the country develops a robust urban sector, there will be additional demands on the farm sector. With added demand for fruits, vegetables, chickens, eggs, milk, and the plethora of other things demanded by even lower middle-income consumers, the farming landscape will adapt by producing a far greater variety of goods, even at some distance from major cities. In terms of risk, if the farms increase their scope (number of products), the farm's risk will fall due to diversification.[5] However, to make diversification work well, the farm enterprise needs to be larger than a single hectare so that there is room for many activities all within the same enterprise. The policy implication here is the importance of land tenure, secure property rights, and land markets. With those reforms and better roads, the agricultural landscape will have less inherent production risk and be much better positioned to withstand the stresses of climate change.

Sloping land: an example of how growth interacts with barriers to adaptation

A main takeaway is that economic development increases financial flexibility and stability for farmers, allowing them to more easily adopt the right agricultural packages for their situation. One such package is sloping land management (SLM) – a measure adaptive to climate change.

Assuming that climate change brings heavier and more sporadic rain, SLM measures become even more crucial. SLM measures protect a stock resource, the soil. These measures are not technically demanding, requiring largely physical labor to construct bunds or plant buffer strips. However, the barrier to adoption and maintenance is that the user does not see a very large immediate benefit (Kassie et al., 2008). Failure to take on these measures now will result in less soil later. Soil holds water for plant growth, so shallow soils will exacerbate the problem of more variable weather or of dry spells. In so far as credit institutions reduce effective interest rates and education increases the targeting of measures to land where it has positive benefits, the adoption of these expensive measures will be easier in a more developed East Africa.

Adoption success stories

Education, infrastructure, and institutions define what policies a country can adopt. We offer three success stories to elucidate what the art of the possible has been. We don't think that it is merely coincidental that large-scale agricultural projects in very poor countries are not among them. Countries with low national incomes cannot afford $5 billion programs (see the example from Mexico ahead) that reach millions of households. While countries are still in the low-income phase, they need to make do with interventions that cost considerably less per capita and require much less of the participants.

Deworm the world

Miguel and Kremer (2004) dewormed children in Kenya in a randomized control trial where some schools were treated and some were not. The treatment was to administer a single dose of an antiparasitic medicine. They found that ridding students of intestinal worms had the often overlooked benefit of increased school attendance in addition to the health benefits from removing worms. They also found that there were spillover benefits – there were fewer infections even among the untreated in nearby schools because the overall worm pool decreased. Deworming is a perfect intervention for a less developed country because it requires the subjects to take one pill, no more and no less. Cost is minimal, educational requirements among the treated are minimal, and one size pretty much fits all – none of which describes the requirements for adoption of new agricultural technology. With the cooperation of national governments, Evidence Action (2017) has scaled this up considerably. Deworming campaigns in India, Kenya, Ethiopia, Vietnam, and Nigeria treated 190 million students in the 2015/2016 school year.

One wonders, however, whether a drug-related intervention can remain effective in the face of alternative hosts for the worms, in this case farm animals. Also, will the worms develop resistance? One also wonders whether, 20 years from now, Ethiopians will still eat infected beef. By then, disease eradication in cattle or treatment of beef may well be the norm. Similarly, will they swim in infected waters or walk without shoes on infected ground? Most developed and some developing countries have eliminated worms at the source, rather than medicating afterwards. Likely, this sort of intervention will end with further development. The right intervention this year probably won't be the same in a generation.

Chlorine

Point-of-use water purification stops disease, though not as perfectly as a municipal water supply that both filters and chlorinates, and not as effectively as boiling. However, cost, user effort, and water flavor limit the adoption of point-of-use purification. Luoto et al. (2011) tried four different methods in Bangladesh and, despite the intervention being free, only 30% adopted a method. Albert et al. (2010) did the same in Kenya and found less than perfect uptake. The problem with point-of-use water purification is that it is either expensive or time-consuming. Evidence Action's solution is interesting: free chlorine-based bleach dispensers near water sources. Apart from the taste of the bleach, little is required of the user. Because bleach avoids boiling and therefore avoids fuel use, the scheme is partly funded with carbon credits. What is not required of the users is a great degree of education – the dispensers drop the right amount of bleach each time without the user having to measure it. 4.7 million people in Kenya, Malawi, and Uganda use Evidence Action's bleach dispensers. One would hope that with a doubling of income every decade would come a more elaborate filter and chlorine system. Again, the measure, chlorine, is an easily adopted measure, but it is again a stop-gap measure, unlikely to still be used after a couple of decades. Again, this is a simple fix, whereas agricultural technology adoption is far more complicated.

Progressa Oportunidades

Unlike the previous examples, where adoption was driven by convenience and not hindered by lack of education, is the Mexican government's far-reaching Progressa Oportunidades program. The program incentivizes compliance with health and school measures by offering cash transfers. The cash transfers are significant (in some cases, one-third of family income) and are conditional on participating in the medical system (e.g., obtaining childhood vaccinations) and achieving good school attendance. Mexico is a much richer country than the East African countries and was thus able to devote $5.3 billion to this program in 2012 (Prospera, 2017).Of course, it has levels of education, infrastructure, and institutions to match its income. For instance, banking is well enough developed in Mexico that the target population for Oportunidades, though low-income and low in education, can be part of the banking sector. All of the beneficiary families receive their incentives by means of a bank account, thus providing access to financial services. Mexico's gross national income (GNI) per capita today is what sub-Saharan Africa's GNI will be in several decades, at current growth rates. By the time global climate change has significantly altered the weather, sub-Saharan Africa may well have enough income to run full-scale anti-poverty programs like Progressa Oportunidades.

Conclusions

Adaptation to climate change requires the adoption of modern agricultural techniques. Most of these techniques are needed even without climate change and have not been readily adopted by farmers in SSA. The barriers to adoption today are many and well-studied. However, climate change is a process that takes decades. During those decades, many of the SSA countries will experience economic growth along with development of infrastructure, increasing educational attainment, and institutional reform. The change in education, infrastructure, and institutions will make adaptation to climate change 50 years from now very different from the lack of adoption of modern practices today.

Modern agriculture depends on finding the package of techniques suitable for a given farm. It is not a one-size-fits-all prescription, nor are the packages easy to deploy without education. SSA today has too little education per farm, but educational level is rising quickly in many of these countries. Modern agriculture involves considerable trade in both inputs (e.g., fertilizer) and outputs. Some regions specialize in corn and soy and some in livestock. This type of specialization requires good roads. The needed roads are now being built and so trade in inputs and outputs will be much easier in coming decades than it is today. Finally, institutions such as land titles and credit are needed to make it possible for a farm enterprise to finance the required packages of technology. Some SSA countries are making real progress in land titling. None of them yet have credit institutions as effective as those the United States had 100 years ago.

Given the current infrastructure, institutional, and educational conditions, what are the widely touted development programs and why are these successful? To work in the field, a program needs to demand little of the user. Deworming and chlorine water purification fit this model. To some degree, rapid land titling also fits these requirements. Ambitious programs like Progressa Oportunidades are carried out in middle-income countries, not in most of SSA. But will that be true in the time frame of climate change? Probably not. Development will have proceeded far enough for more ambitious programs.

Climate change adaptation is about the future. It is about future weather and it needs to be about future economic conditions as well. While it is hard to be optimistic about climate change itself, there is some reason for optimism about the economic development needed to cope with climate change in SSA.

Notes

1 At the average US farm size of 434 acres, it would take 86,000 farms to cover all the arable land in Ethiopia.
2 Author's observation in 1978, Guadalajara, Mexico.
3 To be successful on an area-wide basis, groundwater abstraction needs to be limited to recharge. So, as a way of smoothing temporal variance, it is sustainable; however, as a way of increasing water supply, it may not be sustainable.
4 Personal communication with Sidney Hoos.
5 Berck (1981) examined cotton farms and found that the demand for off-farm risk products (futures) shrank with on-farm diversification.

References

Adelman, I. 1984. Beyond Export-Led Growth. *World Development* 12(9): 937–949.
African Union Commission, African Development Bank, Economic Commission for Africa. 2012. Programme for Infrastructure Development in Africa. www.icafrica.org/fileadmin/documents/PIDA/PIDA%20Executive%20Summary%20-%20English_re.pdf
Albert, J., J. Luoto and D. Levine. 2010. End-User Preferences for and Performance of Competing POU Water Treatment Technologies among the Rural Poor of Kenya. *Environmental Science and Technology* 44(12): 4426–4432.
Berck, P. 1981. Portfolio Theory and the Demand for Futures: Case of California Cotton. *American Journal of Agricultural Economics* 63(3): 466–474.
Bezabih, M., S. Di Falco and A. Mekonnen. 2014. On the Impact of Weather Variability and Climate Change on Agriculture: Evidence from Ethiopia. Environment for Development Discussion Paper Series No. 14–15. Gothenburg, Sweden: Environment for Development and Resources for the Future.

Brenton, P., A. Portugal-Perez and J. Régolo. 2014. Food Prices, Road Infrastructure, and Market Integration in Central and Eastern Africa. World Bank Policy Working Paper 7003. Washington, DC: World Bank.

Congressional Budget Office (US). 2013. Options for Reducing the Deficit 2014–2023. Washington, DC, pp. 16–17. www.cbo.gov/budget-options/2013/44738

Di Falco, S., P. Berck, M. Bezabih and G. Köhlin. 2014. Rain and Impatience: Climatic Factors and Investment in the Highlands of Ethiopia. Presented at 14th International BIOECON Conference. www.bioecon-network.org/pages/16th_2014/DIFALCO.pdf

Ekbom, A. 2009. Determinants of Soil Capital. Working Paper No. 339. Gothenburg, Sweden: Department of Economics, University of Gothenburg.

Evidence Action. 2017. www.evidenceaction.org/dewormtheworld/#impact-deworming

Farm Credit Administration. 2016. History of FCA and the FCS. www.fca.gov/about/history/historyFCA_FCS.html

Gil, J. 1987. Improved Stoves in Developing Countries: A Critique. *Energy Policy* 15(2): 135–144.

Giné, X., and D. Yang. 2009. Insurance, Credit, and Technology Adoption: Field Experimental Evidence from Malawi. *Journal of Development Economics* 89(1): 1–11.

Gutman, J., A. Sy and S. Chattopadhyay. 2015. Financing African Infrastructure: Can the World Deliver? Washington, DC: Brookings. www.brookings.edu/wp-content/uploads/2016/07/AGIFinancingAfricanInfrastructure_FinalWebv2.pdf

International Monetary Fund. 2017. Regional Economic Outlook Sub-Saharan Africa. www.imf.org/en/Publications/REO/SSA/Issues/2017/05/03/sreo0517

Kassie, M., J. Pender, M. Yesuf, G. Kohlin, R. Bluffstone and E. Mulugeta. 2008. Estimating Returns to Soil Conservation Adoption in the Northern Ethiopian Highlands. *Agricultural Economics* March(2008). doi:10.1111/j.1574–0862.2008.00295.x.

Luoto, J., N. Najnin, M. Mahmud, J. Albert, M.S. Islam, S. Luby, L. Unicomb and D. Levineet. 2011. What Point-of-Use Water Treatment Products Do Consumers Use? Evidence from a Randomized Controlled Trial among the Urban Poor in Bangladesh. *PLoS ONE* 6(10): e26132. https://doi.org/10.1371/journal.pone.0026132

Miguel, E., and M. Kremer. 2004. Worms: Identifying Impacts on Education and Health in the Presence of Treatment Externalities. *Econometrica* 72(1): 159–217.

Ministry of Agriculture and Rural Development (MOARD), Department of Agricultural Technical and Vocational Training and Education. 2009. Data on ATVET Colleges Graduates. Addis Ababa, Ethiopia: MOARD. www.worldwide-extension.org/africa/ethiopia/s-ethiopia

Porteous, O.C. 2016. Essays on Agricultural Trade in Sub-Saharan Africa (Order No. 10150917). Available from Dissertations & Theses @ University of California; ProQuest Dissertations & Theses A&I; ProQuest Dissertations & Theses Global. (1834922543). https://search.proquest.com/docview/1834922543?accountid=14496

Prospera. 2017. www.prospera.gob.mx/Portal/work/sites/Web/resources/ArchivoContent/2109/BAJA%20Oportunidades%2015yrs%20of%20results.pdf

Rehfuess, E.A., E. Puzzolo, D. Stanistreet, D. Pope and N.G. Bruce. 2014. Enablers and Barriers to Large-Scale Uptake of Improved Solid Fuel Stoves: A Systematic Review. *Environmental Health Perspectives* 122(2): 120–130.

Schlenker, W., and M. Roberts. 2009. Nonlinear Temperature Effects Indicate Severe Damages to U.S. Crop Yields under Climate Change. *PNAS* 106(37): 15594–15598. doi: 10.1073/pnas.0906865106

World Bank. 2017. GDP per Capita. http://data.worldbank.org/indicator/NY.GDP.PCAP.PP.CD?locations=ZG&view=chart

Part VI

Conclusion and policy implications

18 Conclusion

Cyndi Spindell Berck, Peter Berck,
Salvatore Di Falco and Poojan Thakrar

The chapters in this book propose policies to reduce barriers that prevent farmers from adapting to climate change. These can be roughly grouped into policies to manage risk and increase access to financial resources; expand access to education and increase the knowledge base; increase gender equity; build physical infrastructure; develop institutions at both national and local levels; and promote national structural transformations of economies that are now heavily dependent on rain-fed subsistence agriculture.

Manage risk by increasing access to credit, insurance and safety nets

Risk and financial constraints are closely related. Two of the most common reasons for farmers not adopting a new technology are lack of credit (Chapters 5, 7, 9, etc.) and risk avoidance (Chapters 6, 7, 8, etc.). Certain types of improved seeds, for instance, offer higher average yield but greater variance, leading to greater downside risk; see, for example, Chapter 6.

One set of policies addresses the lack of financial resources, such as credit, that are necessary for effective adaptation. Closely related are policies that manage the risk that farmers face if new technology fails to produce desirable results, either because the technology is risky or because uncontrollable events, such as drought, cause the crops to fail. Thus, credit and insurance are examples of financial institutions that can promote adoption of adaptive technologies.

Reducing credit constraints

Not all adaptive strategies require upfront investment, but many do. These include improved seeds and fertilizer (Chapter 9), borehole wells (Chapter 2), airtight bags that reduce postharvest losses (Chapter 12) and other agricultural intensification strategies. Poor smallholder farmers are unlikely to have cash available to invest in new technologies. In many cases, wealthier and better-educated farmers are more likely to adopt new technologies.

Further, some investments, such as soil conservation, may not become profitable for several years (Chapter 9). Even in developed countries, farmers routinely borrow money to finance a season's planting and are expected to pay it back at the next harvest time, not several years later.

Access to credit allows farmers to finance investments before they reap the benefits. In communities with very little capital, it is often the only way to significantly increase the profitability of farms and businesses. For instance, in Chapter 7, Komba and Muchapondwa discovered that 99% of farmers in Tanzania perceive climate change, but only a fraction of those farmers have adopted adaptation methods. Similarly, in Chapter 5, Di Falco observed that 83% of farmers perceive variation in temperature in Ethiopia, but only 44% of them use adaptation strategies. Farmers often cited lack of credit as the reason that they failed to adapt to the perceived threat of climate change. Thus, policy makers should increase access to credit in rural areas so that farmers can invest in adaptation strategies.

Risk management

Programs that protect against downside risk offer farmers some security in case their risky decisions do not come to fruition. For instance, Ethiopia's Productive Safety Net Program allows farmers to work for food or cash in the event of their own crop's failure. As discussed in Chapter 9, Ethiopian respondents who believed that they could rely on the program's help were 20% more likely to adopt more than two sustainable agricultural intensification practices, thus increasing their expected income.

Formal crop insurance is widespread in developed countries, taking two forms: insuring against certain weather events, such as drought, and compensating the farmer if yield falls below a certain level. Such insurance markets are underdeveloped in much of SSA, and, where crop insurance is available, uptake is low. As Chapter 17 argues, a rational farmer in either a developed or developing country may opt not to buy crop insurance at an actuarially fair price, and government subsidies are often needed to encourage uptake. Chapter 17 cautions against allowing a long-term political constituency to lock subsidies in place, but suggests that short-term use of subsidies for insurance can be justified in order to encourage technology adoption by addressing risk.

Social networks play many roles in rural communities in developing countries, including a role in risk management (see, e.g., Chapter 3), as well as responding to cash and credit constraints. Extended family networks, for example, may provide remittances from work away from the farm, which can be used to invest in productivity-enhancing technologies or to meet household needs in case of crop failure.

Increase access to education and knowledge

General education, agricultural extension and a stronger knowledge base are all essential to adaptation. Other avenues to increase information include media, technology spillovers and social networks.

General education

In general, better-educated farmers are more likely to adopt new technologies that are suitable for their specific conditions. Many poor subsistence farmers are illiterate or lack education beyond the primary grades. As Chapter 17 discusses, modern farming techniques require ability to read package directions, compute appropriate amounts of fertilizer, test soil acidity and so forth. General education also equips members of farm households to earn off-farm income, which can provide capital for on-farm improvements, and can otherwise meet households' consumption needs in the event of crop failure.

Agricultural extension

Agricultural extension services are a main source of information about adaptive technologies. Because each farm has specific needs, farmers need information so that they can make the right decision for individual plots and crops. As Chapter 9 points out, both quality and quantity of extension agents are important. Farmers who have confidence in extension agents are more likely to adopt new technologies. Without enough agents, however, information may not reach farmers. Chapter 9 notes that the ratio of extension agents to farmers differs greatly among countries.

Chapter 17 points to the successful model in Ethiopia, which plans to have 60,000 agricultural extension agents in the field. These agents will spread important agricultural techniques on a local level, giving farmers the tools they need to adapt to the effects of climate change. Similarly, Chapters 4 and 6 encourage the Kenyan government to spread agricultural knowledge, give early warning of impending droughts, and educate farmers about adaptive strategies. For example, agricultural extension agents can teach farmers about good practices to reduce postharvest losses, as discussed in Chapter 12.

Increasing the knowledge base

A repeating theme in this book is that both on-farm decisions and public policies must be tailored to specific agroecological conditions. For instance, Chapter 3 breaks Ethiopia into four different agroecological zones and determines the vulnerability of each particular zone. In that chapter, Gebreegziabher et al. were able to discern disparities in adaptive capabilities, climate exposure and general sensitivity in the local zones, exposing substantial differences between the four regions.

Chapter 10 shows the complexities of adapting Ethiopia's crop-livestock farming systems to climate change. Discussing the farmers' choices – whether to keep livestock, and which varieties of animals to keep – the authors assert that "adaptation policy should be based on comprehensive knowledge of the structure and dynamics of livestock production systems, including information about indigenous breeding strategies related to animal adaptation and management in climate-sensitive dry-lands. Research and development programs must prepare appropriate technologies, and information must be disseminated to help farmers adapt to climate change."

Similarly, Chapter 4 recommended monitoring climate change, installing early warning systems about projected droughts, and disseminating other relevant information. These policies will help farmers understand what they can anticipate in the future and what strategies will help them adapt accordingly. Such information can be disseminated to farmers through agricultural extension services and other means.

By using a computable general equilibrium model to study the effects of biofuels production on smallholder income and food security, Gebreegziabher et al. (Chapter 11) also illustrate the importance of evidence-based policy making. While rejecting the contention that biofuels production is a "crime against humanity," the analysis shows that biofuels production is beneficial to food security under many, but not all, circumstances. This brings us back to the importance of considering specific agroecological conditions at the local level and the farm level.

Other ways to disseminate information

Media access can provide information, particularly on weather forecasts. In Chapter 7, Komba and Muchapondwa found that farmers with access to media are 7.7% more likely to change their planting dates, presumably because they have access to weather forecasts and information. Media access is a way to disseminate the weather early warning information recommended by Kabubo-Mariara and Kabara in Chapter 4.

As for social networks, even if their intent is unrelated to agriculture (e.g., funeral societies, discussed in Chapter 3), they can be an effective way to share information.

Lastly, the concept of technology spillovers refers to information or techniques gained in one economic activity – for instance, raising crops for biofuels production – which are then used to improve production in another activity – in this case, growing food crops. Biofuels technology spillovers are discussed in Chapter 11 and ahead, where we discuss structural transformation and overall economic development.

Increase gender equity

Female heads of farming households in sub-Saharan Africa face barriers such as unequal endowments of land and capital, cultural taboos against certain forms of labor, and fewer opportunities for education. Gender gaps in household income and food security are likely to increase under climate change. In Chapter 13, the authors note the different experiences of de facto female-headed households (those headed by married women whose husbands are absent) and de jure female-headed households (where the female head is divorced or widowed or never married). They also disaggregate rural and urban experiences with respect to different essential nutrients. One policy implication is that attention should be paid to different household arrangements, rural versus urban residence, crop choices and specific nutritional needs. Another is that removing barriers to women in agriculture could raise farm productivity by 20%–30%, increase food production by 2.5%–4%, and reduce food insecurity by 12%–17%.

Chapter 15, discussed further ahead, presents a successful example of increasing women's access to agricultural resources, through Ethiopia's land certification program, which gives both men and women documents verifying their land rights. One advantage of land tenure security is that a female (or male) land owner can rent a parcel to another person who is in a better position to farm that land, thereby increasing the land owner's income and the productivity of the land.

Chapter 14 models the vulnerability of the income of male- and female-headed households. When climate variability was added to the model, the income of female-headed households decreased significantly more than that of male-headed households. When the author simulated three policies in his model, the disparity between genders lessened. The policies were promotion of short-term credit, a fertilizer subsidy, and information about new crop varieties. Each policy increased household income, with equal or slightly greater results for female-headed households. The combination of all three policies increased the income of male- and female-headed households by 7.1% and 7.4%, respectively. Thus, all three policies could decrease the gender gap in households' responses to climate change.

Chapter 9 suggests that women's education is associated with a greater level of adoption. In a study in Ethiopia, one year of additional education on the part of the spouse increased the likelihood of adopting a package of sustainable intensification practices by 12%. The fact that 92% of household heads were male in that study means that additional education for girls and women can contribute significantly to adaptation.

Physical infrastructure

Roads are emphasized in Chapters 2 and 17 and several chapters in between. Roads connect farmers to markets where they can buy fertilizer and improved seeds, and connect them with extension services (Chapter 9). They facilitate migration to off-farm employment (Chapter 16), allow underutilized land to be planted with biofuel crops so that smallholder farmers of food crops are not displaced (Chapter 11), and permit transport of stored food to markets (Chapter 12). As discussed further ahead, road building is part of the overall development context in which agricultural adaptation takes place.

Institutions

The broad concept of institutions ranges from national-scale legal systems, such as land tenure, to local social networks. We will consider those two examples.

The policy implications of Chapter 15 are clear. That chapter looked at the effectiveness of Ethiopia's land certification program. The study found that the policy was in fact successful. In sum, the program has enhanced tenure security

428 *Cyndi Spindell Berck et al.*

for households/villages participating in the program. Certification was found to reduce border disputes, decrease gaps in gender welfare, and increase investment in soil conservation and tree planting. Chapter 15 thus confirms that effective land reforms do improve the welfare of rural households. Chapter 10 adds that such policies should include security of rights to use land for pasture, because access to pasture plays a role in livestock decisions.

At the local level, social networks play many roles in rural communities in developing countries, including a role in risk management (Chapter 3) and in providing informal solutions to the lack of credit access (Chapter 7). Extended family networks, as described earlier, may provide remittances from work away from the farm. As also mentioned earlier, social networks also play a role in sharing information. Social networks are described as "social capital" and various authors suggest strengthening such local institutions.

Structural transformation

As Chapter 17 stresses, agricultural development takes place in a broader development context. For the nations that receive the most attention in this volume, adaptation to climate change takes place in a context of urbanization, diversification of economic activities (e.g., increases in their industrial sectors), and gains in income.

A common theme is diversification in the broadest sense: diversification of farm products and diversification of income sources.

Diversification of on-farm production in response to climate change includes intercropping (e.g., growing legumes and grain at the same time, to restore nitrogen to the soil and to spread risk among crops), crop rotation (for similar reasons), choosing different varieties of crops (for consumption versus sale), changing the crop-livestock mix and so forth. Each of these options is made more feasible for smallholder farmers through the common themes of credit, education and roads, which also are key to development beyond the agricultural sector.

Biofuels are a controversial example of diversifying into production of nonfood crops. Although biofuels are often seen as competitors to food crops, Gebreegziabher et al. found in Chapter 11 that biofuels production has neutral to positive effects on household welfare in several settings, particularly when technological spillovers are considered. With its caveats regarding the type of biofuel crop and agroecological setting, this study illustrates the complex interactions among traditional and modern economic activities.

Extending the diversification of income sources beyond the farm, in Chapter 16, Alem et al. find that weather variability induces households to send migrants to earn off-farm income. They suggest policy options that will reduce the need for this step, including improving farmers' access to water sources and developing insurance products to cushion the impact of rainfall variability. The authors add, "Because several factors enter into the migration decision, it is also clear that some migration is unavoidable – and even desirable from a household viewpoint – and policy measures should also be directed toward supporting development in the urban sectors of the country that will receive the out-migrants from the rural areas that are subject to high rainfall variability." Once again, adaptation takes place in a broad context.

Closing thoughts

We conclude by emphasizing that adaptation to climate change is no substitute for global efforts to mitigate rising temperatures by controlling greenhouse gas emissions. Adaptation and mitigation must proceed together. National and regional policies, in cooperation with NGOs and international partners, can remove some of the barriers that smallholder farmers face in making the adaptation decisions that are best for their farms and households. As stated before, many of the policies that can help farmers adapt to climate change would be good for agricultural production and income even without a changing climate.

Index

Page numbers in italic indicate a figure and page numbers in bold indicate a table on the corresponding page.